Programming and Computational Thinking in Technology Education

AF720349

International Technology Education Studies

Series Editor

Marc J. de Vries (*Delft University of Technology, The Netherlands*)

Editorial Board

John R. Dakers (*Delft University of Technology, The Netherlands*)
Jonas Hallström (*Linköping University, Sweden*)
Philip Reed (*Old Dominion University, USA*)

VOLUME 20

The titles published in this series are listed at *brill.com/ites*

Programming and Computational Thinking in Technology Education

Swedish and International Perspectives

Edited by

Jonas Hallström and Marc J. de Vries

BRILL

LEIDEN | BOSTON

Cover illustration: iStock.com/gorodenkofff

All chapters in this book have undergone peer review.

The Library of Congress Cataloging-in-Publication Data is available online at https://catalog.loc.gov

Typeface for the Latin, Greek, and Cyrillic scripts: "Brill". See and download: brill.com/brill-typeface.

ISSN 1879-8748
ISBN 978-90-04-68789-9 (paperback)
ISBN 978-90-04-68790-5 (hardback)
ISBN 978-90-04-68791-2 (e-book)
DOI 10.1163/9789004687912

Copyright 2024 by Jonas Hallström and Marc J. de Vries. Published by Koninklijke Brill NV, Leiden, The Netherlands.
Koninklijke Brill NV incorporates the imprints Brill, Brill Nijhoff, Brill Wageningen Academic, Brill Schöningh, Brill Fink, Brill mentis, Vandenhoeck & Ruprecht, Böhlau and V&R unipress.
Koninklijke Brill NV reserves the right to protect this publication against unauthorized use. Requests for re-use and/or translations must be addressed to Koninklijke Brill NV via brill.com or copyright.com.

This book is printed on acid-free paper and produced in a sustainable manner.

Contents

PART 2
Curriculum and Teacher Perspectives on Computational Thinking and Programming in Technology Education

PART 3
Computational Thinking and Programming in Technology Teaching

Preface

This edited book with the title *Programming and Computational Thinking in Technology Education: Swedish and International Perspectives* is part of the Brill book series International Technology Education Studies. As such, it contributes to the international research field of technology education, with a particular focus on Sweden albeit in an international context. Because programming and computational thinking are currently urgent topics for education in many countries, the book also makes a contribution to computer science, policy research and other similar areas that concern the introduction of programming, CT and digital tools in schools.

We as editors would like to thank all the authors who contributed their respective chapters to the book, as well as many good comments at the webinar and in the writing process. We would also like to express our deepest gratitude to John Bennett at Brill who made this project possible and supported us throughout the writing and editing stages.

Figures and Tables

Figures

Tables

Notes on Contributors

Cecilia Axell
is Senior Lecturer and Researcher in Technology Education at the Department of Behavioural Sciences and Learning, Linköping University, Sweden. She holds a PhD in Technology Education. Her research interests include technology education in early years, the implementation of programming as teaching content in primary school technology education, philosophy of technology, gender and technology, how indigenous knowledge is implemented in technology education in Sámi school, and how fictional literature can contribute to developing learners' critical thinking skills in technology education. Cecilia has published articles in scholarly journals, written a book, contributed chapters in national and international books, and presented papers at international conferences. She has also been involved in externally funded research projects in technology education.

Astrid Berg
(Linköping University, Sweden) is PhD in Chemistry and Senior Lecturer in Science Education. Her main research interests are the role of animations in the learning of chemistry-related content in science education, and programming and artificial intelligence in technology education.

Eva Björkholm
is PhD and affiliated with the Department of Learning in Engineering Sciences, KTH Royal Institute of Technology, Sweden. Her research concerns, for example, practical knowledge traditions—especially technology—in primary education.

Lars Björklund
holds a position as a university lecturer at Linköping University, Sweden. There, he is responsible for the education of aspiring engineers and teachers in the fields of natural sciences and technical subjects. He is also a Civil Engineer and has previously worked with both hardware and software. He holds a PhD in Educational Science with a specialisation in neurodidactics and tacit knowledge. His research has been focused on teachers' assessment and grading across various subjects, with a particular emphasis on subconscious assessment criteria.

Kajsa Bråting
(Uppsala University, Sweden) is Professor of Mathematics Education. Her research concerns algebra learning, computational thinking and textbooks.

She leads a research project on the implementation of programming in Swedish school mathematics and is the first author of the article Exploring the intersection of algebraic and computational thinking (*Mathematical Thinking and Learning*, 2021).

Helen Brink
is a PhD student in Pedagogical Work at Karlstad University. She teaches in the teacher education program at that university. She is also the author of an educational book for technology education. Her research interest arises from her own teaching experiences concerning digitalisation.

Anne-Marie Cederqvist
is a Lecturer at Halmstad University, Sweden, and holds a PhD in Technology Education. Her research concerns teaching and learning in relation to technology and programming, and also Science and Technology Education in relation to Education for Sustainable Development (ESD), digital competence and the use of digital tools.

Marc J. de Vries
is Professor of Philosophy of Technology and Professor of Science and Technology Education. He is the Editor-in-Chief of the *International Journal of Technology and Design Education* (Springer) and author of, among other publications, the book *Teaching about Technology: An Introduction to the Philosophy of Technology for Non-philosophers* (Springer, 2016, 2nd ed.).

Susanne Engström
(KTH Royal Institute of Technology, Stockholm, Sweden) is an Associate Professor in Engineering/Technology Education. Her research focuses on educational, sociological and didactical aspects within technology teaching. Two of her latest articles are "From a teacher student's view – how STEM-actors have impact on teacher education and teaching in STEM" and "How physics courses can make highly valued strategies and dispositions visible to physics teacher students".

Jonas Hallström
(Department of Behavioural Sciences and Learning, Linköping University, Sweden) is Professor of Technology Education. His research concerns historical, philosophical, and sociological aspects of technology (education), and his most recent book is *Teaching and Learning about Technological Systems* on Springer (2022, co-edited with P. John Williams).

Fredrik Heintz
(Linköping University, Sweden) is Professor of Computer Science. He leads the Division of Artificial Intelligence and Integrated Computer Systems (AIICS) and the Reasoning and Learning lab (ReaL). His research focus is artificial intelligence, especially Trustworthy AI, autonomous systems, and the intersection between machine reasoning and machine learning. Furthermore, he is the Director of the Wallenberg AI and Transformative Technologies Education Development Program (WASP-ED), Director of the Graduate School of the Wallenberg AI, Autonomous Systems and Software Program (WASP), Coordinator of the TAILOR ICT-48 network developing the scientific foundations of Trustworthy AI, Vice President for AI research of Adra the AI, Data, and Robotics partnership, and President of the Swedish AI Society.

Helena Isaksson Persson
is a lecturer and doctoral student at the Department of Learning in Engineering Sciences, KTH Royal Institute of Technology, Sweden. Her research centers on technology education in both lower and upper secondary school contexts.

Cecilia Kilhamn
(University of Gothenburg, Sweden) is a researcher and editor at the National Center for Mathematics Education (NCM). Her research concerns the teaching and learning of school algebra and how algebraic thinking is related to other aspects of mathematics, such as computational thinking, generalized arithmetic and mathematical discourse.

Andreas Larsson
(Linköping University, Sweden) has a PhD in Technology Education and is currently working as a postdoctoral scholar. His research concerns technology teachers' knowledge and beliefs about their practice. Moreover, his research explores the role of conceptual metaphors in education.

Linda Mannila
is Adjunct Associate Professor in Computer Science Education at Linköping University, Sweden, and Associate Professor in Computer Science at the University of Helsinki, Finland. Her research interests include questions related to computational thinking, digital competence and AI literacy at K-9 level, both from a student, teacher and organisational perspective.

Jörgen Nissen
(Linköping University, Sweden) is Associate Professor of Pedagogic Practices. His research concerns digital technology, youth, and education. His most

recent publications are in the fields of Visual Analytics and AI related to education.

Charlotta Nordlöf
(Linköping University, Sweden) is a PhD in Technology Education. Her research interests concern different aspects of technology education such as teachers views and attitudes, knowledge in technology education and aspects of learning in technology education.

Arnold Pears
(KTH Royal Institute of Technology and Uppsala University) holds professorships as Professor of Engineering Education (KTH) and Professor of Computer Science specialising in Computing Education (Uppsala). His research focusses on the engineering and technical education experience, learner identity and participation and theory of engineering education research. His most recent book contributions include chapters on computing education research.

Anna Perez
(Linnaeus University, Sweden) is a PhD student in Technology Education. Her upcoming dissertation will focus on teacher education in the field of technology education, with particular emphasis on the educational aspects of technology involving programming content. The thesis is expected to be defended in the autumn of 2025.

Lennart Rolandsson
(Uppsala University, Sweden) is a Senior Lecturer of Technology Education. His research concerns primary and secondary school teachers' beliefs, intentions, and transpositions on integrating digital technology into the curriculum, e.g., programming, big data, artificial intelligence, and machine learning.

Maria Sparf
(Linköping University, Sweden) is a university lecturer and doctoral student in pedagogy. Her research interests are teaching and learning with digital tools, and learning in informal environments. Her licentiate thesis (2020) addresses programming education at Science Centres.

Linnéa Stenliden
(Linköping University, Sweden) is Associate Professor of Digital Education. Her research focuses on teaching and learning in relation to Digital Media and Artificial intelligence. She is, among others, director of the WASP-HS project: "AI Literacy in Swedish Teacher Education – A participatory design project"

and the sister project "AI Literacy for Swedish Primary Education: A Co-Design Approach" funded by the Swedish Research Council (SRC). Her most recent book is *Qualitative (digital) competence – a study of multimodality in secondary school* (Linköping University Electronic Press, 2023, co-edited with Anna Martín Bylund).

Karin Stolpe
(Linköping University, Sweden) is Senior Lecturer in Pedagogic Practices and PhD in Science Education. Her research interest is mainly focused on teachers' work in science and technology classrooms. For example, the research concerns teachers' use of models in the design process, and programming education at upper secondary level.

Maria Svensson
(University of Gothenburg, Sweden) is Associate Professor in Technology and Science Education. Her research focus is the field of technology education with a special interest in technological systems, systems thinking, sustainable development and technology teacher education. Her most recent book is *Science and technology teacher education in the Anthropocene – Addressing challenges in the North and South*, on Routledge (2022, co-edited with Miranda Rocksén, Elaosi Vhurumuku, Emmanuel Mushayikwa and Audrey Msimanga).

Peter Vinnervik
(Umeå University) is PhD in Pedagogical Work. His research concerns the digitalisation of the school system with focus on preconditions for and consequences of the introduction of programming in technology education.

CHAPTER 1

Introduction

Programming and Computational Thinking in Technology Education

Jonas Hallström

Abstract

In recent years, digital technology, programming, and computational thinking have been incorporated on a larger scale as curriculum components in technology education in many countries across the globe. Technology education research about programming and computational thinking is needed, and this edited book, *Programming and Computational Thinking in Technology Education: Swedish and International Perspectives*, contributes with new research that has many international applications and grounds for comparison, both in schools and in the research community. This introduction not only serves as a background for the reader as s/he delves into the various chapters of the book, but also introduces some pertinent and sometimes longstanding issues in technology education and its relation to computers and computing. These issues are: 1. Programming and computational thinking in school curricula: the early 21st century wave; 2. Computational thinking and literacy; 3. Teacher competence in programming; and 4. Computational thinking, programming, and learning in technology education.

Keywords

programming – computational thinking – computational literacy – teacher competence – technology education

•••

Computers and computer systems are arguably some of the most significant and iconic technologies to have emerged in the last hundred years. When the Technical Museum (*Tekniska museet*) in Stockholm ran its interactive exhibition "100 Innovations", adults and school students were given the opportunity to vote for what they believed were the 100 most important innovations of all time. Not surprisingly, the computer was selected as one of these, and so

© JONAS HALLSTRÖM, 2024 | DOI:10.1163/9789004687912_001

were related digital technologies such as computer games, robots, the Internet, the mobile phone, and the Global Positioning System, GPS (Tekniska Museet, 2023). This museum investigation was, of course, very subjective though statistically solid. It shows, however, that at least in Sweden—and probably in the whole Western world—people rate the importance of computers highly and think of themselves as affected by computers in their daily lives. School curricula have also to varying degrees included computers, computing, programming, and similar areas for decades, even though it is only recently that there has been a concerted movement motivated by the resurgence of the concept of computational thinking, CT (Denning & Tedre, 2019; Williams, 2018; Wing, 2006).

In spite of the ubiquity of computers and computing in society, they have been rather slow in coming in technology education, at least as more comprehensive curriculum components. The reasons for this may be the historical roots of technology education in crafts, sloyd and vocational subjects in many countries (e.g., Hallström et al., 2014) and the ensuant focus on designing and making of physical products (Hallström in this volume). It is also conceivable that it has to do with the academic division between engineering and computer science (CS) since in many countries technology education has at least partly been modelled on engineering in a broad sense, not computer science (de Vries et al., 2016).

Whatever the reasons for the slow pace of including computing in technology education, in this edited volume we explore historical, philosophical, educational, and pedagogical perspectives on programming and computational thinking in technology education, particularly from a Swedish viewpoint. Sweden followed suit in the early 21st century push for more computing in schools. Swedish—and Nordic—research about computational thinking and programming in general has thereby been extensive, although in technology education it was especially after the compulsory school curriculum reform in 2018 when programming was introduced that research began to be carried out (cf. Hallström, 2018; Mannila, 2023; Nouri et al., 2020). Technology education research is therefore needed, and this volume contributes research perspectives based on Swedish examples which also have many international applications and grounds for comparison, both in schools and in the research community.

The book was written by a team of internationally renowned technology education and computer education researchers, mainly based in Sweden. In the first part, chapters are more general in nature and provide historical, philosophical, and international perspectives on the introduction of computers, programming and computational thinking in schools. In the second and third parts chapters are mainly based on Swedish empirical data, from technology education contexts but also in a couple of cases mathematics because of the close relationship between these two subjects as it concerns computing in school. The problematics that are addressed throughout the book are, however, of a general nature, common to national and international contexts that deal with computing and programming in schools, particularly in technology education, STEM education and related subjects or curricular areas. As such, the book can therefore also be utilized as a resource for research about programming and computational thinking in education more broadly, across countries and school systems. Furthermore, there is a wealth of teaching examples in the chapters of the book, so it can also be used as a resource for teachers and teacher educators when teaching about programming and computational thinking.

This introduction not only serves as a background for the reader as s/he delves into the various chapters of the book, but also introduces some pertinent and sometimes longstanding issues in technology education and its relation to computers and computing. These issues are: 1. Programming and computational thinking in school curricula: the early 21st century wave; 2. Computational thinking and literacy; 3. Teacher competence in programming; and 4. Computational thinking, programming, and learning in technology education.

1 Programming and Computational Thinking in School Curricula: The Early 21st Century Wave

The innovation of automated systems and computers in the first half of the 20th century was groundbreaking and led to demands for improved education by including computing (Denning & Tedre, 2019; Mindell, 2002). Computing and programming in schools thus have a long history, dating back to the 1960s. Although the first part of this book deals with some historical examples of early computer and programming initiatives (see chapters by Nissen & Stenliden, and Rolandsson, Kilhamn & Bråting), the book mainly deals with the latest "wave" of computing in schools. This 21st century wave of programming in schools owes a great deal to Jeannette Wing's influential paper in which she introduced *computational thinking* as a central skill set for everyone to learn, not just computer scientists (Wing, 2006). Despite the fact that the CT concept was not new but was suggested by, for instance, Seymour Papert a decade earlier (Papert, 1996) and can be traced back at least to 1980 (Denning & Tedre, 2019), it really caught on and inspired renewed debate, curriculum innovation and research during the first two decades of the 21st century. Thus, it can be considered an early 21st century iteration or wave of computing in the schools.

When CT was subsequently introduced in schools on a larger scale, a major way of doing this was through programming. In the last decade or so, programming has thus been introduced in school curricula and standards all over the world. In countries such as the UK a new school subject—computing—was created to harbour this partly new subject content, whereas in countries such as Sweden programming was included in existing subjects, notably mathematics and technology education. This inclusion was cheered on by politicians, teachers, and software companies, but it also met with resistance and criticism (cf. Raptopoulou, 2021). Following a long curricular debate, in 2017–2018 the Swedish National Agency for Education (Skolverket) thus updated the syllabuses in technology and mathematics for compulsory school to include a programming component (programming is also mentioned in the social science syllabus). Computing and digital components have also subsequently been strengthened in the curriculum for upper secondary school.

The book addresses the introduction of computers, computing and programming in particularly Swedish but also international curricula and schools, primarily in the early 21st century wave. It is a complex introduction that is interwoven with the general development of society and various political ambitions and arguments. The goal was often for computerized schools to contribute to an industrialized and competitive nation, through the development of digital future citizens. Many actors had an interest in introducing computers and programming in schools, with varying and sometimes failed results, at least if we look at the late 20th century developments (e.g., Machado-Trujillo, 2021; chapter by Nissen & Stenliden).

The introduction of computers, programming, CT, and information and communication technologies (ICT) in schools can be said to often, but not always, have been "hope-based", that is, introduced with little if any prior testing in schools. This was also a specific Swedish way of carrying out school reforms and has perhaps been especially evident when it comes to the 2018 introduction of programming (Raptopoulou, 2021; Mannila & Heintz, and Vinnervik in this volume). The actual impact of computers, programming, and ICT on educational outcomes has, however, proven to be complex, heterogeneous, and dependent on a range of factors such as school subject, learning objective, type of computer technology, and the uses of that technology (e.g., Fernández-Gutiérrez et al., 2020; Saqr et al., 2023). The introduction of programming in Swedish schools is still ongoing and it remains to be seen how well it will be implemented.

2 Computational Thinking and Literacy

Jeannette Wing's (2006) re-introduction and outline of the concept of computational thinking might hide the fact that it is really a contested notion with many different, and often conflicting, definitions. There are more generic perspectives according to which CT is defined in alignment with a more general skill set related to computer science but also so-called 21st century skills such as problem solving and critical thinking skills (e.g. Voogt et al., 2015). Programming plays a role here, although it is more of a background activity compared to acquiring generic skills which could for all purposes be learned in many different situations, not just by a physical computer.

There are, on the other hand, perspectives on CT that strongly advocate computer programming. For example, Grover and Pea (2013) wrote that "programming is not only a fundamental skill of CS and a key tool for supporting the cognitive tasks involved in CT but a demonstration of computational competencies as well" (p. 40). Denning and Tedre (2019) seem to support both generic and concrete approaches in their definition, although for them CT is largely a mental process in a science and engineering context:

> Computational thinking is the mental skills and practices for:
> – *designing* computations that get computers to do jobs for us, and
> – *explaining* and interpreting the world as a complex of information processes (p. 4).

There have also been competing concepts such as diSessa's (2000) *computational literacy*, which resonates more fully with the literacy discussions in particularly the STEM subjects in the last decades (cf. Dakers, 2006; Stolpe & Hallström, this volume). Computing has a strong connection to STEM—science, technology, engineering, and mathematics—and has been conceived of variously as a science, a branch of mathematics and/or a branch of technology and engineering (De Mol & Primiero, 2014).

In response to the focus of this book, what is really needed is a conception of computational thinking and computational literacy adapted for technology education. In this volume, Isaksson Persson & Pears go some way toward developing such a literacy framework. Hallström also attempts to forge a conception of what technology is that does not foreground either concrete, abstract, analog, or digital components of technology but instead see them as a unified whole (cf. Grigoriadis et al., 2022). Thus, computational thinking and literacy could be seen as part of a broader technological literacy, divided into several technological sub-literacies or multiliteracies including CT (Hallström, 2023).

As regards the definition of computational thinking, in this book we will not use just one single definition of computational thinking, but the chapter authors employ definitions appropriate for the particular contexts of their chapters.

3 Teacher Competence in Programming

As several chapters in this book show, technology teachers are largely unprepared for the introduction of programming that is now being implemented. There is generally a lack of competence among teachers, but there is also a corresponding insufficiency of in-service training in programming (see chapters by Engström & Björkholm, and Vinnervik). Existing in-service courses in programming, for example, include the programming language Python which has been considered difficult by participating novice teachers. One result of this lack of programming experience and competence among teachers is that they need to get help from outside of the immediate school context, for example, from science centers which have more competence in providing programming teaching. This is lucidly shown by Sparf, and Engström and Björkholm, in their chapters.

4 Computational Thinking, Programming, and Learning in Technology Education

The introduction of programming in technology education is particularly interesting because it presupposes a particular relationship between programming and technology. Programming is usually performed with technological artefacts—various types of computers—but it can also be used to control other artefacts, for example, sensors and other computers in control systems (Brey & Hartz Søraker, 2009). The context of technology education is therefore in many ways ideal for CT and programming because of the focus on technological artefacts and systems.

This means that programming can be made authentic and related to real-world uses, for example, in engineering and STEM projects which involve programmed solutions. Contemporary software such as Scratch or BBC Micro:bit are also good implements for programming control units in smaller systems, which are prominent in certain technology curricula (cf. Norström, 2022). In this volume, both Perez and Svensson, and Cederqvist address so-called programmed technological solutions (PTS) in which programming is employed to control artefacts or systems. Berg and Axell also address applications of programming in the primary technology classroom, by studying students' views of the human—machine relationship when working with programmable robots.

Furthermore, models and modelling feature more or less explicitly in many technology curricula. In her chapter, Brink expounds on teachers' conceptions of programming as a form of modelling. Larsson also points to the fact that knowledge of how gesture and metaphor play out in programming education could improve learning. Finally, Björklund and Nordlöf bring up the important issue of assessment of programming skills, which has not gained much attention in technology education so far despite its importance in computer science education (e.g., Tang et al., 2020; Weintrop et al., 2021). Björklund and Nordlöf provide a typology for formative assessment of programming skills.

The studies presented in the chapters of this book will hopefully contribute to a better implementation of programming and computational thinking in technology education, STEM education and across a broad array of other school subjects, in Sweden and elsewhere. This is all the more pertinent given the recent advances in generative artificial intelligence (AI). To equip today's young generation with computational thinking skills could thus contribute to their developing even more dynamic and advanced technological multiliteracies that would include even AI literacy (cf. Hallström, 2023; Park & Kwon, 2023).

References

Brey, P., & Hartz Søraker, J. (2009). Philosophy of computing and information technology. In A. Meijers (Ed.), *Philosophy of technology and engineering sciences* (Vol. 9). Elsevier.

Dakers, J. R. (Ed.). (2006). *Defining technological literacy: Towards an epistemological framework* (1st ed.). Palgrave Macmillan.

De Mol, L., & Primiero, G. (2014). Facing computing as technique: Towards a history and philosophy of computing. *Philosophy & Technology*, *27*(3), 321–326.

De Vries, M. J., Gumaelius, L., & Skogh, I.-B. (2016). Pre-university engineering education: An introduction. In M. J. De Vries, L. Gumaelius, & I.-B. Skogh (Eds.), *Pre-university engineering education* (pp. 1–12). Brill Sense.

Denning, P. J., & Tedre, M. (2019). *Computational thinking*. The MIT Press Essential Knowledge Series. The MIT Press.

diSessa, A. A. (2000). *Changing minds: Computers, learning, and literacy*. The MIT Press.

Fernández-Gutiérrez, M., Gimenez, G., & Calero, J. (2020). Is the use of ICT in education leading to higher student outcomes? Analysis from the Spanish Autonomous Communities. *Computers & Education, 157*, 103969.

Grigoriadis, K., Poulsgaard, K. S., & Hua, H. (2022). Digital and physical materiality in contemporary architectural design: A material engagement approach. *Frontiers of Architectural Research, 11*, 591–593.

Grover, S., & Pea, R. (2013). Computational thinking in K–12: A Review of the State of the field. *Educational Researcher, 42*(1), 38–43.

Hallström, J. (2018). Ett forskningsfält i tillväxt: Teman i svensk teknikdidaktisk forskning. In K. Stolpe, G. Höst, & J. Hallström (Eds.), *Teknikdidaktisk forskning för lärare. Bidrag från en forskningsmiljö. Naturvetenskapernas och teknikens didaktik No. 2.* NATDID, The Swedish National Centre for Science and Technology Education.

Hallström, J. (2023). The philosophical and political value of technology education: Fostering technological multiliteracies. In D. Gill, D. Irving-Bell, M. McLain, & D. Wooff (Eds.), *The Bloomsbury handbook of technology education*. Bloomsbury Academic.

Hallström, J., Hultén, M., & Lövheim, D. (2014). The study of technology as a field of knowledge in general education: Historical insights and methodological considerations from a Swedish case study, 1842–2010. *International Journal of Technology and Design Education, 24*, 121–139.

Machado-Trujillo, C. (2021). *Education and historical evolution of information and communication technologies: Background, international influences and their development in Spain in the 1980s* (Diss). Universidad de la Laguna, Tenerife.

Mannila, L. (2023). Integrating programming in other subjects at primary level: Tool, glue or ideation. In T. Keane & A.E. Fluck (Eds.), *Teaching coding in K-12 schools: Research and application* (pp. 127–144). Springer International Publishing.

Mindell, D. A. (2002). *Between human and machine: Feedback, control, and computing before cybernetics.* The Johns Hopkins University Press.

Norström, P. (2022). Technological systems in national standards and curricula. In J. Hallström & P. J. Williams (Eds.), *Teaching and learning about technological systems: Philosophical, curriculum and classroom perspectives* (pp. 35–51). Springer.

Nouri, J., Zhang, L., Mannila, L., & Norén, E. (2020). Development of computational thinking, digital competence and 21st century skills when learning programming in K-9. *Education Inquiry, 11*(1), 1–17. https://doi.org/10.1080/20004508.2019.1627844

Papert, S. (1996). An exploration in the space of mathematics educations. *International Journal of Computers for Mathematical Learning, 1*(1), 95–123.

Park, W., & Kwon, H. (2023). Bringing computational thinking to technology education classrooms: Hacking car activity for middle schools in the Republic of Korea. *International Journal of Technology and Design Education, 33*, 863–881. https://doi.org/10.1007/s10798–022-09750-5

Raptopoulou, A. (2021). *Politics of contemporary education policy: The case of programming in the Swedish Curriculum* (Diss.). Department of Education, Stockholm University.

Saqr, M., López-Pernas, S., Helske, S., & Hrastinski, S. (2023). The longitudinal association between engagement and achievement varies by time, students' subgroups, and achievement state: A full program study. *Computers & Education*, 104787. https://doi.org/10.1016/j.compedu.2023.104787

Tang, X., Yin, Y., Lin, Q., Hadad, R., & Zhai, X. (2020). Assessing computational thinking: A systematic review of empirical studies. *Computers & Education*, *148*, 103798. https://doi.org/10.1016/j.compedu.2019.103798

Tekniska Museet. (2023). *100 innovationer*. https://www.tekniskamuseet.se/lar-dig-mer/100-innovationer/

Voogt, J., Fisser, P., Good, J., Mishra, P., & Yadav, A. (2015). Computational thinking in compulsory education: Towards an agenda for research and practice. *Education and Information Technologies*, *20*, 715–728. https://doi.org/10.1007/s10639-015-9412-6

Weintrop, D., Wise Rutstein, D., Bienkowski, M., & McGee, S. (2021). Assessing computational thinking: An overview of the field. *Computer Science Education*, *31*(2), 113–116.

Williams, P. J. (2018). Technology education: history of research. In M. J. de Vries (Ed.), *Handbook of technology education* (pp. 85–99). Springer.

Wing, J. M. (2006). Computational thinking. *Communications of the ACM*, *49*(3), 33–35.

PART 1

Definition, Philosophy and History of Programming and Computational Thinking, in Relation to Technology Education

CHAPTER 2

How Computers Entered Swedish Classrooms

The Importance of Educating Digital Citizens

Jörgen Nissen and Linnéa Stenliden

Abstract

The current effort to establish computational thinking in compulsory school is not a new agenda. In Sweden, the history starts in the late 1960s with the introduction of computer technology in schools. This chapter explores how the arguments for this endeavour have varied over time and what preceded the initiation of the concepts of digital competence and programming in the subject of Technology and the Swedish national curriculum of 2017.

The foundation for this is, on one hand, a scrutiny of four large-scale national campaigns and, on the other hand a review of the Swedish curricula from 1969, 1980, 1994 and 2011. It was found that an underlying political agenda has tightly connected the introduction of "new" technology in schools with the competitiveness of Sweden as an industrial nation.

Keywords

computational thinking – digital competence – history – programming – Sweden – school

1 Introduction

Dear Bill,

Apart from testing this connection on the global Internet system, I want to congratulate you on your decision to end the trade embargo on Vietnam. I am planning to go to Vietnam in April and will certainly use the occasion to take up the question of the MIA*s. From the Swedish side we have tried to be helpful on this issue in the past, and we will continue to use the contacts we might have.*

© JÖRGEN NISSEN AND LINNÉA STENLIDEN, 2024 | DOI:10.1163/9789004687912_002

Sweden is—as you know—one of the leading countries in the world in the field of telecommunications, and it is only appropriate that we should be among the first to use the Internet also for political contacts and communications around the globe.

Yours,
CARL

This message was transmitted on 5 February 1994, from the Swedish Prime Minister Carl Bildt to the US President Bill Clinton (The American Presidency Project, 1994). It has been claimed as the first email exchange between two statesmen. In it, the Swedish Prime Minister confirms his support for the USA, at the same time as he positions Sweden as *a nation at the forefront of communication technologies and as a future-oriented digital society*.

At that time, we were seeing the beginnings of "the world wide web", and electronic communication would soon become increasingly common to many. In 1994, the general interest in Information technologies (IT) accelerated and often that year is regarded as the beginning of the systematic digitalisation of Swedish schools (Riis et al., 2000, p. 1). Computers in school were viewed as important for a society increasingly permeated by this technology, and the development of schools became part of the vision to position Sweden as a digitalised, leading knowledge nation characterised by high-quality education and lifelong learning (Swedish Government Official Report [SGOR], 1994). This was all in line with the Prime Minister's statement in the email above, which declared Sweden to be a world-leading digital nation.

However, in this text we will show that the endeavour to introduce computers into Swedish schools had started much earlier. The aim is to illustrate how computers entered Swedish classrooms through a description of four large-scale national campaigns, some other projects, and changes in Swedish curricula and syllabuses between 1969, 1980, 1994, and 2011.[1] Altogether, this is an attempt to provide a historical background to the renewed syllabus of 2017, where *digital competence* and *programming* were reinforced. The National Agency for Education (NAE) [Skolverket] argued that these concepts covered the same meaning as *computational thinking*, and therefore this concept was left out of the governing document (NAE, 2016, p. 7).

We conclude this chapter by discussing the processes of introducing computers into schools and the efforts to develop citizens' digital competence through the concepts of *pull* and *push* factors (Schoen, 1967). These concepts contribute to looking at progress from a demand perspective (opportunities pulling from people's needs and the market) and a supply perspective (technological opportunities pushing forward from scientific discoveries and various other forces). In this case, has the technology and educational content been *pulled* into schools by students, teachers, and school principals, or has it been *pushed* there in the interests of others?

1.1 *Technological Progress, Politics, and Education*

The process of introducing computer technology into Swedish schools has been ongoing for more than 50 years; as mentioned above, it began in the late 1960s and 1970s. In the Swedish context, technological progress, politics, and education have all been salient features that have contributed to digital competence and programming now being part of the compulsory school curriculum.

During the first 25 post-war years, Sweden had an exceptionally long and cohesive period governed by the Social Democratic Party. Their policy was coloured by a positive attitude to technological development. Sweden as a nation had also flourished in terms of technological advancement in the years following the end of World War II, not only owing to a hegemonic attitude towards technology among both politicians and the population in general, but also thanks to the facts that: the country had stayed out of the two world wars, and it was rich in resources like iron ore and wood products. However, during the 1970s, something happened; the well-established trust in increasing prosperity connected to technological advancement was no longer a matter of course. For example, the country became divided in attitudes towards nuclear power technology, and in 1980 there was a referendum (which is rare in Sweden) about its future. Computer technology, being relatively new, also came to be met with public mistrust relating to automation, rationalisation, unemployment, and risks regarding personal integrity (Johansson, 1997).

At this time, the educational system was suggested as an instrument in attempts to both counteract this mistrust and, at the same time, promote the use and development of computer technology, IT/ICT, the Internet, and numerous other applications and devices. Two events, initiated by the Social Democratic Party in 1984 and the Conservative government ten years later, illuminate this. In January 1984, at a political seminar, the journalist leading the debate pressed the Deputy Prime Minister, Ingvar Carlsson—who was also "Future Minister"—on how the government would handle the development of computer technology. Confronted several times by this journalist, he finally said: "Well, now, you must accept that our strategy is to attack the issues of education first, because it is basic and we have prioritised it /…/ we will certainly manage the young generation, upper secondary school will manage it"

(National Board of Education [NBE], 1985, p. 21) [Skolöverstyrelsen].[2] Thus, Ingvar Carlsson connected the "handling" of computer technology directly with education and schools, and worried about its use and consequences.

These aspects were also highlighted ten years later when the Conservative Prime Minister, Carl Bildt, followed up the email exchange with President Bill Clinton by profiling himself, and his government, as enthusiastic about IT. Carl Bildt also launched an IT commission in 1994, to realise his vision of Sweden as a digitalised country. This resulted in the Swedish Government Official Report: *Wings to human ability* (1994). One of the prominent promotors of such a society, which was emphasised in the report, was the introduction of IT in schools. This was supposed to contribute not only to the development of new, effective teaching methods, both within schools and for distance education, but also to citizens' digital abilities and lifelong learning. If school and teaching made a joint effort to position this as a *primary task*, it was presumed that Sweden would become a prominent nation of IT users, the report enthusiastically claimed. The line of thought was that the development and application of IT concerns everyone: "Swedish competence, know-how, and information environments must be attractive and dynamic in international terms" (SGOR, 1994, p. 7), and that, by introducing IT, the school system would build not only young people's knowledge, but also their habits and digital behaviours. The argument was that new ways of utilising the knowledge that IT ultimately constituted must grow out of—and be disseminated by—school environments and their working methods.

With a ten-year interval, these two ministers articulated the importance of computers in society and declared strong beliefs in education. It is likely that their political views affected how computer technology and related content was brought into the Swedish compulsory school. Hence, the next section thoroughly examines the process employed to "computerise" the Swedish school.

2 National Reforms and Campaigns to Enhance Computer Knowledge in School, 1969–2001

The first national campaign to equip compulsory schools with computers took place during the mid-1980s. However, the first activities started back in the 1960s. Thereafter, during the period 1984–2001, four crucial national campaigns were launched and evaluated. Other national efforts have also been carried out, but we judge these four campaigns to be decisive.[3] Changes in national curricula and the efforts made between 1969 and 2001 are described in the following.

2.1 *The First Steps: 1969–1983*

The first parliamentary proposition was presented at the end of the 1960s. In the national curriculum of 1969, established that year and in force from 1970, computer-related content was mentioned in three subjects: Mathematics (orientation about computers), Technology (data processing), and Art (ICT and the interpretation of images) (NBE, 1969). In 1971, the NBE was given an assignment by the government to initiate experimental activities related to "teaching computer technology" (Riis, 1991, p. 19). This project was labelled "The computer in school" [Datorn I Skolan, DIS] (NBE, 1980a). Other limited (in relation to what would follow) initiatives were PRODIS (Software and Computer Equipment in School) and PRINCESS (Project for Research on Interactive Computer-based Education Systems). The latter was co-financed with the National Swedish Board for Technical Development [Styrelsen för teknisk utveckling, STU]. These projects were carried out at the Royal Institute of Technology in cooperation with the NBE (Riis, 1991).

As a result of these activities, in 1980 the NBE adopted its first action plan concerning "the computer in schools", where three school activities were differentiated:

- Teaching about computers and the use of computers in society—"datalära"
- Using computers in schools with the purpose of modernising teaching, for example as a tool for arithmetic—subject-related computer use
- Computers as learning aid—computer-supported teaching (NBE, 1980a; Riis, 1991, pp. 19–20).

The Swedish expression "datalära" is a play on words. In Swedish, "data-", comes from "data machine" and "-lära" can be teaching, learning, or a branch of knowledge. As a comparison, the expression "datalära" was also used early on in Denmark, with inspiration from Peter Naur, who coined "datalogi" in a protest against "computer science", which he thought wrongly emphasised the computer. "Datalogi" was the creed of data, and that was a matter of human understanding, not the machine. When experimental activities were conducted in Danish compulsory schools during the 1970s and 1980s, the Ministry of Education chose the label "datalära", which was thought of as even broader than "datalogi" (Caeli & Bundsgaard, 2019, 2020).

In the Swedish action plan, it was suggested that, in compulsory school:

1. teaching about computers and their use in society should be part of social science and mathematics and should give students adequate knowledge such that they should:
    a. be *able to encounter and affect* the use of computers in society
    b. be *willing and dare* to exercise such influence

2. students would be able to use computers in their future working life
3. "datalära" should be conducted without computers (Riis, 1991, 2019).

Altogether, this was the position of the NBE in 1980. In the same year, a new national curriculum was published, according to which compulsory schools were required to implement from 1982 (NBE, 1980b). In this, "datalära" was previously an area of knowledge in Mathematics, but our analysis of the syllabuses shows that more content was included than in the earlier curriculum from 1969 (such as: orientation about the use of computers, that the computer is a tool governed by humankind, computer programming tasks, and methods of problem solving). The topic was also introduced into two more subjects: Natural Sciences, now including Technology (the computer, its development and consequences for humankind and society), and Social Sciences (computers and information and data processing, computer law, computerisation and its consequences for the individual and society, in particular for working conditions and employment). Art was no longer a subject that addressed ICT and images.

Note that computer-related teaching was still expected to be conducted *without* computers. Despite this fact, a national effort to produce a Swedish school computer was soon introduced (see below). At this time, the NBE saw no way of redistributing resources for the purchase of hardware, nor any way to impel the municipalities to shoulder such costs (Riis, 2019).

2.1.1 The "Mate" Computer, 1981

At this time, another initiative, well worth relating, was the government-run technical project to develop a Swedish school computer: Compis (short for "Computer in school"; however, "compis" also means "mate" in Swedish). The project was initiated in 1981 by the National Board of Technical Development as an assisting industrial development enterprise, even though it was a relatively small project (Kaiserfeld, 1996). The Board often successfully dealt with larger project, as for example manufacturing of fighter aircrafts or electronics. Both the Ministry of Industry and the Ministry of Education were involved in setting up the development of the Compis computer, and the initiative was related to the NBE's action plan involving "Datalära" and PRINCESS (see above).

The "mate" computer has been considered an "unequivocal failure", at least in respect to schools (Karlsohn, 2018, p. 24). An educational aid company, Esselte, won the bid to produce the computer and four municipalities signed up to test it in 15 upper secondary schools in 1983. In the procurement made within the framework of the first national campaign, directed towards ages 13–15 in compulsory school, Compis was one of six approved computers, at the same time as a PC from IBM was rejected (Kaiserfeld, 1996, p. 274). But, due to technical problems and rapid development within personal computer technology, in reality Compis was abandoned by schools in Sweden.

2.2 *"Datalära" in Compulsory School, 1984–1987*

In March 1981, with only a couple of months until the next election, the Conservative government presented the bill Coordinated Computer Policy [Samordnad datapolitik] (Swedish Government, 1982). It singled out five political areas for special action, one being education. It was stated that all students should "be oriented about the use of computers in society and the rapid digital development" (NBE, 1980a, p. 107) and that they should develop "insights that computers are technical aids controlled by humans" (Swedish Government, 1982, Appendix 3, p. 2). This was nicely in line with the new curriculum of 1980. Nothing was mentioned about whether or not computers were to be used in compulsory school, but it was explicitly stated that computers must be procured for upper secondary schools.

In the autumn of 1983, the new Social Democratic government had recently come to power. The new government surprised the NBE with a three-year assignment to support municipalities to obtain computers, combined with an economic subsidy targeted at the municipalities for the purchase of hardware. From this point onwards, the idea of "datalära" without computers was abandoned. In 1984, the NBE released two documents: An orientation document, *Education for the Computer Society* [Utbildningen inför datasamhället] (NBE, 1984a), and a study plan, *"Datalära" in Compulsory School* [Datalära i grundskolan] (NBE, 1984b).

As an outcome of this assignment, the first of the four national campaigns, "Datalära" in compulsory school, was introduced in 1984. This campaign was directed towards 13–15-year-olds in compulsory school. It was planned for three years duration and represented the first national effort to introduce hardware into general and mandatory education. The main components were:

- Every pupil should have 80 hours of education in "datalära"
- Every school should formulate a plan for these 80 hours and its distribution across the subjects of Mathematics, Natural Sciences (including Technology), and Social Sciences. The hours for "free activities" could also be used for this teaching, even though this was not encouraged
- At least one teacher per school should be given a 10-week course as preparation for teaching "datalära"
- The municipality should finance the computers to at least the same extent as the government, to a maximum of 60,000 SEK per school (the total sum of 120,000 SEK per school was sufficient for eight computers)
- Only computers certified by the NBE could be financed in this way (Riis, 2019).

The main conclusions from an evaluation of this campaign raised issues about the 80 hours and their content:

- Pupils seldom received the stipulated 80 hours of "datalära"; rather, they only received 8 or perhaps 16 hours. This teaching was mostly conducted within hours stipulated for selectable content by students, "free activities". Consequently, the students' options for choosing content became significantly narrowed.
- The main content was related to Natural Sciences and mostly consisted of simple, not advanced, programming exercises (Riis, 2019).

To sum up, the main results from the evaluation of "Datalära" were that the key content and the 80 hours of teaching were not achieved, although the schools were equipped with computers. Nor were stipulated content concerning contemporary issues (for example, personal integrity or unemployment due to automation) that coloured the political debate, part of the actual teaching (Riis, 1991).

2.3 *The Computer as an Educational Aid, 1988–1991*

In 1988, the Swedish parliament decided to conduct a second national campaign, targeting both compulsory schools and upper secondary schools: "The computer as an educational aid".

The background to the launching of this campaign was that "information technology affects all parts of our society" (Riis, 1991, p. 10), which leads to demands for new knowledge and increased competence for at least two reasons: (a) providing a qualified workforce for industry and business, and (b) civic influence (democratic control) should be based upon knowledge. Concerning education in schools both *about* and *with* computer technology, the minister noted that: technology has become a necessary aspect of vocational education, since it is a prominent part of working life; it could also be a tool for pupils with special needs; and it could be a tool for individualisation, but it should not be used for routine drill exercises (or programmed instruction) (Riis, 1991, pp. 10–11).

So, this time the focus was much more on computers as a pedagogical tool. The campaign was carried out at three levels: experimental activities at *a local level* (in schools) and development work *at a regional* and *a national level*. The NBE was responsible for initiating activities at all three levels. At the national level, the NBE also performed the development work. In the end, 160 local and 11 regional projects were undertaken. At the NBE, 15 "advanced" educational programmes were developed.

In their final report, the evaluators were rather critical, making the following points:

- The most common use of computers is skills training and other rather simple drill exercises, the advantage being individualised feedback and one more pedagogical tool (as variation)
- Advanced pedagogical software developed at the national level is so far *not* being used by ordinary teachers
- For students with special needs, the positive influence of the new tool is invaluable and deemed difficult to overestimate
- The largest increase in computer use during the campaign was word-processing in Swedish and foreign-language teaching
- In most of the projects, the teachers worked with the same content as previously. The method of conducting the teaching was also quite similar, despite the computers. Changes that were found were mostly within Mathematics and Social Sciences (Riis, 1991, pp. 75–77).

This campaign was seen as a direct continuation of the previous one. The next campaign was to be given a different character in several ways.

2.4 *The Lighthouse Project, 1996–2000*

The political situation was different when the third national campaign was put into motion. Previous computer technology had featured in the public debate to a high degree in relation to concerns about risks, unemployment due to automation, rationalising, and issues of threatened integrity. The Prime Minister since 1991, Carl Bildt, represented a Conservative government. He profiled himself as viewing IT as a crucial issue for the nation, and he sent the positioning email to Bill Clinton in 1994. As Prime Minister, he himself chaired an IT commission established in 1994. Several ministers were also members of that commission (SGOR, 1994, p. 5). Such a composition for a commission, including the Prime Minister and several other ministers, was unique for Sweden. The commission considered IT as a means to improve the quality of life, and primarily as a way to increase the nation's international competitiveness (SGOR, 1994).

Simultaneously, in 1991, the Swedish National Agency for Education (NAE) [Skolverket] replaced the NBE, and a new compendious curriculum for compulsory school was launched in 1994 (NAE, 1994). In 1996, this was supplemented with subject syllabuses and grade criteria (NAE, 1996b). According to our analysis, content related to computers was still given the most attention in Mathematics (related to mini calculators and the computing capacity of computers). But now this content was also mentioned in the following subjects: Swedish, English and Modern Languages, usually French or German (dictionaries, word processing, information searches, and communication). It was also touched upon in History (databases), Crafts (as computer support), Music (media technology), Art (computer-generated images), and Technology (electronic information processing). In any case, the relatively sparse content (note that there was almost nothing mentioned in Technology or subjects within Natural and Social Sciences) compared to previous curricula, was counteracted by clear instructions concerning the principal's responsibilities. This duty extended across "access to computers so that the students themselves can seek and develop knowledge" and "the students should be able to orient themselves within a complex reality, with a large flow of information and a rapid rate of change" (NAE, 1994, pp. 5, 17).

The NAE emphasised that changed living conditions were affecting schools:

> The function of school has changed from simply communicating information, to teaching students how to locate information and knowledge. What is needed today, more than ever, is the ability to sort out, interpret, and use information. Changes in living conditions for children and young people mean that schools must now deal with issues and information outside their control. It is important to accept that young people have knowledge gained outside the school and then use it. (NAE, 1996a, p. 4)

The great importance of students' relation to information and knowledge can also be linked to another aspect of the new curriculum: the emphasis on the active learner. This is explicitly spelled out; for example, that the teacher shall "strengthen each pupil's desire to learn", and "organise and carry out the work so that each pupil develops according to his or her capacity". The active learner was to become a significant element in the rhetoric of the third campaign, which was *en route* (Jedeskog, 2005, p. 16).

The Bildt government set up and financed a special foundation, the Foundation for the Development of Knowledge, Competence, and Skills—and thereby made it independent of political control, unlike the NBE and NAE, which were government agencies. Although independent in terms of *how* to do it, the foundation's statutes, established by the government, clearly stated that one of its primary missions was to "spread the use of IT in schools". The effort to carry this out was embodied in the Lighthouse Project, also called the "IT billion" for school purposes (Karlsohn, 2018, p. 8). The Foundation's financing of its mission toward schools became huge: SEK 1 billion, equivalent to around EUR 100 million.

Unlike the earlier ones, this third campaign did not target all schools. The strategy was to provide a few schools with well-financed conditions to develop good examples, which could later inspire others, so-called lighthouse projects.

It was directed towards education from preschool to grade 12. The two previous campaigns, and some minor ventures, were financed by the government, costing a total of 220 million SEK (Riis, 1991, p. 19).

The main part of the Foundation's "IT billion for schools" was 27 lighthouse projects. These were financed with SEK 336 million from the Foundation and SEK 495 million from participating municipalities, a total of SEK 831 million (Nissen et al., 2002, p. 194). Other parts were 65 smaller school projects (Grensjö, 1999), the development of teaching materials, and an extensive information effort. In addition to the campaign being large, even by international comparison, by not primarily directing its efforts towards all schools, but instead investing in just a few units, the model meant additional concentrations of funds:

> What perhaps first strikes a viewer is the concentration of development funds. Never before have such large sums been invested in school development in one single area. Neither have such large sums been invested in so few units. (Riis & Jedeskog, 1997, p. 28)

The projects differed in character from each other, but roughly they had the same financial conditions regardless of factors such as the number of schools, teachers, or students encompassed. Some covered entire municipalities, while a few projects concerned only one school. Early on, the latter became "one-to-one" schools where each student was equipped with a computer. One of those schools was for students aged 13–15. Another was a brand-new secondary school, with the ambition to combine problem-based learning and one-to-one equipment throughout the teaching.

Enormous expectations were attached to the overall school investment. From an efficiency point of view, students' time for learning and teachers' time for work with teaching, planning, and follow-up was expected to be improved (Riis et al., 1997, p. VI, 70). At the same time, the objectives of the various projects were very imprecise and vague (Riis et al., 2000, p. 101). Yet, in most cases, but not all, an overall picture was presented that the central conditions in the school would be revolutionised:

> In the descriptions of the different projects, it appears that IT's strength is considered to lie in the promise of solving most of the school's problems; teaching becomes more fun, truer, more effective and just. (Riis et al., 1997, pp. 24–25)

Like the previous campaign, this one was extended by one year. When evaluators summed up the efforts made, they found that the effects of the work did not match the high ambitions or widely touted goals. It was stated that "it is clear that IT is not a 'magic wand' that can be waved over a school to evoke development" (Nissen et al., 2002, p. 12). The conclusions were that the projects had promoted the use of IT in schools, but in a more general and overall way than through convincing examples (lighthouses) (Nissen et al., 2002, p. 193). A broad acceptance of the technology could be seen. Its use was commonly connected to fairly general areas, such as writing, searching for information via the internet, and communication via email and chat. In other words, IT emerged as a versatile tool that was being used in a relatively singular way. However, many of the participating teachers made important contributions (Riis & Jedeskog, 1997).

Other changes relating to the entire school system, such as the new curriculum and a new agreement between trade unions and employers concerning teachers' working conditions, probably affected the project schools to at least the same extent as the lighthouse project. Moreover, the project's investments also meant that attention was directed towards IT and schools by the media, politics, and the public. In turn, this led to a general increase in attention given to schools. To sum up, the evaluators took an international perspective:

> By international standards, the investment in 27 well-funded projects is extremely extensive (in relation, for example, to population or number of school pupils), something that our neighbouring countries in particular have clearly shown that they want to learn from. (Riis, 2000, p. IV, see also Riis & Jedeskog, 1997, p. 28)

It is likely that this interest may have contributed significantly to the fourth campaign.

2.5 *Tools for Learning, 1999–2001*

The third campaign had not even ended when, during the parliamentary elections of 1998, the Social Democratic Party proposed one more crusade: "Tools for Learning" (Ministry of Education and Science, 1989). As the winner of the election, the Social Democratic Party launched this fourth national campaign under its official title: "The National Programme for ICT in School" (ITiS), but it was colloquially perceived as "one computer for each teacher". This campaign saw a return to earlier strategies in its ambition to direct the measures towards as many school units and teachers as possible. In terms of economic value, it was even bigger than that of the Foundation, with a budget of SEK 1.9 billion (Jedeskog, 2005, p. 66).

In other respects, as well, the basic structure was different. Despite its name, which hails IT as a tool for learning, the central element of ITiS was competence development in work teams. The teachers were made an offer of in-service training in teacher teams. Each team should implement a specific project coloured by *an interdisciplinary, problem-based, and student-oriented approach together with their pupils*. The Ministry of Education led the venture, and such an approach was in line with the national curriculum of 1994, so it was considered a means to implement those traits (Jedeskog, 2005). The evaluators concluded that the initiative was a conscious government strategy to generate school development with the help of a new educational tool (Tebelius et al., 2003, p. 28). IT was used as an incentive for school development and to increase teachers' digital competence. The initiative was broad and almost 70,000 teachers, corresponding to 60% of all teachers in Swedish schools, took part in these projects (Chaib et al., 2004, p. 165).

Teachers in the teacher teams were provided with one computer each as a "working tool". The municipalities were also given financial support from the government to improve schools' connections to the internet and to create conditions for all students and teachers to have their own email address (Tebelius et al., 2003). In addition to these four elements, the venture also consisted of resources to develop both the Swedish and the European Schoolnet, special measures for pupils with special needs, and awards for excellent pedagogical contributions (Jedeskog, 2005, p. 65).

The fourth campaign came so soon after the previous one that it became difficult to delimit traces of the different school development initiatives and truly show where the Foundation's initiative ended and the new one took over (Riis, 2000). The Foundation also made genuine efforts to support the transition (Ministry of Education and Science, 2002).

Similarly to the previous ones, this campaign was also prolonged for one year. One of the final evaluation reports stated that a great deal of development work had been conducted by the teachers. For many, it was the first time they had cooperated in teacher teams (although organising teachers in teams had already been recommended in the 1980 curriculum, Jedeskog, 2005, p. 70). Approaches involving interdisciplinarity and problem-based learning were also a new experience for many teachers. Experiences of student-oriented methodology were more common and were further deepened through the project. The working teams had mentors, who were highly appreciated by teachers. Concerning schools and competence development, a great deal was experienced and achieved. The evaluators also discussed the role of the technology. Officially, ITiS was about school development with support from IT. In the end, groups of both students and teachers were disappointed that, to a very large extent, the initiative turned out to be about other factors than IT as a pedagogical tool:

> The overall purpose of developing the school may well have been noble and right, but the initiative would have served its purpose even better if the teachers could have been offered more IT training than they have received. (Chaib et al., 2004, p. 175)

We think it is fair to conclude that this campaign was more directed towards school development, using IT, than to strengthening IT-related content. This means that it was not only a large-scale investment in IT and schools, but also applied to the large field of continuing education:

> [This] is the most comprehensive investment to date in the competence development of teachers in schools that has ever been made, both in financial terms and measured in the number of participating teachers. (Tebelius et al., 2003, p. 174)

Concerns about what would happen after the project ended were raised by the evaluators. There were worries both about whether the school was "mature" enough for changes in the spirit of the initiative, and whether the technology would continue to be good enough and remain functional. What happened after 2002 is another question. Nevertheless, we can state that, to date, "Tools for learning" is the last time that issues of IT and school have been part of a national election, and the last national campaign with ambitions to strengthen both hardware access and IT-related content.

2.6 *Evaluations*

Under the leadership of Ulla Riis, the ELOïS group (at Uppsala University and Linköping University) conducted evaluations and research regarding the first three initiatives (Jedeskog, 2000; Nissen et al., 2002; Riis, 1991, 2000, 2019; Riis, Holmstrand, & Jedeskog, 1997). In Swedish, ELOïS is short for: students [elever], teachers [lärare], organizations and information technology in [i] school [skolan]. The fourth campaign was evaluated by Mohamed Chaib (2004) and Ulla Tebelius (2003). Repeatedly, the ELOïS group, as well as other researchers and evaluators involved in similar efforts, concluded that the targets of these campaigns were very seldom reached, and that the goals—which during certain periods claimed that computers (or the internet, etc.) would "revolutionise" education—were too ambitious. Nevertheless, more new reforms were still launched. It was suggested that the strong belief in the importance of Sweden becoming a world-leading nation within information technology was an interpretive pattern that overshaded negative results. To describe this, the concept of "IT-ism" was introduced, with reference to the notion of "Orientalism", according to which conceptions about the "Orient" become a hindrance to perceiving it in a nuanced way (Said, 1978; Johansson et al., 1998). A comparable notion today would perhaps be "the digital imperative" (Clark, 2010).

Jedeskog (2005) metaphorically described the campaigns as four waves:

> "Waves" is, on the one hand, an attempt in a general way to describe administrative and political reasoning for concrete campaigns. On the other hand "waves" is used as something irresistible not possible to defend oneself against from teachers' and principals' perspectives. (Jedeskog, 2005, p. 27)

3 Thereafter: 2002–2011 (2017)

Up until now, this chapter has primarily concerned the campaigns, and has presented an account of changes in national curricula. For the sake of completeness, we now provide a brief account of developments during the period 2002–2017.

After 2002, there were no further national campaigns, no fifth wave. There might be no final answer as to why there have not been any more, but there are several factors that are probably important. One reason is that the bursting of the dot.com bubble struck Sweden hard in the 1990s. The rampant IT euphoria was transformed into its opposite, and IT ventures vanished from the political agenda.

Concurrently, the Swedish school system was gradually becoming decentralised, from the state to municipal ownership, and the volume of the private school sector grew rapidly. Also, the number of national school authorities grew and their assignments shifted. In 2003, the National Authority for School Development [Myndigheten för skolutveckling] was established, but after just five years, in 2008, this authority was replaced by the Swedish Schools Inspectorate [Skolinspektionen]. Additionally, Sweden, which by this time had become part of the European Union, was also affected by several strong policy organisations which became engaged in school issues to a much greater extent than before. Together, these aspects created uncertainty concerning leadership. Was it the state or the municipality that was responsible for school development? Should it be conducted mainly by guidance, regulation, or inspection? Should policy build on national or international expertise?

Another related aspect is that the school system internationally had started to attract researchers from more disciplines than previously. In accordance with the increasing accessibility of digital technology for all citizens they brought with them two different schools of thought. Both of them emphasised the need to develop literacy skills beyond what had previously been established. On the one hand, The New London Group, consisting of researchers within Social Science, Linguistics, and Educational Science, argued that success in a digital, information-oriented society requires *multiliteracies*, that is, competence in an even more diverse set of functional, academic, critical, and electronic skills than before (New London Group, 2000). To be considered multiliterate, one must acquire a battery of skills that enables one to take advantage of the diverse modes of communication made possible by new technologies, and to participate in global learning communities (Kress, 2003). On the other hand, computer scientists, led by Jeanette Wing (2006), argued that the new literacy of the 21st century was *computational thinking*, which was another take on the issue (Heintz & Mannila, 2018, p. 66). This kind of competence would enable everyone to shape computation to their own needs. Wing (2006, p. 33) claimed that "to reading, writing, and arithmetic, we should add computational thinking to every child's analytical ability". In this respect, it was viewed as a "fundamental analytical skill that 'everyone /…/ can use to help solve problems, design systems, and understand human behavior'" (Committee for the Workshop of Computational Thinking, 2011, p. 3). Denning and Tedre (2019, p. 4) defined the concept as the mental skills and practices for designing computations that get computers to do jobs for "us", explaining and interpreting the world as a complex of information processes.

At this time, post-2006, arguments were raised highlighting the importance of creating a curriculum that reflected the current world, rather than the 1970s, looked upon as the "exciting opportunities and challenges of IT" (Patterson, 2006, p. 27; Committee for the Workshop of Computational Thinking, 2011).

Returning to a national perspective, in Sweden the period after the fourth wave can be seen as the aftermath to an intensive period that had featured several extensive campaigns spread across 25 years. It also seems to have left a void and challenges to take advantage of Sweden's relatively early expansion in the field. Moreover, the period after 2001 can be compared with, for example, Norway, which in 2000, perhaps inspired by the early and extensive Swedish efforts, took a strong national leadership and established a national unit for IT and education. This was based at the University of Oslo, and in 2006 digital competence was introduced as a core competence on a par with spoken skills, reading, writing, and numeracy (Godhe et al., 2020, p. 78; Kløvstad & Kristiansen, 2003; Olofsson et al., 2021).

In Sweden, there were no similar initiatives at that time, and the curriculum from 1994 still regulated digital activities in schools. However, a new curriculum was introduced in 2011 (NAE, 2011). This covers compulsory school, and the

IT-related aspects are pointed out more clearly in the descriptions of schools' overall responsibilities and objectives. Compared to previously, these aspects are explicitly declared in many more subjects, specifically in 11 subjects (out of a total of 19). To some extent the increase follows from the fact that the ability to use the technology, or the effect of using it, is much more present than in previous curricula. In five other subjects, the aspects are only implicitly referred to, covering for example: information searching, source criticism, or similar formulations. So, in total there are 16 subjects with content related to IT. There are thus only three subjects, Mother-Tongue Tuition, Crafts, and Sign Language for the Deaf or Hard of Hearing, in which none of the content is related to IT.

However, in 2016, the NAE was tasked with proposing a national strategy for the digitisation of the education system (NAE, 2016).[4] A decisive factor for this assignment was the EU's definition of digital competence as one of the key competencies for lifelong learning, which has been influential in educational policy documents (Godhe et al., 2020, p. 76). Other important influences, albeit to varying degrees, were the two lines of thought that emphasised either multiliteracy or computational thinking. However, neither of these concepts was explicitly mentioned in the new curriculum.

Computational thinking was considered in the revision process for the 2017 curriculum, although, according to the NAE, this concept refers to skills and approaches that were already covered in the 2011 curriculum (NAE, 2016, p. 7). Furthermore, the argument was that those skills were also reinforced by the NAE's proposed changes. The revision implemented in 2018 instead included the concept of "digital competence", and the aims and core content of several subjects were altered (NAE, 2017; Godhe et al., 2020, p. 76). The revisions concern six specific areas:

- to include programming primarily in Technology and Mathematics
- to strengthen the ability to critically evaluate sources
- to strengthen the ability to solve problems and translate ideas into action in a creative manner
- to work with digital texts, media, and tools
- to use and understand digital systems and services
- to develop an understanding of the impact of digitisation on individuals and society (Swedish Government, 2017, p. 1).

One could say that, instead of using the notion of "computational thinking", the NAE chose what the agency understood as a broader concept, "digital competence", and made programming a part of this (Raptopoulou, 2021). This is in line with Denning and Tedre (2019), who argue that policymakers in particular interpreted the concept of computational thinking quite narrowly compared to the scientific notions of computational thinking developed previously. As mentioned, the concept of multiliteracy/multimodality was not included in the curriculum at all (Godhe et al., 2020).

In summary, regarding the various curricula, it can be said that the changes since 1969 brought with them an expanded number of subjects with IT-related content and assigned increased importance to what is today called digital competence. In addition, the degree to which digital technology is used has broadened in scope through the years, to emphasise aspects such as word processing, geographical information systems (GIS), and computer games (remember that, at the beginning it was not even assumed that computers were to be *used*). Likewise, this field of knowledge has increasingly been accentuated in schools' overall goals and areas of responsibility.

The motives for these ventures will be further discussed in the final section, below.

4 Discussion

The results of the evaluation and research concerning the four campaigns are very much in line with Cuban's work (Cuban, 1990). Just as he recognised, it is actors from outside the school system who launch the reforms to improve schools, with the support of new technology. This is also in line with what happened in several other European countries (Flury & Geis, 2022; Machado-Trujillo, 2021). At the beginning, during the 1960s and 1970s, the neighbouring country of Denmark saw similar development to that in Sweden. But the efforts during the 1980s and 1990s were not so forceful in Denmark (Caeli & Bundsgaard, 2019). Until the middle of the 2000s, it was, as explained above about the same in Norway.

In the Swedish case, it is evident that politicians at a national level, and under their direction, state agencies, were important actors. It is they who inaugurated the four major campaigns, including the hardware and software project to develop and implement the use of a Swedish school computer. It is also apparent that, during the period 1980–2001, the computerisation of schools is consistent with Cuban's (1986) finding that technology was mainly pushed out (Schoen, 1967) by forces outside the school. Of course, there were some signs of computers being pulled into schools; for example, by individual teachers who took the initiative to introduce computers, both due to their own personal interest and in the belief that the technology could be an effective educational aid. They formed a long-lived professional association and a magazine to support each other (*The Computer in Education*). Of course, there were also influences from other actors, such as computer hardware and software producers, telecommunications companies, publishers of teaching materials, and business associations, which could also be further studied (Raptopoulou, 2021). Nevertheless, it is clear that "computers entered the classrooms" mainly by being pushed there.

The question is: why? Jedeskog (2005) has identified four main motives:

- The *democratic argument,* a compensatory function. Children with no computer at home should have access to computers at school and all pupils, irrespective of where they live in the country or to which social class they belong, have a right to become computer literate in the Swedish school. Later on, a fifth basic skill, digital literacy, positioned alongside spoken skills, reading, writing, and numeracy, was equated with the ability to search, manage, assign meaning to, and evaluate information in the form of facts, and all students should have the same opportunities to develop all five basic skills.
- *Preparation for their future in working life and as citizens in the future society.* Pupils have to manage the new technology and live in a society impregnated by digital technology. As already described, this motive was being actualised even before 1984!
- *Improving pupils' learning,* using IT as a pedagogical tool for enhanced learning.
- *ICT as a force for change of education and the school system* which came to the fore in the 1990s.

These motives were important in the decision-makers' rhetoric. This is understandable since these motives relate to what should take place in teaching practice and schools' everyday life. A possible reasoning might also be that perhaps some of the arguments were "wrong", but at the same time the expectations may now have been proven correct. Wrong in the sense that, for example, the strong expectations tied to IT as an educational aid or as a lever for learning and school development were not fulfilled. But perhaps accurate in the sense that, today, our existence is impregnated by IT and Artificial Intelligence (AI). It can therefore be seen as reasonable that schools had to consider the then new technology. On the other hand, does this justify the relatively extensive investment and rather unrealistic hopes attached to the computer as a game-changer for schools' working methods and teaching?

In relation to questions like these, the current 20-year distance from the end of the fourth wave offers a belated insight considering a fifth motive—the prosperity of the nation. The importance of maintaining the country's competitiveness and hence the strong welfare state, and how this is related to the new technology and schools was already vital in the first campaign, and most clearly articulated by Prime Minister Bildt in the arguments presented for the third campaign (SGOR, 1994). But then, perhaps it was not IT *competence* that was the most crucial? Was it perhaps more important that the population was to become IT savvy and IT willing, or as the report *Wings to human ability* phrased it: school also builds young people's "habits and behaviour" (SGOR, 1994, p. 8)? Could it be that the primary contribution of the four campaigns, as they seamlessly meshed with each other, was not advanced knowledge, improved learning, or a revolutionised school, but the Swedish population's willingness to become digital citizens, by first adopting IT and now AI? Which of course would be a good breeding ground for technical and industrial success. Modern comparisons between different countries, from which it appears that Sweden exhibits a very high degree of IT use, indicate that, in this respect, the politicians' goals since the end of the 1970s have been successfully met. Would we even have ended up in today's situation, as a "leading digitalised nation" without several large, pushed out, and in our view nothing less than, full-scale experiments? The enthusiasm of Swedish politicians was important for the creation of these experiments, an enthusiasm that at the time was also pushed forward on the international political arena:

> I share your enthusiasm for the potential of emerging communications technologies. This demonstration of electronic communications is an important step toward building a global information superhighway. (*From US President Clinton's reply to Prime Minister Bildt*, The American Presidency Project, 1994)

This email exchange between Bildt and Clinton, which also introduced this chapter, took place during the preparations for the third Swedish school campaign. Almost 30 years later, the information superhighway is now truly global, and schools around the globe world are part of an ongoing process of digitalisation. This may correspond to influences of *IT-ism* in ongoing implementation of digital competence, including programming (i.e., computational thinking), in Swedish compulsory schools (Johansson et al., 1998). How strong is todays' conviction that all children must learn programming and AI literacy, and thereby become digital citizens contributing to a flourishing nation? Do a simplified digital imperative, or lucid considerations, still constitute a hindrance to perceive contemporary efforts in a nuanced way? The future will tell.

Acknowledgements

The authors thank Professor Ulla Riis, Uppsala University, for sharing her invaluable experience and insights on the topic of this chapter, and her tireless contributions to both the details and overall conclusions.

Notes

1 The syllabuses were analysed using the following keywords: ADB (short for automatic data processing [automatisk databehandling]), Automation [automatisering], Data (in Swedish a computer is a "dator" or "datamaskin (data machine) ("data" also catches "datalära, datavetenskap, databehandling"), Digital, Information, Internet, Integrity [integritet], Register, Programming [programmering], and Technology [teknik].
2 All translations from Swedish have been carried out by the authors.
3 See Riis (1991, pp. 21–23) for other relevant campaigns.
4 In 2022 NAE suggested a new version of the National digitalization strategy for the Swedish education system, this time covering 2023-2027. The new government, elected in the autumn of 2022, sent the proposal on a re-referral (Ministry of Education and Research, 2023).

References

Caeli, E. N., & Bundsgaard, J. (2019). Datalogic thinking and the comprehension of technology in primary school. A return journey [Datalogiskt taenking og teknologiforståelse i folkeskolen tur-retur]. *Laering & Medier, 19*, 1–30. https://tidsskrift.dk/lom/article/view/110919/162230

Caeli, E. N., & Bundsgaard, J. (2020). Computational thinking in compulsory education: A survey study on initiatives and conceptions. *Education Tech Research Dev, 68*, 551–573. https://doi.org/10.1007/s11423-019-09694-z

Chaib, C., Bäckström, Å., & Chaib, M. (2004). *To live with ITiS. National evaluation* [*Leva med ITiS. Nationell utvärdering av IT i Skolan*]. National Centre for Lifelong Learning, Rapport 2204:1. Högskolan för lärande och kommunikation, Jönköping University. http://hj.diva-portal.org/smash/get/diva2:4381/FULLTEXT01.pdf

Clark, E. J. (2010). The digital imperative: Making the case for a 21st-century pedagogy. *Computers and Composition, 27*(1), 27–35. https://doi.org/10.1016/j.compcom.2009.12.004

Committee for the Workshops on Computational Thinking. (2011). *Report of a workshop of pedagogical aspects of computational thinking*. National Academies Press. https://nap.nationalacademies.org/catalog/13170/report-of-a-workshop-on-the-pedagogical-aspects-of-computational-thinking

Cuban, L. (1986). *Teachers and Machines: The classroom of technology since 1920*. Teachers College Press.

Cuban, L. (1990). Reforming again, again, and again. *Educational Researcher, 19*(1), 3–13. https://doi.org/10.3102/0013189X019001003

Denning, P. J., & Tedre, M. (2019). *Computational thinking*. The MIT Press.

Flury, C., & Geiss, M. (Eds.). (2023). *How computers entered the classroom, 1960–2000 historical perspectives*. De Gruyter Oldenbourg.

Godhe, A.-L., Magnusson, P., & Sofkova Hashemi, S. S. (2020). Adequate digital competence: Exploring revisions in the Swedish National Curriculum. *Educare, 2*, 74–91. https://doi.org/10.24834/educare.2020.2.4

Grensjö, B. (1999). *IT and school development? Survey study among smaller school development projects supported by the KK Foundation* [*IT och skolutveckling? Enkätstudie bland KK-stödda mindre skolutvecklingsprojekt*]. Educational Reports: 61. Umeå University.

Heintz, F., & Mannila, L. (2018). Computational thinking for all: An experience report on scaling up teaching computational thinking to all students in a major city in Sweden. *Proceedings from ACM SIGCSE Conference*. https://dl.acm.org/doi/pdf/10.1145/3210553

Jedeskog, G. (2000). *New in the cl@ss. The relation between the role as a teacher and the use of computers described in international research* [*Ny I kl@ssen. Förhållandet mellan lärarroll och datoranvändning beskrivet i internationell forskning*]. Ekelunds.

Jedeskog, G. (2005). *Ch@nging school: Impementation of ICT in Swedish School, campaigns and experiences 1984–2004*. Uppsala University.

Johansson, M. (1997). *Smart, fast and beautiful: On rhetoric of technology and computing discourse in Sweden 1955–1995* (Diss). Linköping University.

Johansson, M., Nissen, J., & Sturesson, L. (1998). *"IT-ism": Information technology as vision an reality* [*"IT-ism": Informationstekniken som vision och verklighet*]. KFB-rapport 1998:11. Kommunikationsforskningsberedningen.

Kaiserfeld, T. (1996). Computerizing the Swedish Welfare State: The middle way of technological success and failure. *Technology and Culture, 37*(2), 249–279.

Karlsohn, T. (2018). *Digital technology and rhetoric in Swedish educational history*. Uppsala papers in the History of ideas: 18. http://uu.diva-portal.org/smash/get/diva2:1248424/FULLTEXT01.pdf

Kløvstad, V., & Kristiansen T. (2004). *The ITU monitor. The status of digitalisation in school in 2003* [*ITU Monitor, Skolens digitale tilstand 2003*]. Rapport 1. University of Oslo. https://www.udir.no/tall-og-forskning/finn-forskning/rapporter/itu-monitor-2003/

Kress, G. R. (2003). *Literacy in the new media age*. Routledge.

Machado-Trujillo, C. (2021). *Education and historical evolution of information and communication technologies: Background, international influences and their development in Spain in the 1980s*. Universidad de La Laguna.

Ministry of Education and Science, Sweden [Utbildningsdepartementet]. (1989). *Tools for learning. A National program for ICT in Schools*. Delegation for ICT in schools.

Ministry of Education and Science, Sweden [Utbildningsdepartementet]. (2002). Working group for a new national IT strategy for schools, interim report [Arbetsgruppen för en ny nationell IT-strategi för skolan, delrapport]. *Next step* [*Nästa steg*]. Utbildningsdepartementet https://www.regeringen.se/rattsliga-dokument/departementsserien-och-promemorior/2002/05/ds-200219-/

Ministry of Education and Research, Sweden [Utbildningsdepartementet]. (2023). Referral of the National Education Agency's proposal for a national digitalization strategy for the school system 2023–2027 [Remiss av Statens skolverks förslag till nationell digitaliseringsstrategi för skolväsendet 2023–2027]. U2022/03951.

National Agency for Education, Sweden [Skolverket]. (1994) *Curriculum for the compulsory school system, the pre-school class and the leisure-time centre* [*Läroplan för det obligatoriska skolväsendet, förskoleklassen och fritidshemmet, Lpo 94*]. Fritzes.

National Agency for Education, Sweden [Skolverket]. (1996a). *An overview of the school 1996: Summary*. Skolverket.

National Agency for Education, Sweden [Skolverket]. (1996b). *The compulsory school: Course syllabus, grade criteria* [*Grundskolan: kursplaner, betygskriterier*]. Skolverket.

National Agency for Education, Sweden [Skolverket]. (2011). *Curriculum for the compulsory school, pre-school class and school-age educare 2011* [*Läroplan för grundskolan, förskoleklassen, förskoleklassen och fritidshemmet 2011*]. Skolverket.

National Agency for Education, Sweden [Skolverket]. (2016). *Report on the assignment to propose national IT strategies for the school system - changes in curricula, course plans, subject syllabuses and degree goals* [*Redovisning av uppdraget om att föreslå nationella it strategier för skolväsendet – förändringar i läroplaner, kursplaner, ämnesplaner och examensmål*]. Dnr: 6.1.1-2015:1608, 2016-06-16. Skolverket.

National Agency for Education, Sweden [Skolverket]. (2017). *Curriculum for the compulsory school, pre-school class and school-age educare 2011 (revised 2017)* [*Läroplan för grundskolan, förskoleklassen och fritidshemmet 2011 (Reviderad 2017)*]. Skolverket.

National Board of Education, Sweden [Skolöverstyrelsen]. (1969). *Curriculum for the comprehensive school* [*Läroplan för grundskolan*]. Utbildningsförlag Liber AB.

National Board of Education, Sweden [Skolöverstyrelsen]. (1980a). *CIS – The computer in the school: NBE's action program and final report* [*DIS – Datorn i skolan: SÖ:s handlingsprogram och slutrapport*]. SÖ-projekt 628. Skolöverstyrelsen.

National Board of Education, Sweden [Skolöverstyrelsen]. (1980b). *Curriculum for the comprehensive school: General part* [*Läroplan för grundskolan: Allmän del*]. Liber UtbildningsFörlaget.

National Board of Education, Sweden [Skolöverstyrelsen]. (1984a). *Education for the computer society. Starting points and directions* [*Utbildningen inför datasamhället: Utgångspunkter och inriktning*]. 1984-02-03, Skolöverstyrelsen.

National Board of Education, Sweden [Skolöverstyrelsen]. (1984b). *"Datalära" in the compulsory school. Syllabus* [*Datalära i grundskolan. Studieplan*]. Skolöverstyrelsen.

National Board of Education, Sweden [Skolöverstyrelsen]. (1985). *Computer education for all adults* [*Datautbildning för alla vuxna*]. Forum för vuxenutbildningsdebatt. Rapporter - Planering, uppföljning, utvärdering, R 85:23. Skolöverstyrelsen. ISSN: 0282-7522, ISBN: 91-7662-129-4.

New London Group. (2000). A pedagogy of multiliteracies: Designing social futures. In B. Cope & M. Kalantzis (Eds.), *Multiliteracies: Literacy learning and the design of social futures* (pp. 9–38). Macmillan.

Nissen, J. (Ed.). (2002). *"Say IT - that's enough": To develop the school with some luminous IT projects. Evaluation of the KK Foundation's investment in major school development projects* [*"Säg IT – det räcker". Att utveckla skolan med några lysande IT-projekt. Utvärdering av KK-stiftelsens satsning på större skolutvecklingsprojekt*]. The Knowledge Foundation [Stiftelsen för kunskaps- och kompetensutveckling].

Olofsson, A. D., Lindberg, J. O., Young Pedersen, A., Arstorp, A.-T., Dalsgaard, C., Einum, E., Caviglia, F., Ilomäki, L., Veermans, M., Häkkinen, P., & Willermark, S. (2021). Digital competence across boundaries – beyond a common Nordic model of the digitalisation of K-12 schools? *Education Inquiry, 12*(4), 317–328. https://doi.org/10.1080/20004508.2021.1976454

Patterson, D. A. (2006). Computer science education in the 21st century. *Communications of the ACM, 49*(3), 27–30. https://doi.org/10.1145/1118178.1118212

Raptopoulou, A. (2021). *Politics of contemporary education policy: The case of programming in the Swedish Curriculum* (Diss). Stockholm University.

Riis, U. (1991). *The school and the computer. The campaign The computer as a pedagogical aid 1988–1991* [*Skolan och datorn. Satsningen Datorn som pedagogiskt hjälpmedel 1988–1991*]. Tema T Rapport 24, Linköping University.

Riis, U. (2019). To implement educational policies in a shifting world – early computer education in the Swedish compulsory school [Realiserad skolpolitik i en föränderlig värld – tidig datalära i grundskolan]. *Vägval i skolans historia, 1*, https://undervisningshistoria.se/realiserad-skolpolitik-i-en-foranderlig-varld-tidig-datalara-i-grundskolan/

Riis, U. (2000). *The KK-Foundation's campaign on 27 large school development projects and evalutation and research in connection to the campaign: A report to the Foundation's board meeting September 1999* [*KK-stiftelsens satsning på 27 stora skolutvecklingsprojekt och utvärdering och forskning i anslutning till satsningen: En rapport till KK-stiftelsens styrelse i september 1999*]. Rapport 222. Uppsala University.

Riis, U., Holmstrand, L., & Jedeskog, G. (2000). *Visionary enthusiasm and realistic thoughtfulness. The KK Foundation's investment in 27 "lighthouse projects" 1996–1999* [*Visionär entusiasm och realistisk eftertänksamhet* [*KK-stiftelsens satsning på 27 "fyrtornsprojekt" 1996–1999*]. Uppsala University.

Riis, U., & Jedeskog, G. (Eds.). (1997). *Pedagogy, technology or economics? A baseline fixation of the KK Foundation's municipal school development project* [*Pedagogik, teknik eller ekonomi? En baslinjebestämning av KK-stiftelsens kommunbaserade skolutvecklingsprojekt*]. Uppsala University.

Said, E. (1978). *Orientalism*. Pantheon.

Schoen, D. A. (1967). *Technology and change: The new Heraclitus*. Delacorte Press.

Swedish Government [Proposition]. (1982). *Coordinated data policy* [*Samordnad datapolitik*]. Prop. 1981/82:123.

Swedish Government [Regeringen]. (2017). Strengthening digital competence in school's policy documents [Stärkt digital kompetens i skolans styrdokument]. 2017-10-03). https://medarbetarportalen.gu.se/digitalAssets/1621/1621603_regingens_beslut_digital_kompetens_170309.pdf

Swedish Government Official Report. (1994). *Information Technology - Wings to Human Ability* [*Informationsteknologin - Vingar åt människans förmåga, Betänkande av IT-kommissionen*]. SOU 1994:118. Statsrådsberedningen. http://www.itkommissionen.se/dynamaster/file_archive/020124/39d2d69a7f6f35b7ec7cad5a63208c58/Rapport%20Vingar%20%e5t%20m%e4nniskans%20f%f6rm%e5ga%20(1995).pdf

Tebelius, U., Aderklou, C., & Fritzdorf, L. (2003). *ITiS as an incentive for school development: The national evaluation of IT in schools. The survey study* [*ITiS som incitament till skolutveckling: Den nationella utvärderingen av IT i skolan. Surveystudien, 2000–2003*]. Forskning i Halmstad nr 7. ISBN 91-972579-7-4. Halmstad University. http://hh.diva-portal.org/smash/get/diva2:237363/FULLTEXT01.pdf

The American Presidency Project. (1994). *Electronic Mail Message to Prime Minister Carl Bildt of Sweden*. https://www.presidency.ucsb.edu/node/218492

The Computer in the Education. (n.d.). *Datorn i Utbildningen*. https://www.diu.se/diu.asp?val=fdiu

Wing, J. M. (2006). Computational Thinking. *Communications of the ACM*, *49*(3), 33–35. https://doi.org/10.1145/1118178.1118215

CHAPTER 3

Transposition of Computing and Programming Knowledge

The Swedish Upper Secondary School during the 1970s and 1980s

Lennart Rolandsson, Cecilia Kilhamn and Kajsa Bråting

Abstract

This chapter unravels how the teaching of computing and programming is discussed in two Swedish teachers' journals from the 1970s and 1980s, when programming in school was initiated, explored, and consolidated by different actors. These two decades are of special interest as computer technology evolved significantly during this period, and it became possible to teach programming knowledge in school. The aim of the chapter is to understand how programming knowledge was shaped by experts and early adopters who were involved when computers were first integrated into the Swedish upper secondary school. Articles from the journals are analysed based on Chevallard's praxeology describing programming knowledge in terms of praxis: how and what to teach, and logos: why to teach specific elements of programming. During these two decades, experimental work was done to offer programming knowledge to a broad group at upper secondary level. The chapter highlights the transposition of programming knowledge when it moves from experts outside school to early adopters within the school subject of mathematics.

Keywords

praxeology – computing – programming – transposition of knowledge – school mathematics

1 Introduction

The chapter describes a period when programming languages were detached from specific computer hardware so that humans could pay more attention to the application and how the order of code instructions influences our thinking

© LENNART ROLANDSSON ET AL., 2024 | DOI:10.1163/9789004687912_003

or vice versa. During the 1980s, teaching and learning of programming were heavily debated among computer scientists, and one of them, Donald Knuth searched for a way to move from how to instruct computers, towards the logic and the flow of human thinking incorporating much more than simply "explaining to human beings what we want a computer to do" (Knuth, 1984, p. 99). As one step towards that project, he proposed structured programming languages. This is a step that transforms programming languages and algorithms into epistemic objects (Nofre et al., 2014) as a consequence of code being used for something more than solely to instruct computer machines in the most efficient and appropriate way. Partly as a consequence of this expansion, computers and programming started to make their way into education. In the 1980s there was a famous debate in which Papert (1987) argued that students can improve their thinking with specific programming languages (e.g. Logo), while Pea (1987) argued for a careful study of the impact of such languages in education. In the following, we will go into that expansion to study how experts and early adopters in Sweden argued when approaching teaching for mathematical thinking with different programming languages.

1.1 *The Research Aim*

In many ways, programming overlaps and intersects with algorithmic thinking (Kilhamn & Bråting, 2019). However, as the concept of an algorithm is sometimes represented differently in mathematics and computer science (Ensmenger, 2010), and many teachers today consider it problematic to teach programming (Bråting et al., 2022), this chapter approaches different perspectives with the didactical questions of what, how, why, and who. We use Chevallard's (2006) theory of knowledge transposition in order to analyse the voices of experts and teachers as they appear in two Swedish teachers' journals from the 1970s and 1980s.

Historically, many have heralded computers as the solution to learning issues in school (e.g. Segal, 1996). This is problematic because such a technology acceptance influences how computing and programming are taught even today, with deterministic annotations for best educational practices (Rolandsson, 2013). We argue for a careful study of teachers' pedagogy and how it is shaped by computer technology and programming languages. The aim of this chapter is to understand how programming knowledge was shaped by experts and early adopters who were involved when computers were first integrated into the Swedish upper secondary school. In the final part of the chapter, we describe this knowledge in terms of the praxeologies that came to light in pedagogical debates from the 1970s and 1980s.

2 Theoretical Framework

The chapter is theoretically embedded in the Anthropological Theory of the Didactic (ATD), in which knowledge is seen as something changeable and related to a specific institutional context. Transposition of knowledge described in terms of praxeologies is used as an analytical tool (Chevallard, 2006; Chevallard & Sensevy, 2014). According to Chevallard, knowledge moves from professional and scholarly institutions into schools where decisions are made about what to teach, when and how, why, and by whom. All such decisions effect the very nature of the knowledge to be taught, and eventually also the knowledge actually taught. Chevallard calls this a transposition between institutional settings (Figure 3.1).

The theory provides a frame for investigating didactical activities, focusing on their components, dynamics, and the conditions that enable their existence and development. This chapter brings the teachers' work into focus, unravelling how a pedagogical practice is shaped and how the scholarly knowledge is transposed when integrated into the curriculum. This had implications on how computing and programming were debated in the 1970s and 1980s when it moved from *scholarly knowledge* to *knowledge to be taught* in upper secondary school (first two institution levels; see Figure 3.1).

Scholarly knowledge refers to the knowledge of professional programmers and technology developers, as well as academics within both programming and mathematics. In this study, we call them experts. Knowledge to be taught is the knowledge of those who make decisions about what is to be taught in school and how to legitimize the new subject. Here we have, for example, curriculum developers and textbook authors, but also teachers who take a great interest in shaping the new subject. As programming was new in the school setting during the 1970s and 1980s, we use the term *early adopters* for teachers who engaged in the transposition of knowledge to be taught during this period.

Within ATD, knowledge is described as a praxeology (Chevallard, 2006), with tasks and techniques, referred to as knowing-what and knowing-how, constituting the *praxis* elements of knowledge. In programming, *techniques* concern, for example, what programming language to use and what computer hardware is available, but also detailed elements of coding and computing

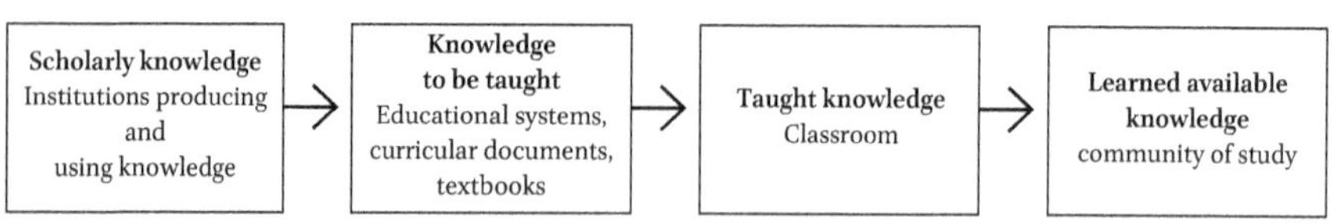

FIGURE 3.1 The arrows describe the transposition between different praxeologies

TABLE 3.1 A praxeology, with praxis and logos blocks for programming in school

Praxis	Didactical question
Tasks (what)	What types of tasks?
Techniques (how)	What type of hardware, language, code and practices?
Logos	**Didactical question**
Technology (why)	Why these techniques for the tasks?
Theory (why)	What are students supposed to learn?

practices. Praxis is legitimized by *logos*, referred to as knowing-why. Logos, in terms of *technology* and *theory*, justifies why certain tasks and techniques are introduced or employed in praxis. A praxeology of programming knowledge in school is described in Table 3.1.

3 Background

The 1970s and 1980s were an experimental time in many ways, as computer technology and operating systems were in constant flux. The market, with IBM, Microsoft, and Digital Research, heavily influenced what computers could offer. In fact, there were different operating systems—like the USCD p-system, IBM PC-DOS, and CP/M—pre-packaged with different programming languages. Computer hardware and memory were major concerns, as the files could be on different hard drives, cassettes, or floppy drives. There were also multi-user computer systems that allowed teachers to monitor their students' work from one computer. In short, the available computer technology had an impact on the development of a pedagogical practice.

3.1 *Computers and Computing in Swedish Schools*

During the 1970s, some Swedish upper secondary schools collaborated with higher vocational education institutions, and at some schools there were teachers with both the resources and the will to explore programming education for the first time. These teachers became early adopters of computer technology. The Swedish Board of Education (Skolöverstyrelsen) decided not to involve the experts from within the higher vocational educational institutions (Rolandsson & Skogh, 2014) and instead suppressed these experts' voices, due mainly to two rationales: firstly, it was not considered realistic to create a new subject because of the continuous changes in society; and secondly, the computer was perceived as a tool for problem-solving rather than as an artifact requiring expert knowledge. In Sweden, this was a time when didactical ambitions and various projects were initiated to approach computing education systematically, e.g. DISK (1971–1973: Fagerström, 1973), DIS (1974–1981: Swedish Board of Education, 1980), and PRODIS (1979–1981: Swedish Board of Education, 1981). The period between 1971 and 1985 was experimental in many ways. If a school could not afford to buy the necessary computer technology, teachers sometimes used industrial mainframe computers, typically at night. These early adopters had to work hard to prepare and debug code such that the mathematical algorithms translated nicely into code (Rolandsson, 2011).

The revival of science education was also in focus during the 1970s, and the Swedish Board of Education advocated an impact of computer technology and programming as part of the educational activities, to encourage more students to pursue a career in natural sciences and mathematics (Rolandsson & Skogh, 2014). In the 1980s, the earliest signs of programming in Swedish curricula appear in two subjects: *Computing literacy* (Swedish: Datalära) in Grades 7–10 and *Computing* (Swedish: Datakunskap) in Grades 10–12 (Swedish Board of Education, 1983, 1984). In the syllabus for both subjects, societal aspects were emphasized while technical details were more or less suppressed (Rolandsson & Skogh, 2014). The subject *Computing* was taught at upper secondary school on its own merits and comprised a diverse palette of modules, including fundamental knowledge about computers and computer programming principles, software development, systemization, numerical methods and statistics, as well as the application of these modules for society.

At the beginning of the 1980s, a Swedish project called "TUDIS 1985" began developing and manufacturing a school computer (Compis), along with a unique programming language (Comal). The Swedish Board of Education offered recommendations regarding the programming language, claiming that "the traditional language, BASIC, would imply bad coding habits, and programs with difficulties in education such that the transfer to structured programming would be difficult" and that with Comal "programming in Swedish schools would build on the basic logical structures".[1] According to Kaiserfeld (1996), this step toward one specific programming language was pivotal in the project and was later criticized among teachers; Comal was an interpretive language that did not follow the standards of the time. This made the authors of this chapter aware of how pedagogy was shaped by implicit values when computer technology and mathematics were merged in a school context, something we can witness again in today's development of curriculum with different programming languages in use.

3.2 *Interacting with Computers*

During the 1970s computer technology improved significantly in society, enhancing the benefits of working in many iterations. However, hardware constraints made the user wait, and the pre-processing and compilation of code were slow. Graphical user interfaces would not exist until the 1990s; instead, upper secondary students had to go through a number of iterations before the code worked as expected. Besides this, there was a drawback when working with some (structured) programming languages: there were no feedback references from the program to the source code, making the debugging process tedious as one had to go through the logic, instruction by instruction, to figure out what was wrong (Habermann, 1973). According to Rogers (1984), this was devastating for the learners, who presumably preferred a quick feedback loop when interacting with computers. Thus, the time at the computer was crucial, and the programs were commonly divided into subroutines for better management when learning how to read and write the code. The time spent on details and the underlying hardware was distracting, as the overall aim involved solving problems. In short, the programming activity had to be student-centric, but still authentic, when experiencing the power of computation through computers.[2] Ekman and Fröberg (1969) argued that merely reading code solutions in a book is a waste of time, as a programming activity has to be experienced in a dynamic process whereby students can learn the limitations of the media (which at the time were code instructions on punched cards/tapes). They also described how typos as well as grammatical and runtime errors emerge when using a programming language, and noted that students therefore rely on feedback from the computer.

In literature from the 1970s (Dahlquist & Fröberg, 1971), we notice that efficiency and interactivity were issues of significance for how both experts and early adopters approached programming, which brought important values and major pedagogical ideas into why, what, and how to teach about these matters. While higher-level programming languages introduced structure to facilitate the coding process, teachers still had to create activities that motivated computation through computer technology such that students could work in an exploratory way with code. We interpret this as early signs of a pedagogical practice relying on students' tinkering actions following the scheme "try something, think about it, make modifications, try again, etc." (Cope & Walsh, 1990, p. 213).[3] Students were to experience how their thinking evolved when experiencing and handling bugs before the code instructions as a whole worked as intended.

3.3 *Structure in Programming*

The question of programming structure was raised in the late 1960s (Dijkstra, 1968; Dahl et al., 1972) and was still debated 20 years later (Rubin, 1987). For many years, structured languages demanded extra computer memory or extension boards in the computer. Therefore, in 1964, the Beginners' All-purpose Symbolic Instruction Code (BASIC) was developed for education, followed in 1975 by the Common Algorithmic Language (Comal). BASIC was heavily criticized, as it was believed that students would develop bad habits when coding. There were guidelines and principles that could help make the code structured (see Lundgren et al., 1980/1982),[4] but problems could often be avoided by using flow charts and avoiding GOTO instructions (Sveinsdottir & Frøkjær, 1988).

In a structured programming language, one expects the code to follow the separation of concerns (Dijkstra, 1982; Laplante, 2007), a design principle that makes the code readable and facilitates a debugging process. In BASIC, however, the programmer does not have to follow this design principle, which makes the code harder to maintain and communicate. In the story of Comal (Christensen, 1989), we come to learn how one Danish teacher trainer tried to integrate this design principle in a programing language in order to help students experience how computers respond almost instantaneously. Christensen writes: "We took what we had of BASIC, and added the structure of Pascal" (p. 2), including the procedures and benefits of reading code with indentation. They also added long variable names, which the "students appreciated while teachers were more hesitant" (p. 3). Other programming languages at the time did not include eight characters in variable names, and did not offer the option to make the code readable. The story eloquently describes how Christensen, a teacher trainer in mathematics,[5] recalls some necessary steps in the transposition of scholarly knowledge while preserving the separation of concerns.

4 Method

Inspired by Breen's "Imagining the Past" (1996) which employs a historical ethnographic approach, our research examines historical debates (as events) that not only narrate the continuous process of knowledge transposition, induced by experts and early adopters, but also explore how they conceptualised themselves within the context of an emerging field of study during the 1970s and 1980s. When scanning existing journals of that time, we identified two journals that gave room for such debates. Obviously, there were different journals and voices about teaching and learning programming in upper secondary school, but not many that brought different perspectives to questions related to what, how, and why in teaching programming (see Table 3.1). In the following, we describe the selected journals, how the debates were found in articles and later analysed.

4.1 *Two Journals: 'Elementa' and 'Skolan och Datorn'*

The two Swedish journals *Elementa* and *Skolan och Datorn* explicitly encouraged teachers (and experts) to write and debate what they considered relevant for teaching about programming in secondary school. The journal *Elementa* existed from 1938 to 2004, as a foundation to facilitate education in mathematics, physics, and chemistry. Its number of subscriptions peaked at 3000, and the journal appeared in almost every Swedish upper secondary school library. Its editorial board originated from three different associations for education in mathematics and the natural sciences, and its authors embraced mainly people from academia and teachers within the natural sciences. The journal *Skolan och Datorn* was a forum for the user association with the same name, and was published from 1980 to 1988 by Esselte Studium, with three to four issues each year. The journal was mainly for the teachers and schools who used computers Esselte 100/1000 and Compis. Its editorial board embraced a diversity of people with interest in pedagogical issues in relation to computer technology, including people from Esselte Studium and professional programmers as well as teacher trainers, upper secondary teachers, textbook authors, and students (Emanuel, 2009; Wennersten, 1980).

4.2 *Identifying Pedagogical Debates*

In our search for differences and similarities during the two decades, we decided to focus on one common task in mathematics education, the extraction of prime numbers. The reason we chose this task was that it had become a benchmarking tool that was widely used during the 1970s and 1980s when evaluating computer hardware and programming language performance (see Gilbreath, 1981; Gilbreath & Gilbreath, 1983). In relation to the purpose of this chapter, that specific task was the prerequisite when searching for articles in the selected journals that describe how the pedagogical debates evolved around different technological values, ideas and epistemological beliefs regarding the pedagogical practice.

The history of programming in school is commonly described in periods that correlate with the computer technology evolution. For instance, Aslan and Reigeluth (2011) distinguish two different periods: the mainframe period (late 1950s to late 1970s) and the microcomputer period (late 1970s to the end of the 1990s), although archive documents, school inventory lists, and contemporary journals depict how different computer technologies co-existed in many schools. We found one debate between experts and early adopters from each of the two periods that provided rich insight into the development of pedagogical ideas at the time, and thus chose these for an in-depth analysis.

4.3 *Analysing the Pedagogical Debates*

The analysis presupposes how experts and early adopters are aware of the significance of what they write for the reader community and how they present pedagogical ideas and technological values as they write their articles. After all, a teacher's pedagogy emerges from "experience, knowledge and discussion in the field of education, not just on the more or less accidental qualities of the technology" (Pedersen, 2001, p. 64).[6]

Early in the analysis, we noticed how the computer science experts pursued authenticity in relation to mathematics, while the early adopters pursued pedagogical benefits in computer technology when learning mathematics. For a more detailed analysis of the knowledge they advocated, we deployed the praxeology from ATD as an analytical tool. For the praxis block (what and how) we analysed how the prime numbers were extracted, and for the logos block we identified why this was done and what the aim and learning goals were. Through this detailed analysis we found how the experts and early adopters disagreed on the *technique, technology* and *theory,* whereof the latter elicits the fact that experts heralded what the National Board of Education suggested early in the process—structural thinking. We return to this in the discussion.

5 Results and Analysis

In the following, we describe and analyze two cases to highlight how programming was debated among experts and early adopters during the 1970s and 1980s, and how the characteristics of computer technology and mathematics evolved in the development of a new pedagogical practice.

5.1 *First Case—A Debate from the Mainframe Period*

The first case is a debate between an expert and an early adopter, found in the journal *Elementa* in 1972–1973. The expert, Hans Riesel, worked as a lecturer and associate professor at the Department of Numerical Analysis at the KTH Royal Institute of Technology, and for many years held the record for discovering the largest known prime number. In 1969, he received his PhD with the thesis "Contributions to numerical number theory".

The early adopter Lars-Erik Björk, taught mathematics and physics at the upper secondary school Sunnerboskolan, and was eager to incorporate computers and programming in school. He was heavily engaged in many projects along with the Swedish Board of Education, and did considerable work in the integration of the subject *Computing* into the natural sciences program. He authored several textbooks on mathematics, with special emphasis on programming in mathematics. Much of his work was anchored in experiences as a teacher trainer. He also published many articles in teaching journals.

5.1.1 Debate between Riesel and Björk

In an article from 1972, upper secondary teacher Björk describes how he used programming as a tool in mathematics to teach about the extraction of prime numbers (see Figure 3.2). He and his students connected with a mainframe computer, via a terminal and a modem.

The article initiated a debate between Björk and the expert Riesel, who elegantly illustrates how the content is shaped by values, ideas and outcomes in computer technology. In following issues of the journal, Riesel suggests how the code can be modified such that the execution time will decrease (see Figure 3.3). He writes:

> It could be of interest to the students to see how these modifications decrease the execution time. The execution time decreased from 0.76 sec to 0.38 sec (excluding the output)[7] for n = 1000. One-half of the execution time was thus saved. (Riesel, 1973, p. 43, translation from Swedish)

In the same issue, Björk replies with the following comment:

```
3 PRINT "ANGE DET STÖRSTA TALET SOM SKALL UNDERSÖKAS"
4 PRINT
5 INPUT N
6 PRINT
7 PRINT "DETTA ÄR EN FÖLJD AV PRIMTAL"
8 PRINT
10 LET T = 2
20 LET D = 2
30 IF D <= SQR (T) THEN 70
40 PRINT T
50 LET T = T + 1
60 GOTO 20
70 IF INT(T/D) = T/D THEN 100
80 LET D = D + 1
90 GOTO 30
100 LET T = T + 1
110 GOTO 20
120 PRINT "VILL DU HA EN NY SERIE?"
130 INPUT A$
140 IF A$ = "JA" THEN 3
150 END
```

FIGURE 3.2 The initial code, by Björk (1972, p. 142). The code extracts a series of prime numbers

```
10 LET T = 2
11 PRINT T
12 LET T = 3
13 PRINT T
14 LET T = 5
15 PRINT T
16 LET T = 6
18 IF INT (T/2) = t/2 THEN 100
19 IF INT (T/3) = T/3 THEN 100
20 LET D = 5
21 LET V = 4
29 LET V = 6 - V
30 IF D <= T/D THEN 70
40 PRINT T
50 T = T + 1
60 GOTO 18
70 IF INT (T/D) = T/D THEN 100
80 LET D = D + V
90 GOTO 29
100 LET T = T + 1
105 IF T <= N THEN 18
```

FIGURE 3.3 The modified code, by Riesel (1973, p. 43)

> In a comment by Hans Riesel, he demonstrates some immediate time savings in the execution of the program. In principle, the remark is interesting, but since the modified version [...] loses significantly in simplicity when it comes to the flow chart, it can hardly be recommended as an exemplary case for the introduction of programming. (Björk, 1973, p. 44, translation from Swedish)

The debate embraces three articles, pinpointing how an early adopter and an expert approach the same algorithm translated into code from different perspectives. Björk holds high standards in simplicity and pedagogy, while Riesel is searching for an educational activity in which the code becomes efficient and interesting to learn, suggesting that it is valuable to optimize the code. The same task is approached with different *techniques* and backed up by different *theoretical* justifications. In addition, their institutional settings influenced them differently, with Björk and his students working persistently with a terminal connected from a distance to a mainframe computer, and Riesel, at KTH, drawing from a computer next door (BESK or ALWAC).

The turn taking between the expert and the early adopter is interesting, as it mirrors the constraints one encounters when working with a specific computer technology in mind. The debate depicts pedagogical ideas about how a learning activity can be authentic and justified in a school setting. It ends abruptly when Björk returns to the original question of how computers (programming) can offer something new to curriculum:

> After this side debate about marginal time savings in simple programs, let us hope we can discuss the pedagogical, organizational and economic issues connected to the integration of computers in school. (Björk, 1973, p. 44, translation from Swedish)

In short, the technicalities and optimization of code were of minor significance to Björk. Aa a teacher trainer and mathematics teacher he embraced a different perspective on the learning and teaching of programming. Today, some of the issues he mentioned, like economy and organization, are routine, while the pedagogical issues remain.

The debate depicts the transposition of knowledge as a teaching practice evolves. In praxeological terms, according to the expert, the technological aspect of the logos (why to choose a specific code) is that the sequence of instructions matter for efficiency and execution time when extracting prime numbers using BASIC, whereas pedagogical issues in mathematics override technical details for the early adopter.

5.2 *Second Case—A Debate from the Microcomputer Period*

The second case is a debate between an expert and an early adopter, found in the teachers' journal *Skolan och Datorn* in 1985, ten years after the first case. The activity and issue at stake are the same: how to extract prime numbers and the execution time. Here the expert is Hans Svensson, who earned a licentiate degree (about half of a PhD) at the Faculty of Engineering at Lund University in the 1960s where he worked as an amanuensis, together with a well-known Swedish computer scientist (Dahlstrand, 2009). Later, he worked as a lecturer in upper secondary school for about 25 years. Svensson was a frequent author in the journal, publishing in total 23 articles, peaking at seven in 1985. In one of his articles, he explains that he is in the process of writing a book about Comal on a Compis computer (Svensson, 1983). In short, he was an expert who brought his scholarly knowledge from academia into the upper secondary school.

The early adopter Roland Engdahl worked from 1954 to 1991 at the Department of Education at Kalmar University (now Linnaeus University), as a teacher trainer in mathematics and physics. According to historical documents (Linnaeus University's archive), the department did not offer any courses on digital computers until the beginning of the 1980s. Engdahl attended a national in-service training program in 1982 on computers in school (within the DIS project), and a year later the department offered courses on the application of computing with terminals, connected to Lund's computer center. In 1983, they requested money from Lund University to buy some personal computers (ZX Spectrum, Commodore 64, and Esselte 100) and a cassette recorder, including 20 floppy disks, so that they could demonstrate and educate pre-service teachers with "state-of-the-art computer technology".

5.2.1 Debate between Svensson and Engdahl

In different articles by Svensson (e.g. 1983 and 1984), we come to know how programming languages are valued in terms of being structured and non-structured. This is clear in an interview by Wettstam (1981), where Svensson suggests teaching programming languages in a progression, from BASIC in lower secondary school to Pascal in upper secondary school. However, two years later, Svensson (1983) changed his opinion about structural aspects, due to the advent of a third language, Comal. In fact, many of his articles revolve around the benefits of that structured language for upper secondary students' thinking when programming.

In the initial article, early adopter Engdahl (1985a) expands on an article from another journal (Wettstam, 1984/1985) in which the code (written in Comal) and its corresponding algorithm describes the sieve of Eratosthenes and its procedure. According to him, the article "is too short in explanation and can be misunderstood by those who are ignorant of the sieve algorithm". Besides, he considers the code to embrace some mathematical misunderstandings and claims that the outcome of the sieve "is actually non-genuine" because of its limit of numbers ($n \leq 20000$). This is the incentive for Engdahl to expand on the initial article. According to him, the explanation of the algorithm would be better off with different code examples in comparison (see Figures 3.4A and 3.4B). He suggests three different examples in which the prime numbers are stored differently: floating point list, integer list, and text string (not discussed in this chapter), to allow students to experience how computer technology carries limitations as to what, and how, numbers are stored in the computer memory in different types of lists. However, in the article's last paragraph he also questions the upcoming school programming language, Comal, and how the language is commonly promoted among the "structural worshipers" (Swedish: strukturdyrkarna), claiming that "one will not gain bad coding habits because of BASIC, as 'some state'".

In the article, Engdahl compares the outcomes of code in BASIC, Pascal, and Comal, all representing the same mathematical algorithm. The examples were executed on different computers (Apple II and ZX Spectrum, see Figure 3.4A and 3.4B), and according to him, the execution time for BASIC and Comal were equivalent. In his experimental work with BASIC, all examples exceeded seven minutes, while they would take approximately seven seconds in Pascal. Therefore, he argues, BASIC is adequate for programming novices. Pascal is mainly for more advanced users as it is "fast and structured, but hardly simple and easy to understand". He ends the article with some open questions: "Where does Comal fit in?"; "Why did they choose Comal as a school language?"; and finally, "Who said Caesar's wife must not even be suspected?" In short, Engdahl questions the benefits of Comal, as BASIC and Pascal are sufficient for programing in school. However, his final question is provocative, and in the following issue of the journal, the expert Svensson is triggered and the pedagogical debate is initiated (although he does not actually address the provocative question).[8]

In his refute, Svensson (1985b) claims that Engdahl's comparison is deficient, as the computer hardware constraints have a huge impact on the execution time with BASIC. The critique is major, and we can identify at least four threads in his argumentation: (1) the three examples could be expanded to make them mathematically interesting; (2) there are faster BASIC dialects; (3) the sieve as a benchmarking tool, and finally (4) the expert and the speed of code. Svensson considers the three examples (with different code) to be of minor interest, but sees a potential interest if one builds a code that traverses the odd numbers, revealing the existence of only one even prime number. In short, Svensson's arguments include theoretical values, ideas and beliefs from computer science and mathematics.

In the following quote, Svensson is concerned about the double loops (Rows 40 to 60, Figures 3.4A and 3.4B) that will be processed only if the conditional statement is false in row 40.

```
Dimensionerad lista med flyttal. Spectrum

10 REM ERATOSTENES SÅLL
20 INPUT "ANGE ÖVRE GRÄNS (<8000) "; N
30 DIM P(N)
40 FOR I=2 TO SQR N: IF P(I)=1 THEN GO TO 60
50 FOR J=I*I TO N STEP I: LET P(J)=1: NEXT J
60 NEXT I
70 LET tab=0: REM UTSKRIFT
80 FOR K=2 TO N: IF P(K)<>1 THEN PRINT
   TAB tab; K;: LET tab=tab+8
90 NEXT K
```

FIGURE 3.4A First example on a ZX Spectrum, with prime numbers in a floating point list

```
Dimensionerad lista med heltal. APPLE

10 REM ERATOSTENES SÅLL
20 INPUT "ANGE ÖVRE GRÄNS (<20000) "; N
30 DIM P%(N)
40 FOR I=2 TO SQR (N): IF P%(I)=1 THEN 60
50 FOR J=I*I TO N STEP I: LET P%(J)=1: NEXT J
60 NEXT I
70 REM UTSKRIFT
80 FOR K=2 TO N: IF P%(K)<>1 THEN PRINT K,
90 NEXT K
```

FIGURE 3.4B Second example on an Apple computer, with prime numbers in an integer list

> Engdahl's programs are simple, but are they easy to understand? There is an IF-THEN jump in row 40, in the Apple program, such that one can skip the next instruction if P% is a prime number. The double loops in rows 40–60 embrace two conditional statements, both FOR and IF, a [coding] technique that is pedagogically devastating. (Svensson, 1985b, p. 8, translation from Swedish)

According to Svensson, the functionality of the sieve becomes hard to understand with Engdahl's code examples. In another article, Svensson (1985a) criticizes teachers' competency in BASIC, saying it belongs to the 1950s, with machine language routines like flow charts, iterations, and jumps. According to him, many teachers are not serious as they are commonly autodidact in programming; hence, they are in need of in-service training in the integration of structured programming. In praxeological terms, Svensson focuses on the technology part of logos, highlighting details that teachers need to know in order to teach.

In the same issue, Engdahl (1985b) defends his three code examples and comparison, as he actually deployed the structure used in the original code (Wettstam, 1984/1985). He suggests that Svensson should "present what a better coding technique would look like in BASIC". At the end of the article, he refers to a professional programmer who emphasizes the benefits of BASIC. After all, Comal as a language demanded an expansion of a computer's memory, which at the time was quite expensive. In contrast to Svensson, Engdahl focuses more on the translation from the mathematical algorithm into code—a praxis in which the code examples are compared and discussed in relation to the mathematical algorithm. He looks more to the theory aspect of logos, focusing on what students learn about mathematics.

Finally, after three turns in different issues, the journal's editor closes the debate because the two seem to have different foci.

6 Discussion

In the two debates, we have found how the experts Riesel and Svensson showed an interest in the accuracy and efficiency of code, while the two early adopters Björk and Engdahl had an ambition to make programming relevant as a subject in school mathematics. In relation to the theoretical framework, it is clear that the task (what) involves the extraction of prime numbers, and that the techniques (how) largely concern technical details in one or several programming languages, and the underpinning hardware. In both cases, the experts push for how the code needs to be structured or exchanged such that the execution time and computer memory are efficiently used, and follow the contemporary knowledge body in computer science. Conversely, the early adopters answer the didactical questions from a mathematics teacher's

TABLE 3.2 The transposition of knowledge described as praxis and logos; first on the level of *scholarly knowledge* expressed by experts, then as *knowledge to be taught* expressed by early adopters

		Experts	Early adopters
Praxis	Task (what)	Extracting prime numbers using code (1970s & 1980s) Interacting with a mainframe, next door (1970s)	Extracting prime numbers using code (1970s & 1980s) Interacting with mainframe, at a distance (1970s)
	Technique (how)	Interacting with a personal computer. Programming lang. Comal (1980s)	Interacting with a personal computer. Programming lang. BASIC (1980s)
Logos	Technology (why)	Optimizing code and memory use (1970s) programming efficiency (1980s)	Transition from algorithm to code (1970s & 1980s)
	Theory (why)	Authentic computing (1970s) Structural thinking (1980s)	Simplicity and clarity, learning mathematics (1970s & 1980s)

perspective, with implications on how the code instructions solve the task and enhance mathematical learning, see a summary in Table 3.2.

From the above examples of pedagogical debates concerning programming in schools we can see that the scholarly knowledge brought forward by experts is transposed when it enters a school context where early adopters are the agents. Firstly, the transposition is influenced by the fact that the content appears within the subject of mathematics. Secondly, the practical conditions differ in that early adopters in schools have less access to efficient computer technology, compared to what experts are used to.

Despite the fact that both experts and early adopters refer to the same task, there is a difference in what they see as the learning aims and outcomes of the activity. In terms of Chevallard's (2006) idea of praxeology, the transposition of knowledge can be described as in Table 3.2. The experts and the early adopters argued differently about the *knowledge to be taught*. In principle, Table 3.2 and the debates described above picture how the experts argued based on the technical details in the programming language, as they knew them very well and strived for authentic and structural thinking, while the early adopters' arguments more concerned the actual transition from algorithms to code in interaction with a computer machine, and how this was made authentic in relation to computational and algorithmic thinking and the underlying mathematical ideas. Although the experts also used mathematical arguments, they approached computation differently with the code's syntactical rules in mind. In the transposition from the experts (representing scholarly knowledge) to the early adopters (representing knowledge to be taught), the chapter reveals how different preferences contributed to the debates. In short, the individuals in the two debates argued from perspectives in both mathematics and computer science, and not solely from a technological perspective, which is of significance even today as teachers develop their teaching practices with different programming languages. Therefore, we would like to suggest how the debates described above picture a transposition process of scholarly knowledge that could inform the transposition of knowledge in the current integration of programming in a school context.

7 Epilogue

Today, on social media, teachers are experiencing debates where they ask questions and receive answers from experts and teacher colleagues. Yet the challenges of the transposition into knowledge to be taught remain. It should be remembered that much of the logos is embedded and black-boxed in the computer technology itself, making it intriguing to create learning activities. In fact, there are moments when teachers need to understand the implicit technological arguments that constrain the praxis, in order to be able to appropriately transpose the scholarly knowledge to the knowledge to be taught, as the work of translating even the simplest mathematical algorithm into code could evolve into a challenging teaching- and learning activity.

Notes

1 Swedish National Archives, SE/RA/420664/F 5/20 "Underlag i Specifikationsarbetet".
2 There were programming environments with microworlds, e.g. Logo, that made the technical details less dominant when learning programming. However, Logo was seldom used in Swedish schools (Hedrén, 1990).
3 According to Cope and Walsh (1990), students in schools commonly approach a programming task with Run-Understand-Debug-Edit (RUDE) while a computer scientist approach programming as a tool somewhat differently with Specify-Prove-Implement-Verify (SPIV).
4 Divide and conquer the original problem into subproblems and comment the code for each section.
5 Christensen was a math teacher at Tønder Seminarium, a teacher's college that received its first minicomputer in 1972.
6 The shift from technology to pedagogy is addressed by many educational researchers (Becker, 1998; Cuban, 1986; Fullan & Hargreaves, 1991).
7 The output was slow in comparison to the execution of the program. This is why Riesel writes "excluding the output" to distinguish between the calculation that happens on the computer and the result that is printed on the teletypes.
8 As Esselte Studium was heavily involved in marketing the Compis computer and was producing educational material, the issue of language makes Engdahl's questions intriguing for the purpose of this chapter.

References

Aslan, S., & Reigeluth, C. M. (2011). A trip to the past and future of educational computing: Understanding its evolution. *Contemporary Educational Technology, 2*(1), 1–17.

Becker, H. J. (1998) Running to catch a moving train: Schools and information technologies. *Theory into Practice, 37*(1), 20–30.

Breen, T. H. (1996). Imagining the Past - East Hampton Histories. University of Georgia Press.

Bråting, K., Kilhamn, C., & Rolandsson, L. (2023). Programming and mathematical competencies: the case of Sweden. In E. Geraniou & U. T. Jankvist (Eds.), *Mathematical competencies in the digital era from a networking of theories point of view*. Springer.

Björk, L.-E. (1972). Elevförsök med dator. *Elementa, 55*(3), 141–146.

Björk, L.-E. (1973). Kommentar till kommentaren. *Elementa, 56*(1), 43–44.

Chevallard, Y. (2006). Steps towards a new epistemology in mathematics education. In M. Bosch (Ed.), *Proceedings of the Fourth Congress of the European Society for Research in Mathematics Education*, CERME 4, 21–30. FUNDEMI IQS-Universitat Ramon Llull.

Chevallard, Y., & Sensevy, G. (2014). Anthropological approaches in mathematics education: French perspectives. In S. Lerman (Ed.), *Encyclopedia of mathematics education* (pp. 38–43). Springer.

Christensen, B. (1989). The story of COMAL. *Comal today, 25*, 1–10. Published by Comal user group.

Cope, P., & Walsh, T. (1990). Programming in schools: 10 y on. *Journal of Computer Assisted Learning, 6*, 119–127.

Cuban, L. (1986). *Teachers and machines: The classroom use of technology since 1920*. Teacher College Press.

Dahl, O., Dijkstra, E. W., & Hoare, C. A. R. (1972). *Structured programming*. Academic Press.

Dahlquist, G., & Fröberg, C.-E. (1971). Datamaskinutvecklingen i Sverige - ett försök till historiebeskrivning. In *Statens naturvetenskapliga forskningsråd. Svensk naturvetenskap: Statens naturvetenskapliga forskningsråds årsbok. Årg. 24*. Statens naturvetenskapliga forskningsråd.

Dahlstrand, I. (2009). To sort things out. Ingemar Dahlstrands egen minnesberättelse. Tekniska museets IT-minnen. Retrieved August 2, 2022, from web.archive.org

Dijkstra, E. W. (1968). Letters to the editor: Go to statement considered harmful. *Communications of the ACM, 11*(3), 147–148.

Dijkstra, E. W. (1982). *On the role of scientific thought. Selected writings on computing: A personal perspective*. Springer-Verlag.

Ekman, T., & Fröberg, C. (1969). *Data för skolan: numeriska metoder, datamaskiner, programmering*. Studentlitteratur.

Elementa: matematik, fysik, kemi. (1938–2004). Elementa.

Emanuel, M. (2009). *ABC 80 i pedagogikens tjänst: exempel på tidig användning av mikrodatorer i den svenska skolan: transkript av ett vittnesseminarium vid Cloetta Center i Linköping den 23 september 2008*. KTH, Avdelningen för teknik- och vetenskapshistoria.

Emanuel, M. (2009). *Datorn i skolan: Skolöverstyrelsens och andra aktörers insatser, 1970- och 80-tal: transkript av ett vittnesseminarium vid Tekniska museet i Stockholm den 30 oktober 2008*. KTH, Avdelningen för teknik- och vetenskapshistoria.

Engdahl, R. (1985a). Eratosthenes' encore! *Skolan och Datorn*, 2, 10–11.

Engdahl, R. (1985b). Debatt, Debatt och Debatt - Finis Eratosthenes. *Skolan och Datorn, 1985*(3), 9, 21.

Ensmenger, N. (2010). *The computer boys take over: Computers, programmers, and the politics of technical expertise*. MIT Press.

Fagerström, P. (1973). *Datorn i skolkommunen (DISK): Kartläggning, beskrivning, förslag till åtgärder*. Linköping.

Fullan, M. G., & Hargreaves, A. (1991). *What's worth fighting for? Working together for your school*. Ontario Public School teachers' Federation.

Gilbreath, J. (1981). A high-level language benchmark. *Byte*, September, 180–198.

Gilbreath, J., & Gilbreath, G. (1983). Eratosthenes revisited: once more through the sieve. *Byte*, January, 283–325.

Goldstein, G. (1968). Computing for mathematics pupils. *Mathematics Teaching, 45*, 22–28.

Goldstein, G. (1971). *Computers and the teaching of numerical mathematics in the upper secondary school: A report prepared for the Mathematical association by a subcommittee of the teaching committee*. Bell.

Grady, M. T., & Poirot, J. L. (1983). Teacher competence: What is needed. In M. T. Grady & J. D. Garonski (Eds.), *Computers in curriculum and instruction* (pp. 78–82). Association for Supervision and Curriculum Development.

Habermann, A. N. (1973). Critical comments on the programming language PASCAL. *Acta Informatica, 3*, 47–57.

Hedrén, R. (1990). *Logoprogrammering på mellanstadiet: en studie av fördelar och nackdelar med användning av Logo i matematikundervisningen under årskurserna 5 och 6 i grundskolan*. Linköpings universitet.

Kaiserfeld, T. (1996). Computerizing the Swedish welfare state: The middle way of technological success and failure. *Technology and Culture, 37*(2), 249–279.

Kilhamn, C., & Bråting, K. (2019). Algebraic thinking in the shadow of programming. In U. T. Jankvist, M. van den Heuvel-Panhuizen, & M. Veldhuis (Eds.), *Proceedings of the Eleventh Congress of the European Society for Research in Mathematics Education, CERME11* (pp. 566–573). Freudenthal Group & Freudenthal Institute, Utrecht University and ERME.

Knuth, D. E. (1969). *The art of computer programming volume 2: Seminumerical Algorithms*. Addison-Wesley.

Knuth, D. E. (1984). Literate programming. *The Computer Journal, 27*(2), 97–111.

Lacina, L. J. (1984). *The determination of computer competencies needed by classroom teachers*. Distributed by ERIC Clearinghouse.

Laplante, P. A. (2007). *What every engineer should know about software engineering* [Electronic resource]. Taylor & Francis.

Lundgren, J., Lundin, B., & Thornell, S. (1980/82). *ABC om programmering och dokumentation*. Emmdata.

Molnar, A. (1982). Summary of the conference. In national computer literacy goals for 1985 conference (1982). *Computer Literacy: Issues and Directions for 1985*. Academic Press.

Nofre, D., Priestley, M., & Alberts, G. (2014). When technology became language: The origins of the linguistic conception of computer programming, 1950–1960. *Technology and Culture, 55*(1), 40–75.

Papert, S. (1987). Information technology and education: Computer criticism vs. technocentric thinking. *Educational Researcher, 16*(1), 22–30.

Pea, R. D. (1987). The aims of software criticism: Reply to professor Papert. *Educational Researcher, 16*(5), 4–8.

Pedersen, J. (2001). Technological determinism and the school. *Journal of Educational Enquiry, 2*(1), 61–65.

Riesel, H. (1973). Kommentar till "Elevförsök med dator". *Elementa, 56*(1), 42–43.

Rogers, W. J. (1984). A new program structure to improve accuracy and readability of Pascal software. *ICSE '84: Proceedings of the 7th International Conference on Software Engineering, 1984*, 454–462.

Rolandsson, L. (2011). Teacher pioneers in the introduction of computing technology in the Swedish upper secondary school. In J. Impagliazzo, P. Lundin, & B. Wangler (Eds.), *History of Nordic Computing 3. HiNC 2010. IFIP advances in information and communication technology* (Vol. 350). Springer.

Rolandsson, L. (2013). Changing computer programming education; The dinosaur that survived in school – an explorative study about educational issues based on teachers' beliefs and curriculum development in secondary school. In *Learning and Teaching In Computing and Engineering* (*LATICE 2013*). Macau, Macao.

Rolandsson, L., & Skogh, I.-B. (2014). Programming in school: Look back to move forward. *ACM Transactions on Computing Education, 14*(2), Article 12.

Rubin, F. (1987). GOTO considered harmful. *Communications of the ACM, 30*(3), 195–196.

Segal, H. P. (1996). The American ideology of technical progress: Historical perspectives. In S. T. Kerr (Ed.), *Technology and the future of schooling*. University of Chicago Press.

Skolan och Datorn. (1980–1988). Esselte studium, Datorgruppen.

Svensson, H. (1983). Skoldatorns operativsystem. *Skolan och Datorn, 1983*(3), 21–23.

Svensson, H. (1984a). Vi besöker IFIP' 83 i Paris. *Skolan och Datorn, 1984*(1), 10–11.

Svensson, H. (1984b). Compas på Compis dvs Pascal på skoldatorn. *Skolan och Datorn, 1984*(1), 24–26.

Svensson, H. (1985a). He who can does, he who cannot teaches Basic. *Skolan och Datorn, 1985*(2), 14–15.

Svensson, H. (1985b). Debatt, Debatt och Debatt - Red. För Skolan och Datorn. *Skolan och Datorn, 1985*(3), 8.

Swedish Board of Education [Skolöverstyrelsen]. (1980). *Datorn i Skolan: DIS: SÖ:s handlingsprogram och slutrapport SÖ-projekt 628*. Skolöverstyrelsen.

Swedish Board of Education [Skolöverstyrelsen]. (1981). *PRODIS: Programvara och datorutrustning i skolan: slutrapport SÖ-projekt 6205*. Skolöverstyrelsen.

Swedish Board of Education [Skolöverstyrelsen]. (1983). *Läroplan för gymnasieskolan 2 Supplement, 95 Datakunskap [Curriculum for upper secondary school 2 Supplement, 95 Computing]*. Liber Utbildningsförlaget.

Swedish Board of Education [Skolöverstyrelsen]. (1984). *Läroplan för grundskolan. Datalära i grundskolan: studieplan [Curriculum for compulsory school. Computer Literacy: syllabus]*. Liber Läromedel/Utbildningsförlaget.

Sveinsdottir, E., & Frøkjær, E. (1988). Datalogy—The Copenhagen tradition of computer science. *BIT Numerical Mathematics, 28*, 450–472.

TUDIS. (1985). *En svensk skoldator. Presentation av teknikupphandlingsprojektet Datorn i skolan*. Styrelsen för teknisk utveckling.

Wennersten, B. G. (1980). Esselte storsatsar för att möta skolans dataoffensiv. *Dagens industri*, 1980-11-27, 28.

Wettstam, L. (1981). Avancerat fleranvändarsystem i Kristianstad, *Skolan och Datorn, 1981*(1), 6.

Wettstam, L. (1984/1985). Uppslaget. *Nämnaren – matematiken i skolan, 11*(2), 35.

CHAPTER 4

Introducing Programming and Computational Thinking in Grades 1–9

Sweden in an International Context

Linda Mannila and Fredrik Heintz

Abstract

In this chapter, we situate Sweden in an international context focusing on how programming and computational thinking have been introduced into primary and lower-secondary education (grades 1–9 in the Swedish system). Our review shows that the strategies used in different countries have their own pros and cons, and there is no clear evidence establishing that one method is preferable. Moreover, due to a lack of clear guidelines, decisions on how programming is taught, by whom, and when, are commonly made at school level, also in Sweden. This freedom, or burden, to locally decide on how to implement the curriculum has left teachers in a difficult position, where they are to fulfil the requirements of the curriculum without proper training, time, and competence needed. This has naturally had a negative impact on how programming and computational thinking have been and are introduced at schools. Based on the review we provide six recommendations, which posit that to succeed, a much more systematic and holistic approach is needed, addressing the needs of teachers, students, and schools.

Keywords

programming – computational thinking – K-12 education – international perspective

1 Introduction

In the 1980s, programming was taught at primary level in many countries. Towards the end of the decade, it became more common to have a computer at home, the machines became increasingly user-friendly, and the amount of ready-to-use software grew. Consequently, in the 1990s, it was considered more important to learn how to use technology as a tool, rather than focusing on the

© LINDA MANNILA AND FREDRIK HEINTZ, 2024 | DOI:10.1163/9789004687912_004 underlying principles of digital technology (Kawai & Burke, 2013). Programming was removed from school curricula, also in Sweden (Utbildningsdepartementet, 1994).

In the 2000s, as a consequence of computer technology's transformational effects in all societal areas, focus again changed as voices were increasingly highlighting the importance of programming and computing for everyone. For instance, in 2013, Informatics Europe and ACM Europe published a report (Gander et al., 2013) emphasizing the importance of helping all children (1) develop skills to use the technology, (2) understand the science behind the technology as well as (3) get insight into the possibilities and challenges it brings.

In 2012, the Swedish government established the Digitalization Committee (Digitaliseringskommissionen) with the task of providing guidelines for the future of work related to digitalization in Sweden. One of the committee's reports (Digitaliseringskommissionen, 2014) highlighted the need for the school system to put a larger focus on digital competence and explicitly pointed out the need to include programming in the curriculum. In 2017, the National Agency for Education (Skolverket) published a revised national curriculum including programming starting from first grade. The revisions were to be implemented in schools by Fall 2018.

This development can be seen globally, as computing-related topics have been reintroduced in an increasing number of countries. For instance, already in 2015, 17 out of 21 European countries surveyed, reported doing so (European Schoolnet, 2015). In 2022, a similar survey (Bocconi et al., 2022) showed that out of 29 European countries analysed, 25 had already introduced computing topics to some degree in their curriculum. The term used to describe the new content varies, including, for instance, computer science, computing, programming, computational thinking, informatics, and digital competence. Regardless of the term used, programming, or computational thinking, is central to all curricula. As we will see below, programming is commonly used as a tool to cover CT, while there are also initiatives where CT is introduced without programming. In order not to end up in a situation where something would need to be left out due to the choice of wording, we, therefore, choose to use the acronym PCT (Programming and Computational Thinking).

In this chapter, we situate Sweden in an international context focusing on how PCT has been introduced in primary and lower-secondary education (grades 1–9 in the Swedish system). The chapter is organized as follows. We first briefly discuss PCT at K-12 level and the international discussion that led to many countries deciding to introduce PCT in their national curricula. Next, we discuss the Swedish situation and the background that resulted in the revised curriculum. This is then related to a corresponding review of the situation in

Finland, Norway, Estonia, England, and South Korea. After the review, we summarize the key findings and make some recommendations for the future.

2 PCT in K-12 Education

Kafai and Proctor (2022) highlight three core questions related to introducing computing topics at K-12 level: *who* should learn, *what* should they learn, and *how* should they learn it?

The most common rationale for introducing PCT in K-12 education both in and outside Europe is to foster the skills and competences needed to participate in the digital world (Bocconi et al., 2016). The answer to the first question, *who*, is hence quite easy: PCT is included in core K-12 curricula to give all citizens agency in terms of computational systems, their operation, structure, construction, and use. Without this knowledge, future generations risk disempowerment in terms of exercising their democratic rights. This, naturally, raises questions of how to achieve this in practice to guarantee inclusion and equal opportunity. For instance, if PCT is only offered in elective courses, the goal of preparing everyone for life and work in the digital world will not be met (Storte et al., 2019).

As computing has traditionally been primarily a university-level discipline, there are no widely accepted general standards for what PCT at K-12 level entails (Kafai et al., 2020; Denning & Tedre, 2021). There is hence no consensus around the question *what*. While some may interpret CT as a synonym for computing, or even only programming, others approach CT from a very general perspective thus ignoring the need for a strong connection with the computing discipline (Storte et al., 2019). In her seminal article, Wing (2006) emphasized that CT and programming are not overlapping sets, as "thinking as a computer scientist means more than being able to program a computer" (p. 33). In practice, however, CT in K-12 education is commonly taught mainly using programming tools (e.g., Lye & Koh, 2014; Hsu et al., 2018; Tang et al., 2020; Bocconi et al., 2022).

In 2016, Bocconi et al. (2016) concluded that "establishing a shared understanding of what CT is and how it is contextualized may facilitate the process of curriculum integration", but still in a somewhat corresponding report from six years later, they concur that it is still not clear what concepts should be taught, when and how (Bocconi et al., 2022). Nevertheless, major stakeholders have founded initiatives supporting curricula development and teacher certification (Barr & Stephenson, 2011). For instance, in 2016, the K-12 CS Framework (k12cs.org) was published as the result of a collaborative effort by the ACM (Association of Computing Machinery), CSTA (Computer Science Teachers Association), Code.org, CIC (Cyber Innovation Center) and the NMSI (National Math + Science Initiative), offering a baseline document that could be used to align standards and local curricula. Kafai and Proctor (2022) suggest that there may not be one single answer to the question of what to teach, but that the answer may differ based on community and culture.

Finally, there is the question of how. PCT has been introduced in curricula in different ways and the question how has been addressed from various perspectives, for instance by reviewing current practice (Mannila et al., 2014), studying how teachers transform curricular content into classroom practice (Vinnervik, 2021), or focusing on how CT is framed in the curriculum (Ottestad & Gudmundsdottir, 2018; Olofsson et al., 2021). Another perspective looks to compare the situation across multiple countries (e.g., Heintz et al., 2016; Dagienė et al., 2019; Niemelä et al., 2022). While some countries have introduced a new subject, others, such as Sweden, have integrated the new content into existing subjects, for example by introducing it as an interdisciplinary element to appear throughout the curriculum or within a few subjects (Heintz, Mannila & Färnqvist, 2016; Heintz et al., 2017; Bocconi et al., 2022). While the former approach seems like the most straightforward alternative, there are several reasons why some countries are opting for the latter alternative: (1) lack of space for introducing a new subject in the curriculum, (2) letting children see and experience the use of programming in different subjects is seen as a way of raising interest (Sands et al., 2018) also among previously underrepresented groups, and (3) computational thinking is considered to be important in all areas of society, thereby also all subjects. In their review of PCT education in 29 European countries, Bocconi et al. (2022) found that the most common model at primary level is to integrate the content into existing subjects, while disciplinary specialization becomes more common at lower secondary level.

3 PCT in Grades 1–9 in Sweden

In Sweden, a revised curriculum was presented in 2017. Preparations started already in 2015 as the Swedish government gave Skolverket (the Swedish National Agency for Education) the task to develop a new curriculum for primary education (grades 1–9) and update the curriculum for secondary education (grades 10–12). The explicit requirement was for the curriculum to (1) strengthen students' digital competence and (2) introduce programming.

3.1 *Curriculum*

The revised curriculum came into force in 2018, introducing digital competence, defined as building on four central aspects: (1) understanding how digitalization affects individuals and society, (2) understanding and knowing how to use digital tools and media, (3) using digital tools and resources critically and responsibly, and (4) being able to solve problems and implement ideas in practice. The curriculum was further revised in 2022 (Skolverket, 2022a).

PCT is seen as part of digital competence, and in supplemental material (Skolverket, 2022b), Skolverket stresses that helping students develop these skills is a cross-curricular responsibility. The material highlights that the focus of teaching PCT is not on developing students' coding skills, but rather on the problem-solving process. In addition, programming should be seen "in a broader perspective that includes creative activities, control and regulation, simulation and democratic dimensions" (Skolverket, 2022a, pp. 9–10).

Apart from being part of digital competence, PCT is also explicitly included in the course plans for two subjects—mathematics and technology—in grades 1–9. In mathematics, PCT is introduced mainly under the topic "Algebra" (Skolverket, 2022a, pp. 55–58):

- Grades 1–3: How unambiguous stepwise instructions can be constructed, described and followed as a basis for programming. The use of symbols as stepwise instructions.
- Grades 4–6: Programming in visual programming environments. How algorithms are created and used when programming.
- Grades 7–9: Programming in visual and textual programming environments. How algorithms are created, tested and used when programming.

The subject technology was introduced into the Swedish curriculum in 1994, to help students develop the knowledge and skills needed in a technology-intensive world. The revision emphasizes the understanding of digital technology and includes programming from two perspectives: understanding the role of programming in technological solutions and using programming to develop technological solutions:

- Grades 1–3: Artifacts in students' everyday life that are controlled using programming, e.g., household appliances and smartphones. Controlling artifacts using programming.
- Grades 4–6: How computers are controlled by programs and can be interconnected in networks. Controlling self-made constructions or other artifacts using programming.
- Grades 7–9: How technical solutions that use electronics can be programmed. Terminology used in this context. Controlling and regulating self-made constructions with the use of programming.

Programming is not explicitly mentioned in the criteria for evaluation of students' knowledge and is thus omitted from what students should know to pass a course or get a certain grade.

3.2 *Implementation*

Skolverket has organized training in PCT for in-service teachers through national conferences and online initiatives since the beginning of the revision process. Teacher education programs do not include mandatory courses, but many universities organize commissioned PCT education. Non-profit and other private organizations have actively supported the introduction of PCT even before the revisions were initiated (Heintz et al., 2015).

Nevertheless, studies show that teachers still lack the subject knowledge and pedagogical approaches needed to teach programming compared to the established practices which exist in other subjects (Nouri, 2020). In addition, programming is not mentioned in the knowledge requirements and the curricular guidelines are perceived as unclear (Vinnervik, 2021). This contributes to teachers being unsure of what they are expected to teach, how, when, and to what extent (Stigberg & Stigberg, 2020), which in turn can lead to programming instruction being sporadic and thus fragmented (Mannila, et al., 2022), superficial (Cederqvist, 2021), unequal (Vinnervik, 2021) as well as boring and irrelevant (Mannila et al., 2022). This is problematic as positive experiences are central to increased and maintained interest (Sun et al., 2022; Gerson et al., 2022). These shortcomings set limits to how programming is introduced into the classroom and thus also to the skills and knowledge students can develop (Nouri et al., 2020). This naturally raises questions about what effect the curriculum changes can have in practice (Vinnervik, 2022), especially when research suggests that clear learning goals and suitable teaching methods are central when programming is integrated into teaching (Popat & Starkey, 2019).

4 Models for Introducing PCT in the Curriculum

In this section, we review the introduction of PCT focusing on primary and lower secondary education (grades 1–9) in five countries: Finland, Norway, Estonia, England, and South Korea. There were several reasons for choosing these countries: we wanted to (1) include countries, whose educational systems share similarities with the Swedish one (Finland, Norway, Estonia) as well as (2) examples where PCT has been introduced in a stand-alone subject (England and South Korea). Finally, we also wanted to (3) include a country, whose traditions, cultures, and educational systems are quite different from those in Sweden (South Korea). As such, this selection of countries lets us relate the strategies and curricular reforms made in Sweden to a variety of other contexts. To identify similarities and differences, we have reviewed relevant documentation such as official national curricula, research articles, policy documents and reports for each country.

We start with Finland, where PCT, similarly to Sweden, has been introduced framed by digital competence in existing subjects. Next, we review the situation in Norway and Estonia, where PCT has been introduced both across the curriculum and as part of a computing-related stand-alone subject. Finally, we review the curricula in two countries that have introduced PCT as part of a stand-alone subject only: England and South Korea.

4.1 *Finland*

Finland was the first Nordic country to introduce programming into the curriculum. While previous Finnish curricula focused on using IT as a tool, in 2012, discussions on a new curriculum were initiated highlighting the need for introducing topics, such as programming, that would also help students learn about the technology and its underlying principles.

4.1.1 Curriculum

The new curriculum was introduced in 2014 (Opetushallitus, 2014) and came into force gradually, starting in 2016. Programming was introduced as part of digital competence, which is described in terms of four areas: (1) practical skills and own production, (2) responsible and safe ways of working, (3) information management as well as exploratory and creative work, and (4) communication and networking.

Digital competence is in turn one of seven dimensions of the so-called transversal competence, which plays a central role in the curriculum and can be seen as a combination of knowledge, skills, values, attitudes and will, and are to be covered throughout the curriculum. The transversal competence is closely related to the 21st century skills, involving competencies such as critical thinking, communication, collaboration, problem-solving, creativity, and digital competence. These are not taught in isolation, but rather, digital competence—and hence also programming—should be integrated into other subjects.

In addition to this implicit integration, programming is also explicitly mentioned and part of the assessment in two subjects (mathematics and sloyd):

- In mathematics, the focus is on logical thinking, problem-solving and algorithms, starting with creating and testing step-by-step instructions in grades 1–2, followed by programming in visual environments in grades 3–6, and deepening of programming skills and practice in grades 7–9. Hence, in

grades 1–2 programming can be introduced completely unplugged, while computers, robots, or other digital devices are needed starting in grade 3. The Finnish curriculum does not state when text-based programming languages should be introduced, leaving the choice of programming environment to the teacher.
- Sloyd is a subject taught in the Nordic countries and can be described as crafts or handicrafts using both soft and hard materials. When introducing programming in this context (grades 3–9), new materials and tools are introduced, such as microcontrollers, electronic components, 3D printers and robots. This is hence closely related to the maker culture; a modern version of the DIY culture integrating technology-based materials, which has become popular in both formal and informal educational settings.

In addition to viewing programming in terms of problem-solving, implementing of ideas and creative expression, it can also be considered from a societal point of view: teachers and students can engage in discussions on, for instance, the role of algorithms, data and programmable devices in our society and everyday life today as well as potential scenarios for the future.

4.1.2 Implementation

Since the curriculum was made official, the Finnish government, private actors and universities have funded development programs and projects providing teachers with training, material, and other support. Examples include MOOCs (e.g., mooc.fi), networks (e.g., Innokas, innokas.fi and PROFFS, proffs.edu.pargas.fi) and programming tasks in mathematics textbooks. Similarly to Sweden, many Finnish teachers, nevertheless, lack the time and resources needed to develop their competence to introduce PCT in their teaching (Tanhua-Piiroinen et al., 2020; Mannila, 2021). For instance, the results from a study among Swedish-speaking schools in Finland showed that only a few teachers in grades 1–6 teach PCT, with even fewer doing so in grades 7–9 (Mannila, 2021). The situation was no better if only considering the subjects where programming is explicitly introduced in the Finnish curriculum (mathematics and sloyd). Programming is also one of the areas in which Finnish teachers wish to receive more training (Mannila, 2021). In addition to sufficient professional development, factors such as collegial learning, organizational support, and a joint vision, have been shown to play a role in the successful introduction of PCT in education (Mannila, 2021, 2023).

Also like Sweden, the Finnish curriculum was perceived as too vague and unclear regarding the expectations for PCT-instruction. To mend this situation, the Finnish Ministry of Education and Culture initiated the "New Literacies

Programme" (uudetlukutaidot.fi) in 2020 aimed at developing more detailed and concrete descriptions of the curriculum objectives in three related areas: digital competence, media literacy and programming. The program covers not only grades 1–9, but also early childhood education and pre-primary education. The goal is to make it easier for schools to know what they are expected to teach and thus guarantee equal learning opportunities for all students. A revised version of the descriptions was published in June 2022.

PCT is described in terms of three main areas: (1) computational thinking, (2) exploratory work and production, and (3) programmed environments. These, in turn, include several sub-areas, for which learning outcomes are specified for each grade level (early childhood, pre-primary, grades 1–2, grades 3–6, grades 7–9). Table 4.1 (Mannila, 2023) lists the learning outcomes for grades 1–6 as an example.

The programme was complemented by projects throughout the country, where municipalities and schools were to plan for implementing the learning outcomes in the classroom.

4.2 *Norway*

Digital competence has been part of the Norwegian curriculum since 2006. A government report from 2013 reviewing the barriers to digital value creation (NOU, 2013), however, showed that students are taught communication and presentation, while development, programming, calculations, analysis or understanding aspects were neglected. This initiated a debate on how to introduce such aspects into the Norwegian curriculum. Sanne et al. (2016) recommended the introduction of a new stand-alone subject, including PCT, as this could make it easier to get qualified and competent teachers. Also, they found a risk of the other model, integrating the new content into other subjects, resulting in the new content being deprioritized in favour of established subjects as it would not be given dedicated time and resources. The Center for ICT in Education (Sevik et al., 2016), on the other hand, argued that the most realistic way of introducing PCT in the curriculum was to integrate it into other subjects, while still acknowledging that having a separate subject should be a long-term goal.

4.2.1 Curriculum

Based on this debate, a pilot project was initiated in Fall 2016 involving 146 Norwegian lower secondary schools, where PCT was introduced as an elective course. The pilot had three main goals, to (1) increase students' competence and interest in technology, (2) get experience from teaching PCT in grades 8–10, and (3) increase teachers' PCT competence (Corneliussen & Tveranger,

TABLE 4.1 Learning outcomes for PCT in grades 1–6 according to the New Literacies Programme

	The students can...	Grades 1–2	Grades 3–6
Computational thinking	Logical thinking and processing of information	– Organize and compare things based on different conditions, such as similar form. – Identify logical operations, such as "and", "or" and "not" – Present the choices and observations made using concepts and concrete tools.	– Organize, compare and present information using technology related concepts and symbols. – Perceive connections between different entities. – Find and describe causal relationships between different phenomena.
	Solving and modelling of problems	– Decompose familiar everyday problems into smaller parts, and try to solve them in different ways. – Tell others about the ways in which a problem has been solved.	– Use different methods and strategies for solving problems and also try create such themselves. – Evaluate solutions using given criteria, such as functionality, readability or effectiveness.
	Basic programming concepts and constructs	– Create step-by-step instructions using simple instructions and loops. – Identify error situations caused by the instructions and try to correct them in different ways.	– Create exact and detailed instructions using repetition and selection. – Look for and correct errors in the instructions and the program code.
Exploratory work and production	Co-creation processes	– Present their ideas, listen to others and collaboratively try solutions. – Alternate between different roles when creating programs as a team.	– Describe, in different ways, their own ways of thinking, take the perspectives of others into account and work persistently to achieve the common goal in a programming project.
	Creative production	– Model different things both according to instructions and creatively implementing their own ideas. – Present and share their ideas with others. – Create a digital product containing storytelling or gamified elements using simple programming under guidance or in co-operation with others.	– Use their own observations, measurements and sensors in their products, and combine these with robotics. – Refine existing solutions in a cyclic manner while practicing iterative work, that is, continuous ideation, doing, testing and revising. – Identity computational features in animations and games.

(cont.)

TABLE 4.1 Learning outcomes for PCT in grades 1–6 according to the New Literacies Programme (*cont.*)

	The students can...	Grades 1–2	Grades 3–6
	Programming as a tool for learning	– Process content from other subjects by playfully using and trying out programming related approaches and tools.	– Use programming related tools and approaches for creative expression and own production as well as to explore and explain phenomena in different subjects and multidisciplinary projects.
	Practical skills	– Control a programmable device or a figure in an application..	– Use a visual programming environment and create a program, animation or game with it.
Programmed environments	Programmed technology in different areas of life	– Identify and name programmed devices and services they have experienced and encountered. – Familiarize themselves with robotics. – Describe the purpose and operational principles of devices.	– Describe how technological applications can be used, their basic underlying principles and their role in the students' own lives. – Observe the presence of programmed elements, such as robotics and AI, in the surrounding society.
	Impacts of programmed technology in everyday life	– Reflect on the data collected based on their activities in digital environments, as well as the connection between programming and this data collection. – Understand that data collected are stored and give at least one example of what data collected about themselves could be used for.	– Give examples of targeted digital content and ways in which this targeting is accomplished. – Reflect on their own activities and how data collected are used in digital environments..

2018). At the same time, a digitalisation strategy for education was prepared. When published in 2017, it stated that the elective PCT course was to become a permanent part of the lower secondary curriculum in 2019, while PCT should be integrated into other subjects at primary level (Kunnskapsdepartementet, 2017).

The revised primary curriculum (Kunnskapsløftet, 2020) came into force in 2020, introducing PCT mainly in mathematics, science, arts and crafts, and music. In some of these subjects, PCT is mentioned explicitly and throughout all grades, whereas in others it is mentioned more sporadically and implicitly. The following lists show examples of learning objectives for PCT in a selection of these subjects: mathematics, science and music. Curricular guidelines specify what the students should have achieved at the end of the given grade.

In mathematics, PCT is mentioned as part of the general core objectives, under the element "Exploration and problem-solving": *Algorithmic thinking is important in the process of developing strategies and procedures to solve problems and involves breaking down a problem into sub-problems that can be solved systematically*:

- Grade 2: recognize and describe repeating units in patterns; create patterns
- Grade 4: explore and describe structures and patterns in play and games; create algorithms and express them using variables, conditions and loops
- Grade 5: create and program algorithms using variables, conditions and loops
- Grade 6: use variables and formulas to express relationships in practical situations; use variables, loops, conditions and functions to explore geometric figures and patterns
- Grade 7: use programming to explore data in tables and datasets
- Grade 8: explore how algorithms can be created, tested and improved using programming
- Grade 9: simulate outcomes in random experiments and calculate the probability that something will occur, using programming
- Grade 10: explore mathematical properties and relationships using programming

In science (naturfag) PCT is explicitly mentioned under the core element "Technology": The students shall understand, create, and use technology, including programming and modelling, in science:

- Grade 2: present own ideas for technological inventions
- Grade 4: explore technological systems that are made up of different parts, and describe how the parts work and work together
- Grade 7: explore, create, and program technological systems consisting of parts that work together
- Grade 10: use programming to explore scientific phenomena

In arts and crafts, as well as music, PCT is not mentioned as explicitly. It is, nevertheless, still part of the learning objectives in some grades:

- Arts and crafts, grade 7: use programming to create interactivity and visual expression
- Music, grade 10: create and program musical sequences by experimenting with sound from different sources

4.2.2 Implementation

In 2017, the National Framework for Teachers' Professional Digital Competence (Kelentric, Helland, & Arstorp, 2017) was published. Subsequently, the Ministry for Education and Research funded digitalisation projects in teacher education to promote teachers' and students' digital competence. In addition, several Massive Open Online Courses (MOOCs) for teachers have received funding based on the framework to enhance in-service teachers' digital competence as part of a new competence development model (OECD, 2019). Non-profit organizations have also supported the introduction of programming, e.g., "Lær Kidsa Koding", which provides teaching material and professional development.

Similar challenges as in Sweden and Finland have been encountered in Norway. While some teachers find programming interesting and exciting, it is still found to be difficult and time-consuming (Stenlund, 2021). Teachers also lack sufficient support, training, subject pedagogics, and preparation time.

4.3 *Estonia*

Since the late 1990s, the national strategy in Estonia has focused on creating an information society. As part of this strategy, the Tiger Leap project (Tiigrihüpe)[1] was founded to develop and expand computer and internet access, particularly in education. As a result, internet access was rolled out to all Estonian schools in the early 2000s. At that time there was no separate subject on computing, rather "IT and media" was taught as a cross-curricular theme, with a heavy focus on the usage aspects.

4.3.1 Curriculum

In 2011, a new national curriculum was introduced including programming both as a topic to be integrated into existing subjects and as elective courses.

PCT is integrated into other subjects through the cross-curricular theme "Technology and Innovation". According to the curriculum (Riigi Teataja, 2022) the theme should help students "be innovative and able to use modern technologies purposefully [and] cope in a rapidly changing technological living, learning and working environment". In appendix 7 of the curriculum,[2] the objectives of the theme are summarized as:

> Various materials and processing methods are used in the classes. By solving tasks and presenting the results, they learn to use computer programs, find ways to use the digital environment in the learning process. Computer-controlled devices and machines are introduced, as working with them creates an opportunity to learn about modern technological possibilities.

In addition, there is a standardized national curriculum for two courses (35 hours in grades 5 and 8), presented in appendix 10.[3] These courses do not, however, include much PCT, if any. Schools can also design their own elective courses.

The curriculum was revised in 2014. As a response to the Estonian lifelong learning strategy 2020, digital competence was added and defined according to the EU DigComp framework, consisting of five main competence areas: information and data, collaboration and communication, content creation, security and problem-solving. PCT only needs to be included in schools where it is taught as a separate subject (Haridus- ja Teadusministeerium, n.d.):

- Grades 1–3: The student creates simpler programs that include variables, loops, conditional statements, and procedures, using a visual programming language.
- Grades 4–6: The student creates a robot control software, an interactive game running in a browser, or a mobile application using a visual programming language.
- Grades 7–9: The student contributes a common theme to the software made in the framework for "Technology and Innovation" as a member of a development team (as a programmer, designer, tester, analyst or team leader).

4.3.2 Implementation

In 2012, the ProgeTiger programme was launched as a response to an active discussion on introducing PCT and robotics in schools. Teacher training and resources were developed, and over the years the programme grew. It was originally based at the HITSA Development Centre of IT Education, funded by the Estonian Ministry of Education and Research. Other partners included the Estonian Ministry of Economic Affairs and Communications, as well as universities, the private sector, and third-sector institutions. At the same time, Codecademy.com was becoming more known and their courses were translated into Estonian as part of the ProgeTiger program.

ProgeTiger is currently coordinated by the Education and Youth Board of Estonia, with support and funding from the Estonian government through the Ministry of Education and the European Union. HITSA also developed an unofficial curriculum for grades 1–6, placing more focus on PCT, which is still used today. The curriculum was complemented with two digital textbooks,[4] one for each grade level, offering material for 30 hours for each of the three respective areas. Teacher material is also available. As digital competence was introduced in the curriculum a support site was developed,[5] providing material, assessment instruments, and other resources for both students and teachers. We have not been able to find any national evaluation reports or research covering the introduction of PCT in the Estonian curriculum.

4.4 *England*

In England, a new national curriculum introducing the subject "Computing" came into force in 2014, in effect replacing the previous ICT programme of study. This change was the result of a debate on the role of computing in UK schools. For instance, a non-profit organization called Computing at School had been established already in 2008, starting as a working group to establish computing as a school subject and build a network supporting teachers in their work on computing in the classroom (Jones et al., 2013). A year later the British Computer Society Academy of Computing was formed and initiated a partnership with Computing at School. In August 2010, the Royal Society initiated a project reviewing the state of computing education in the UK, resulting in a set of recommendations highlighting, for instance, the need for qualified teachers and recognition of computing as an academic discipline in schools (Royal Society, 2012).

4.4.1 Curriculum

The curriculum of 2014 introduced computing as a broad subject composed of three areas: information technology, digital literacy and computer science. PCT is a core component of a computing curriculum and is not only important in coding but can also be an important life skill for solving problems. The main aims of the study programme are to ensure that all students:

- can understand and apply the fundamental principles and concepts of computer science, including abstraction, logic, algorithms and data representation
- can analyse problems in computational terms, and have repeated practical experience of writing computer programs to solve such problems
- can evaluate and apply information technology, including new or unfamiliar technologies, analytically to solve problems
- are responsible, competent, confident and creative users of information and communication technology

How these aims are to be reached is described in four key stages (Department for Education, 2013), covering both primary and secondary education. In the following, we list the main content related to PCT:

Key stage 1: Students should be taught to
- understand what algorithms are, how they are implemented as programs on digital devices, and that programs execute by following precise and unambiguous instructions
- create and debug simple programs
- use logical reasoning to predict the behaviour of simple programs

Key stage 2: Students should be taught to
- design, write and debug programs that accomplish specific goals, including controlling or simulating physical systems
- solve problems by decomposing them into smaller parts
- use sequence, selection, and repetition in programs
- work with variables and various forms of input and output
- use logical reasoning to explain how some simple algorithms work and to detect and correct errors in algorithms and programs

Key stage 3: Students should be taught to
- design, use and evaluate computational abstractions that model the state and behaviour of real-world problems and physical systems
- understand several key algorithms that reflect computational thinking
- use logical reasoning to compare the utility of alternative algorithms for the same problem
- use 2 or more programming languages, at least one of which is textual, to solve a variety of computational problems; make appropriate use of data structures
- design and develop modular programs that use procedures or functions
- understand simple Boolean logic and some of its uses in circuits and programming
- understand how numbers can be represented in binary, and be able to carry out simple operations on binary numbers

Key stage 4: Students should be taught to
- develop their capability, creativity and knowledge in computer science, digital media and information technology
- develop and apply their analytic, problem-solving, design, and computational thinking skills

4.4.2 Implementation

Computing at school (CAS, computingatschool.org.uk) is a grass-root community aimed at supporting computing teachers. In September 2022, the organisation had 11 500 active users in over 330 CAS communities. CAS offers, for instance, events and discussions, as well as resources for school leaders and teachers at different grade levels.

The Network of Excellence is part of CAS and when the new curriculum was introduced, the network was tasked with four key objectives to be fulfilled by 2018 (CAS, 2021). For instance, they were to ensure a good geographical spread of support, recruit and train Master Teachers, and annually provide at least 30 000 hours of teacher training and support at least 8500 teachers from a minimum of 2500 schools. The network contract came to an end on March 31, 2018.

The private sector has also played an important role in the implementation of the new subject, including companies such as Microsoft and BBC, and non-profits such as the Code Club and the Raspberry Pi Foundation. The National Centre for Computing Education (NCCE) is a consortium funded by the Department for Education and partners, which has made a notable investment in promoting computing education in England. The centre offers teachers courses, teaching resources and community support.

Although England, by introducing a stand-alone subject, has been able to focus on training only a subset of teachers, i.e., those responsible for the Computing subject, reports (Royal Society, 2017) show similar challenges related to teacher preparation as, e.g., Finland and Sweden. In a study among 300 teachers, Sentence and Csizmadia (2017) identified three challenge categories from a teacher perspective: challenges related to (1) the teacher (e.g., lack of subject knowledge, time and teaching approaches), (2) the student (e.g., lack of understanding, resilience and engagement) and (3) resources (e.g., technical problems and lack of good resources). Moreover, Larke (2019) described how teachers have "blocked" the computing curriculum, which they "view as narrow, difficult to teach and in conflict with their beliefs and practices as educational professionals" (p. 1137).

4.5 *South Korea*

South Korea has a long tradition of technology development. In the early 2000s, the Ministry of Education, Science and Technology installed computer labs in schools across the country, which made it possible to also invest in IT education. Nevertheless, although students had access to technology, they did not have PCT skills (OECD, 2005) leading the ministry to revise the curriculum to have a stronger emphasis on PCT. Subsequent updates and revisions further strengthened the role of computational thinking in the curriculum and reflect the government's recognition of the importance of computing education.

4.5.1 Curriculum

In the 2015 revised curriculum, computing education was made compulsory in primary and lower secondary education (Ministry of Education, 2015). Students at primary level are taught about problem-solving, algorithms, PCT, and digital ethics in the context of the subject "Practical arts", and primary school teachers are obliged to teach these aspects for a minimum of 17 class hours per year using e.g., simplified programming languages or unplugged materials (Kim & Kim, 2018).

Informatics existed as an elective subject at lower-secondary level before the 2015 revision. As a result of the revisions, informatics was introduced as a compulsory subject in all lower secondary schools (Lee & Lee, 2021; The Korea Ministry of Education, 2015). Students are to get at least 34 class hours, one hour per week, covering CT-based problem-solving, algorithms and programming at a more advanced level (Kim & Kim, 2018). The revised curriculum came into force in 2018.

In November 2021, the Ministry of Education announced the 2022 revised curriculum, which will be introduced in primary schools starting in 2024, and in lower and upper education in 2025 (Hyeon-Kyeong & Jian, 2021). The curriculum includes courses on software and artificial intelligence. The ministry has also announced a plan to by 2024 have trained 5000 AI teachers and developed an AI textbook for primary schools.

4.5.2 Implementation

The Korean government has supported the implementation of the new curriculum through, for instance, large-scale training for in-service teachers. As primary school teachers teach all subjects, all pre-service teachers are to take a course on how to teach PCT and other related content. The government has also partnered with educational associations, companies and teacher groups to help with the implementation (Kim & Kim, 2018).

Textbooks play an important role in the Korean educational system (Kim et al., 2021). Private companies can develop and publish books if they include the content presented in the curriculum. Schools are free to use textbooks that have been accredited. In 2021, for instance, there were 17 informatics textbooks for the lower secondary level. South Korea deals with the same challenges seen above related to training both pre-service and in-service teachers (Kim & Kim, 2018). They also highlight the need for training principals and school leaders, as earlier research (Choi et al., 2016) suggested that their attitudes and views on the need for PCT education are related to, for instance, the amount of support and time provided to teachers.

5 Experiences and Lessons Learned

The above review of the curricular development in six countries highlights differences in the way that PCT has been introduced, not only concerning the model used, but also the preparatory work done, and the implementation support offered.

5.1 *Who?*

The model used affects *who* is reached by the curricula changes. As an example, a pilot introducing PCT as an elective course in Norwegian lower-secondary education was dominated by boys, failing to attract girls (Corneliussen & Tveranger, 2018). One could therefore argue that integrating PCT into existing subjects is to be preferred, but that again brings about other challenges, mainly related to time, resources and teacher competence as seen above for both Sweden and Finland.

Sands et al. (2018) found that many teachers both within and outside the STEM subject areas lack an awareness of what CT skills entail and how these can be supported in the classroom. These teachers also seem to share a similar understanding—and sometimes also the same misunderstandings—of CT, highlighting the need to develop higher level understanding among non-computing teachers. Vinnervik (2021) argues that with limited subject knowledge and time, and an otherwise extensive curriculum to cover, teachers might approach PCT narrowly as an isolated part of the subjects it is integrated into, focusing mainly on coding while overlooking the other aspects of PCT. Teachers have a notable influence over what curricular content reaches students (Larke, 2019). This is particularly true in countries where teachers are autonomous, as it can make top-down reforms a poor fit for the educational culture, calling for more teacher involvement when planning curricular reforms (Korhonen et al., 2022).

Nevertheless, similar challenges related to preparing enough teachers and finding ways to reach all students despite their gender or other background factors, have also been encountered in, for instance, England, where PCT has been introduced as part of a new stand-alone subject (Royal Society, 2017; Ofsted, 2022). Teacher confidence and self-efficacy in PCT are also a challenge (Mannila et al., 2018). Curricular requirements should optimally meet teachers' beliefs, needs and abilities (Larke, 2019), while also promoting positive emotions and attitudes (Korhonen et al., 2022). The need for continuous training for both in- and pre-service teachers thus remains highly relevant (Kong et al., 2020), but as the former group already has established pedagogical practices and learning objectives, the same strategies cannot be used for training in- and pre-service teachers (Sands et al., 2018). In addition, where PCT is to be integrated into existing courses, the training needs to be content-specific to aid teachers in how to accomplish this in practice (Sands et al., 2018). Regardless of the model used, leadership and support are crucial, and Bocconi et al. (2016) highlight the importance of making all stakeholders aware of what CT is and how it is relevant for K-12 education.

5.2 *What?*

The review shows that the curricula seldom state any details regarding the second question, *what*. For instance, research and national evaluations have indicated that teachers and school-leaders need additional detail on what students are expected to learn—and hence teachers are supposed to teach—at different grade levels (Royal Society, 2017; Tanhua-Piiroinen et al., 2020; Ifous, 2020; Vinnervik, 2021; Mannila, 2021). The lack of clear objectives and suitable teaching methods is problematic as these are crucial when integrating PCT into education (Popat & Starkey, 2019). Without such information, it is difficult to guarantee equal learning opportunities for all children, regardless of, for instance, what school they attend. When the curriculum is non-descriptive, many choices and details need to be decided upon at school or even teacher level, although teachers lack resources that could contextualize PCT into their subject areas (Pears et al., 2017).

The lack of details can also lead to the content being covered only superficially (Cederqvist, 2021), which in turn can lead to the content being decided upon by external actors, such as textbook companies who design the material that teachers use in the classroom. As seen in the discussion above, in South Korea, on the other hand, textbooks are evaluated and accredited to ensure that they cover what is intended by the curriculum. In Finland, on the other hand, additional detail has been provided in the "New Literacies Programme". While such clarifications make the learning goals more concrete, there is still also a need for training efforts that point out how PCT can meet learning objectives in specific subjects instead of merely being an add-on in the curriculum (Greenleaf et al., 2011). In the Swedish case, this type of professional development would be needed in mathematics and technology.

In addition to knowing what to teach, the importance of terminology has also been raised in national evaluations. Diethelm and Goschler (2014) point out that PCT terminology includes words with specific meanings within computing, which may have different meanings in other contexts. Therefore, they argue, PCT terminology needs to be taught and learned as completely new words, to prevent students from trying to interpret them based on their experiences from other contexts, which can lead to misconceptions. Munashinghe, Bell, and Robins (2021) suggest that teachers should be given explicit guidance on terminology and opportunities to use it in a relevant PCT context.

5.3 *How?*

Finally, there is the question of *how*. In Denmark, where PCT has not yet been officially introduced in the curriculum, a pilot has been carried out where half of the participating schools have followed an integrated approach while the other half have taught PCT in a stand-alone subject (Børne- og undervisningsministeriet, 2021). The evaluation of the pilot showed no clear preference for either model, as both approaches have their benefits and drawbacks. The introduction of PCT in K-12 education has thus given rise to several tensions, for instance, between the "urgency of curriculum change" and the "need for costly and time-consuming professional development of teachers" (Storte et al., 2019). Despite the model, the challenges tend to be the same and it is thus unclear if it is easier or harder to integrate PCT into other subjects.

National evaluations in Finland show that there are yet no established approaches to teaching PCT, and many students meet PCT mainly in projects, sometimes involving external experts (Tanhua-Piiroinen et al., 2020). In addition to finding suitable pedagogical methods to teach PCT, assessment also plays an important role as it indicates that certain content is important (Bocconi et al., 2022). If PCT is not part of the assessment, teachers may not have the same motivation to teach it (Larke, 2019) and efforts have been made to aid teachers in assessing PCT-related skills (e.g., Roman-Gonzalez, 2015; Grover, 2017; Mannila et al., 2020). Since 2018, the international IEA ICILS study has included an optional part assessing students' CT skills. This can be seen as a breakthrough in terms of making objective assessment of PCT possible, while also allowing the results to be compared and used for national and cross-national policymaking.

6 Conclusions

National evaluations in several countries identify similar challenges such as teacher time, lack of resources, lack of teacher self-efficacy and skills, negative beliefs and attitudes, unclear responsibilities, and conflicting organizational priorities. School reforms downsize the amount of time teachers have to prepare, which may leave them with too little time to learn the new content themselves and become confident with teaching it. As teachers have different needs and ambition levels, professional development needs to come in different forms, ranging from short inspirational workshops to longer courses where the competence is developed over time.

Our review shows that the strategies used in different countries have their own pros and cons, and there is no clear evidence establishing that one method is preferable. We conclude that the freedom, or burden, to locally decide on how to implement the curriculum has left teachers in a difficult position, where they are to fulfil the requirements of the curriculum without proper training and the competence needed. When integrating PCT across the curriculum, more teachers need to be trained and there is an increased need for pedagogical content knowledge. This has naturally had a negative impact on how PCT has been and is introduced at schools. Based on our review, we have identified the following six recommendations:

- The need for large-scale professional development remains, as PCT is an area under continuous development. Training should, however, not only be provided in PCT as such but rather on the pedagogical aspects of how PCT can be used to address learning objectives in those subjects into which it has been integrated. In a Swedish context, this means mathematics and technology, in particular.
- There is a need for standardization in PCT education, for instance, related to terminology. A common vocabulary with precise meanings can give students and teachers a joint language, helping them communicate more confidently in PCT contexts.
- PCT should be made part of the skills and knowledge that are to be assessed; this can raise the perceived level of importance of covering the content in class.
- More specific curricular guidelines are needed to provide a clear progression with learning outcomes for different grade levels. These need not be defined in the curriculum, which often is a static document, but rather be provided in a format that allows for adaptions. Such adaptions should naturally still only be made when needed in order not to place new requirements on teachers too often.
- Textbooks and other teaching resources should be evaluated or accredited to ensure that they correspond to the curricula requirements.
- Organizational support and a joint vision are important to reach long-lasting change and training should be extended to also include those making decisions about curricular priorities and the time and resources allocated to teachers (e.g., principals and school leaders).

Taken together, these could provide the basis for a more systematic and holistic approach to introducing PCT into grades 1–9, addressing the needs of teachers, students, and schools.

Acknowledgments

This work was supported by the Marcus and Amalia Wallenberg Foundation.

Notes

1 en.wikipedia.org/wiki/Tiigrihüpe
2 riigiteataja.ee/aktilisa/1120/4202/2010/1m%20lisa7.pdf
3 riigiteataja.ee/aktilisa/1120/4202/2010/1m%20lisa10.pdf#
4 courses.cs.ut.ee/t/digiopik/
5 digipadevus.ee

References

Barr, V., & Stephenson, C. (2011). Bringing computational thinking to K–12: What is involved and what is the role of the computer science education community? *ACM Inroads*, 2(1), 48–54. https://doi.org/10.1145/1929887.1929905

Bocconi, S., Chioccariello, A., Dettori, G., Ferrari, A., & Engelhardt, K. (2016). *Developing computational thinking in compulsory education – Implications for policy and practice*. European Commission, Joint Research Centre. https://doi.org/10.2791/792158

Bocconi, S., Chioccariello, A., Kampylis, P., Dagienė, V., Wastiau, P., Engelhardt, K., Earp, J., Horvath, M. A., Jasutė, E., Malagoli, C., Masiulionytė-Dagienė, V., & Stupurienė, G. (2022). *Reviewing computational thinking in compulsory education: State of play and practices from computing education*. In A. Inamorato dos Santos, R. Cachia, N. Giannoutsou, & Y. Punie (Eds.), *Publications office of the European Union*. European Commission, Joint Research Centre. https://doi.org/10.2760/126955

Børne- og undervisningsministeriet. (2021). *Forsøg med teknologiforståelse i folkeskolens obligatoriske undervisning*. Report.

CAS. (2021). *Network of teaching excellence in computer science*. https://www.computingatschool.org.uk/resource-library/2013/may/network-of-teaching-excellence-in-computer-science

Cederqvist, A. (2021). *Seeing the parts, understanding the whole – A technology education perspective on teaching and learning in processes of analysing and designing programmed technological solutions* (Diss.). Göteborgs universitet.

Choi, S., Kim, H., Sung, J., Seo, Y., & Choi, Y. (2016). *A study of activating Gyeonggido SW education*. Gyeonggido Institute of Education.

Corneliussen, H. G., & Tveranger, F. (2018). Programming in secondary schools in Norway: A wasted opportunity for inclusion. In *Proceedings of the 4th Conference on Gender & IT (GenderIT '18)* (pp. 175–182). Association for Computing Machinery, New York, NY, USA. https://doi.org/10.1145/3196839.3196867

Dagienė, V., Jevsikova, T., Stupurienė, G., & Juškevičienė, A. (2019). Teaching computational thinking in primary schools: Worldwide trends and teachers' attitudes. *Computer Science and Information Systems, 19*(1), 1–24.

Department for Education. (2013). *National curriculum in England: Computing programmes of study*. https://www.gov.uk/government/publications/national-curriculum-in-england-computing-programmes-of-study/national-curriculum-in-england-computing-programmes-of-study

Denning, P. J., & Tedre, M. (2021). Computational thinking: A disciplinary perspective. *Informatics in Education, 20*(3), 361–390. https://doi.org/10.15388/infedu.2021.21

Diethelm, I., & Goschler, J. (2014). On human language and terminology used for teaching and learning CS/Informatics. In *Proceedings of the 9th Workshop in Primary and Secondary Computing Education (WiPSCE '14)* (pp. 122–123) ACM, New York, NY, USA. https://doi.org/10.1145/2670757.2670765

Digitaliseringskommissionen. (2014). *En digital agenda i människans tjänst – en ljusnande framtid kan bli vår. Delbetänkande av Digitaliseringskommissionen*. Statens offentliga utredningar, SOU 2014:13, Stockholm, Sverige.

European Schoolnet. (2015, October). *Computing our future: Computer programming and coding. Priorities, school curricula and initiatives across Europe*. Report. http://www.eun.org/documents/411753/817341/Computing+our+future_final_2015.pdf/

Gander, W., Petit, A., Berry, G., Demo, B., Vahrenhold, J., McGettrick, A., Boyle, R., Drechsler, M., Mendelson, A., Stephenson, C., Ghezzi, C., & Meyer, B. (2013). *Informatics education: Europe cannot afford to miss the boat*. Report of the joint Informatics Europe & ACM Europe Working Group on Informatics Education. Retrieved February 20, 2022, from https://www.informatics-europe.org/images/documents/informatics-education-acm-ie.pdf

Gerson, S. A., Morey, R. D., & van Schaik, J. E. (2022). Coding in the cot? Factors influencing 0–17s' experiences with technology and coding in the United Kingdom. *Computers & Education, 178*, 104400. https://doi.org/10.1016/j.compedu.2021.104400

Greenleaf, C. L., Litman, C., Hanson, T. L., Rosen, R., Boscardin, C. K., Herman, J., Schneider, S. A., Madden, S., & Jones, B. (2011). Integrating literacy and science in biology: Teaching and learning impacts of reading apprenticeship professional development. *American Educational Research Journal, 48*(3), 647–717. https://doi.org/10.1177/0034637398095003l9

Grover, S. (2017). Assessing algorithmic and computational thinking in K-12: Lessons from a middle school classroom. In P. Rich & C. Hodges (Eds.), *Emerging research, practice, and policy on computational thinking* (pp. 269–288). Springer. https://doi.org/10.1007/978-3-319-52691-1_17

Haridus- ja Teadusministeerium. (n.d.) *Digipädevus Õppekavades* (*eng.* Digital competence in the curriculum). https://www.hm.ee/sites/default/files/digipadevusoppekavades_2016veebi.pdf

Heintz, F., Mannila, L., & Färnqvist, T. (2016). A review of models for introducing computational thinking, computer science and computing in K-12 education. In *Proceedings of the 2016 IEEE Frontiers in Education Conference* (*FIE*) (pp. 1–9). https://doi.org/10.1109/FIE.2016.7757410

Heintz, F., Mannila, L., Nordén, L.-Å., Parnes, P., & Regnell, B. (2017). Introducing programming and digital competence in Swedish K-9 education. In V. Dagienė & A. Hellas (Eds.), *Informatics in schools: Focus on learning programming. ISSEP 2017. Lecture notes in computer science* (Vol. 10696). Springer. https://doi.org/10.1007/978-3-319-71483-7_10

Hyeon-Kyeong, M., & Jian, L. (2021, November 24). Lots of changes proposed for Korea's students. *Korea JoongAng Daily.* https://koreajoongangdaily.joins.com/2021/11/24/national/socialAffairs/education-covid19-schools/20211124174731580.html

Hsu, T.-C., Chang, S.-C., & Hung, Y.-T. (2018). How to learn and how to teach computational thinking: Suggestions based on a review of the literature. *Computers & Education, 126*, 296–310. https://doi.org/10.1016/j.compedu.2018.07.004

iFous. (2020). *Programmering i skolan – vad, hur, när och varför?* Ifous rapportserie 2020:5 – Slutrapport från FoU-programmet Programmering i ämnesundervisningen.

Jones, S. P., Michell, B., & Humphreys, S. (2013). *Computing at school in the UK.* https://microsoft.com/en-us/research/wp-content/uploads/2016/07/ComputingAtSchoolCACM.pdf

Kafai, Y. B., & Proctor, C. (2022). A revaluation of computational thinking in K–12 education: Moving toward computational literacies. *Educational Researcher, 51*(2), 146–151.

Kafai, Y. B., Proctor, C., & Lui, D. (2020). From theory bias to theory dialogue: Embracing cognitive, situated, and critical framings of computational thinking in K-12 CS education. *ACM Inroads, 11*(1), 44–53. https://doi.org/10.1145/3381887

Kawai, Y. B., & Burke, Q. (2013). Computer programming goes back to school. *Phi Delta Kappan, 95*(1), 61–65. https://doi.org/10.1177/003172171309500111

Kelentric, M., Helland, K., & Arstorp, A. (2017). *Professional digital competence framework for teachers.* Center for ICT in Education. https://www.udir.no/in-english/professional-digital-competence-framework-for-teachers/

Kim, S., & Kim, H. Y. (2018). A computational thinking curriculum and teacher professional development in South Korea. In M. Khine (Ed.), *Computational thinking in the STEM disciplines.* Springer. https://doi.org/10.1007/978-3-319-93566-9_9

Kim, J., Woo, H., Kim, J., & Lee, W. (2021). Relationship between Korean Informatics Curriculum and Textbook Learning Element Considering Compound Word. *International Journal on Informatics Visualization, 5*(4), 430–437. http://dx.doi.org/10.30630/joiv.5.4.731

Kong, S.-C., Lai, M., & Sun, D. (2020). Teacher development in computational thinking: Design and learning outcomes of programming concepts, practices and pedagogy. *Computers & Education, 151*, 103872. https://doi.org/10.1016/j.compedu.2020.103872

Korhonen, T., Salo, L., Laakso, N., Seitamaa, A., Sormunen, K., Kukkonen, M., & Forsström, H. (2022). Finnish teachers as adopters of educational innovation: perceptions of programming as a new part of the curriculum. *Computer Science Education*, 1–23. https://doi.org/10.1080/08993408.2022.2095595

Kunnskapsdepartementet. (2017). *Framtid, fornyelse og digitalisering. Digitaliseringsstrategi for grunnopplæringen 2017–2021.* https://www.regjeringen.no/contentassets/dc02a65c18a7464db394766247e5f5fc/kd_framtid_fornyelse_digitalisering_nett.pdf

Larke, L. R. (2019). Agentic neglect: Teachers as gatekeepers of England's national computing curriculum. *British Journal of Educational Technology, 50*(3), 1137–1150.

Lee, M., & Lee, J. (2021). Enhancing computational thinking skills in informatics in secondary education: The case of South Korea. *Education Tech Research Dev, 69*, 2869–2893. https://doi.org/10.1007/s11423-021-10035-2

Lye, S. Y., & Koh, J. H. L. (2014). Review on teaching and learning of computational thinking through programming: What is next for K-12? *Computers in Human Behavior, 41*, 51–61. https://doi.org/10.1016/j.chb.2014.09.012

Mannila, L. (2021). *Digital kompetens i Svenskfinland – nulägesanalys och goda modeller.* Svenska Kulturfonden.

Mannila, L. (2023). Integrating programming in other subjects at primary level – tool, glue or ideation. In T. Keane & A. E. Fluck (Eds.), *Teaching coding in K-12 schools – Research and application.* Springer. https://doi.org/10.1007/978-3-031-21970-2_9

Mannila, L., Åkerfeldt, A., Kjällander, S., & Heintz, F. (2022). Exploring gender differences in primary school programming. In *Proceedings of the 2022 IEEE Frontiers in Education Conference (FIE)* (pp. 1–9). Uppsala, Sweden. https://doi.org/10.1109/FIE56618.2022.9962482

Mannila, L., Dagiene, V., Demo, B., Grgurina, N., Mirolo, C., Rolandsson, L., & Settle, A. (2014). Computational thinking in K-9 education. In *Proceedings of the working group reports of the 2014 on Innovation & Technology in Computer Science Education Conference* (ITiCSE-WGR '14) (pp. 1–29). ACM, New York, NY, USA. https://doi.org/10.1145/2713609.2713610

Mannila, L., Heintz, F., Kjällander, S., & Åkerfeldt, A. (2020). Programming in primary education: towards a research based assessment framework. In *Proceedings of the 15th Workshop on Primary and Secondary Computing Education* (*WiPSCE '20*). Association for Computing Machinery, New York, NY, USA, Article 10, 1–10. https://doi.org/10.1145/3421590.3421598

Mannila, L., Nordén, L.-Å., & Pears, A. (2018). Digital competence, teacher self-efficacy and training needs. In *Proceedings of the 2018 ACM Conference on International Computing Education Research* (*ICER'18*) (pp. 78–85). https://doi.org/10.1145/3230977.3230993

Ministry of Education. (2015). *The National Curriculum for the primary and secondary schools*. Proclamation of the Ministry of Education #2015-74. http://www.koreaneducentreinuk.org/wp-content/uploads/2021/02/The-National-Curriculum-for-the-Primary-and-Secondary-Schools-2015.pdf

Munasinghe, B., Bell, T., & Robins, A. (2021). Teachers' understanding of technical terms in a Computational Thinking curriculum. In *Proceedings of Australasian Computing Education Conference* (*ACE '21*), *February, SA, Australia*. ACM, New York, NY, USA. https://doi.org/10.1145/3441636.3442311

Niemelä, P., Pears, A., Dagiene, V., & Laanpere, M. (2022). Computational thinking – Forces shaping curriculum and policy in Finland, Sweden and the Baltic countries. In D. Passey et al. (Eds.), *OCCE 2021* (IFIP AICT 642, pp. 131–143). https://doi.org/10.1007/978-3-030-97986-7_11

NCCE. (2022). *Programming and algorithms within the computing curriculum*. https://static.teachcomputing.org/Programming+and+Algorithms+Report.pdf

NOU. (2013). *Hindre for digital verdiskaping* (*eng.* Obstacles to digital value making). Norges offentlige utredninger, 2013:2. Oslo. https://www.regjeringen.no/contentassets/e2f0d5676e144305967f21011b715c16/no/pdfs/nou201320130002000dddpdfs.pdf

Nouri, J. (2020). Forskningsinsatsens genomförande och resultat. In A. Jahnke (Ed.), *Programmering i skolan. Vad, hur, när och varför?* Slutrapport 2020:5, FoU-programmet Programmering i ämnesundervisningen, pp. 35–49. https://www.ifous.se/app/uploads/2020/11/202010-ifous-2020-5-f.pdf

Nouri, J., Zhang, L., Mannila, L., & Norén, E. (2020). Development of computational thinking, digital competence and 21st century skills when learning programming in K-9. *Education Inquiry*, *11*(1), 1–17. https://doi.org/10.1080/20004508.2019.1627844

OECD. (2005). *Are students ready for a technology-rich world? What PISA studies tell us*. OECD.

OECD. (2019). *Improving school quality in Norway: The new competence development model, implementing education policies*. OECD. https://doi.org/10.1787/179d4ded-en

Ofsted. (2022). *Research review series: computing*. https://www.gov.uk/government/publications/research-review-series-computing

Olofson, A. O., Lindberg, J. O., Pedersen, A. Y., Arstorp, A.-T., Dalsgaard, C., Einum, E., Caviglia, F., Ilomäki, L., Veermans, M., Häkkinen, P., & Willermark, S. (2021). Digital competence across boundaries – beyond a common Nordic model of the digitalisation of K-12 schools? *Education Inquiry*, *12*(4), 317–328. https://doi.org/10.1080/20004508.2021.1976454

Opetushallitus. (2014). *Perusopetuksen opetussuunnitelman perusteet 2014*. https://www.oph.fi/sites/default/files/documents/perusopetuksen_opetussuunnitelman_perusteet_2014.pdf

Ottestad, G., & Gudmundsdottir, G. B. (2018). Information and communication technology policy in primary and secondary education in Europe. In J. Voogt et al. (Eds.), *Handbook of information technology in primary and secondary education*. Springer International Handbooks of Education. https://doi.org/10.1007/978-3-319-53803-7_92-1

Pears, A., Dagiene, V., & Jasute, E. (2017). Baltic and Nordic K-12 teacher perspectives on computational thinking and computing. In V. Dagienė & A. Hellas (Eds.), *Informatics in schools: Focus on learning programming. ISSEP 2017. Lecture notes in computer science* (Vol. 10696). Springer. https://doi.org/10.1007/978-3-319-71483-7_12

Popat, S., & Starkey, L. (2019). Learning to code or coding to learn? A systematic review. *Computers and Education, 128*, 365–376. https://doi.org/10.1016/j.compedu.2018.10.005

Riigi Teataja. (2022). *Põhikooli riiklik õppekava* [*National curriculum of elementary school*]. https://www.riigiteataja.ee/akt/114012011001?leiaKehtiv

Roman-Gonzalez, M. (2015). Computational thinking test: Design guidelines and content validation. In *Proceedings of the EDULEARN15 conference* (pp. 2436–2444). https://doi.org/10.13140/RG.2.1.4203.4329

Royal Society. (2012). *Shut down or restart? The way forward for computing in UK Schools*. https://royalsociety.org/-/media/education/computing-in-schools/2012-01-12-computing-in-schools.pdf

Royal Society. (2017). *After the reboot: Computing education in UK schools*. https://royalsociety.org/computing-education

Sands, P., Yadav, A., & Good, J. (2018). Computational thinking in K-12: In-service teacher perceptions of computational thinking. In M. S. Khine (Ed.), *Computational thinking in the STEM disciplines*. Springer. https://doi.org/10.1007/978-3-319-93566-9_8

Sanne, A., Berge, O., Bungum, B., Jørgensen, E. C., Kluge, A., Kristensen, T. E., Mørken, K. M., Svorkmo, A.-G., & Voll, L. O. (2016). *Teknologi og programmering for alle. En faggjennomgang med forslag til endringer i grunnopplæringen* [*Technology and programming for everyone. A subject review with proposals for changes in basic education*]. https://www.udir.no/globalassets/filer/tall-og-forskning/forskningsrapporter/teknologi-og-programmering-for-alle.pdf

Sevik, K., et al. (2016). *Programmering i skolen* [*Programming at school*]. Notat fra Senter for IKT i utdanningen.

Sentance, S., & Csizmadia, A. (2017). Computing in the curriculum: Challenges and strategies from a teacher's perspective. *Education and Information Technologies, 22*, 469–495. https://doi.org/10.1007/s10639-016-9482-0

Skolverket. (2022a). *Läroplan för grundskolan samt för förskoleklassen och fritidshemmet*. https://www.skolverket.se/publikationer?id=9718

Skolverket. (2022b). *Få syn på digitaliseringen på grundskolenivå. Ett kommentarmaterial till läroplanerna för förskoleklass, fritidshem och grundskoleutbildning.* https://www.skolverket.se/publikationsserier/kommentarmaterial/2022/fa-syn-pa-digitaliseringen-pa-grundskoleniva

Stenlund, E. (2021). *Programmering og Fagfornyelsen. En casestudie om lærernes nye situasjon og holdninger til programmering i skolen.* Masteroppgave. Universitetet i Oslo.

Stigberg, H., & Stigberg, S. (2020). Teaching programming and mathematics in practice: A case study from a Swedish primary school. *Policy Futures in Education, 18*(4), 483–496. https://doi.org/10.1177/1478210319894785

Storte, D., Webb, M. E., Bottino, R., Passey, D., Kalas, I., Bescherer, C., Smith, J., Angeli, C., Katz, Y. J., Micheuz, P., Røsvik, S., Brinda, T., Fluck, A. E., Magenheim, J., Anderson, B., & Fuschek, G. (2019, June). Coding, programming and the changing curriculum for computing in schools. *Report of UNESCO/IFIP TC3 Meeting at OCCE, Linz, Austria.*

Sun, L., Hu, L., & Zhou, D. (2022). Programming attitudes predict computational thinking: Analysis of differences in gender and programming experience, *Computers & Education, 181*, 104457. https://doi.org/10.1016/j.compedu.2022.104457

Tang, X., Yin, Y., Lin, Q., Hadad, R., & Zhai, X. (2020). Assessing computational thinking: A systematic review of empirical studies.*Computers & Education, 148*, 1–22. https://doi.org/10.1016/j.compedu.2019.103798

Tanhua-Piiroinen, E., Kaarakainen, S.-S., Kaarakainen, M.-T., & Vitel, J. (2020). *Digiajan peruskoulu II*. Opetus- ja kulttuuriministeriön julkaisuja 2020:17. http://urn.fi/URN:ISBN:978-952-263-823-6

Utbildningsdepartementet. (1994). *Kursplaner för grundskolan, Lpo 94.* http://hdl.handle.net/2077/30959

Vinnervik, P. (2021). *När läraren formar ett nytt ämnesinnehåll: intentioner, förutsättningar och utmaningar med att införa programmering i skolan* (Diss.). Umeå universitet.

Vinnervik, P. (2022). An in-depth analysis of programming in the Swedish school curriculum—rationale, knowledge content and teacher guidance.*Journal of Computers in Education*, 1–35. https://doi.org/10.1007/s40692-022-00230-2

CHAPTER 5

Design and Make—and Code?

Technology Education and a Unified Conception of Technology

Jonas Hallström

Abstract

The aim of this chapter is to discuss how a unified theory of technology could be forged philosophically, and suggest some implications for technology education. A post-phenomenological model of human-technology relations was employed as an analytical tool. It is concluded that both digital and analog technologies could be seen as *technical artefacts with a dual nature* and *technologies of representation*. The dual nature of technical artefacts, that is, the functional/intentional and physical dimensions of artefacts and systems, is reflected e.g. in the abstract programming language in conjunction with a specification, which relates to a physical configuration. Representational technologies could include everything from simple control systems to computers to AI systems, and it would be possible to conceive of the concrete and abstract parts of these technologies as different components of their representational capacity; a component could either be seen as representing (concrete) or represented (abstract), but part of the same representational system that makes up the technology. In both these "dual" perspectives on technology, artefacts and systems could be viewed from a common point of view and may consist of both digital, analog, concrete, and abstract components that together make up the technology. One important implication for technology education is that teaching needs to involve both abstract and concrete technological components. When programming, for instance, students need to learn not only about the code or software in itself, but also about what digital technology *does* in terms of solving real-world problems and achieving technical purposes.

Keywords

digital technology – analog technology – programming – philosophy of technology – post-phenomenology – technology education

© JONAS HALLSTRÖM, 2024 | DOI:10.1163/9789004687912_005

1 Introduction

Technology education is a subject with a traditionally strong emphasis on designing, making and using physical artefacts (e.g., de Vries, 2016). This is not surprising given the fact that sloyd, craft and/or vocational subjects—which include the making of products as central components—preceded technology education in many countries over the world (Hallström et al., 2014; Jones et al., 2013). In recent years, however, digital tools, programming, and computational thinking (CT) have also been incorporated on a large scale as curriculum components in technology education in many countries.

When implemented in technology education programming has partly come to be carried out in pure coding tasks (e.g. Bjursten et al., 2022), at the same time as technology education is focused on the designing, making and using of artefacts and systems. This potential mismatch stems from uncertainty about exactly how programming and digital technology should be conceived of in technology education. The aim of this chapter is therefore to discuss how a unified theory of technology could be forged philosophically, and suggest some implications for technology education.[1]

2 Digital, Analog, Abstract, and Concrete Dimensions of Technology

When defining technology, it is important to separate between four of its dimensions: digital and analog on the one hand, and abstract and concrete, on the other. Digital technology in everyday usage is a very broad term that encompasses computers and all manner of digital tools, but in essence digital really denotes technology that relies on or processes information in the form of 0s and 1s, such as in binary code. A digital thermometer can thus show temperature as numbers on an LCD screen, in which case both the gauging of the temperature and the display are digital. Analog, on the other hand, means technology that uses or processes information in the form of *all* values. Examples include a traditional mercury thermometer with a seamless transition between the display of different temperatures due to the expansion or retraction of the mercury, or an oscilloscope showing sinus waves of electronic signals on a tube screen. Concrete technology is usually called artefact but can be all manner of physical devices or physical networks or systems, from coffee cups to windmills to sewer pipes to spaceships. When we talk about computers or information technologies, their concrete manifestations are also referred to as hardware. Abstract technology is essentially information in various forms,

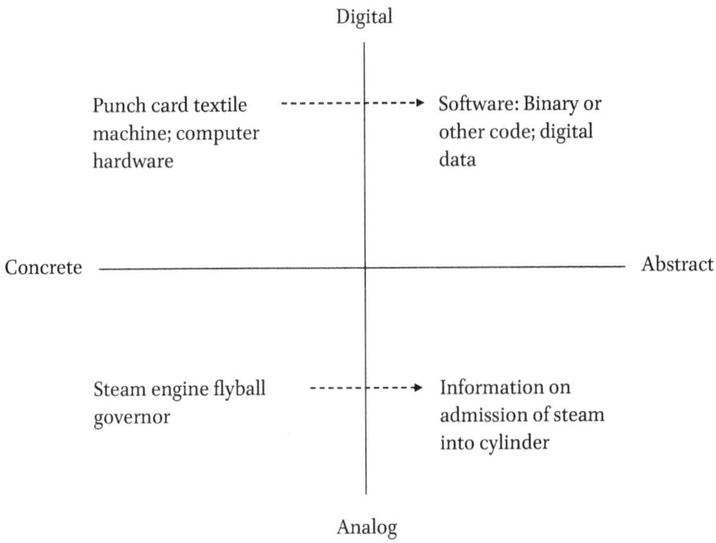

FIGURE 5.1 Schematic outline of digital, analog, abstract, and concrete dimensions of technology

both analog and digital, and in the latter case it is also referred to as software. See Figure 5.1 for an outline of these four dimensions of technology.

The examples in Figure 5.1 show that at least in the case of information-processing technologies, be they analog or digital, there is always a concrete and an abstract component. It might be odd to see a steam engine as an information-processing technology, but even in such early controlled systems information had to be fed back for control purposes (Hallström & Kaijser, 2022). The focus of this chapter, however, is mostly on the upper right-hand corner because coding has often been seen as something not really related to technology but rather to mathematics and information processing in that it takes an abstract, or virtual, form (Brey & Hartz Søraker, 2009). Consequently, computing and information technology have developed their own academic trajectories not always related to technology and engineering (e.g. Denning & Tedre, 2019), but it is also the case that technology education traditionally embraced a view of technology as mainly being physical artefacts, that is, concrete forms of technology.

3 Literature Review: Technology Education and Physical Artefacts

Today, therefore, there is a strong focus on designing and making products in technology education, even though there are different emphases in different countries. For example, in the UK there is emphasis on design and manufacturing of products, while in the USA engineering design in the context of STEM (science, technology, engineering and mathematics) education has gained traction (Buckley et al., 2021; ITEEA, 2020; Rossouw et al., 2011; Williams, 2018). Unsurprisingly, much earlier research shows that students primarily equate technology with artefacts in general and modern products such as TVs, cell phones and computers in particular (e.g. Jones et al., 2013; Solomonidou & Tassios, 2007; Svenningsson, 2020).

The field of philosophy of technology is an important "sister discipline" to technology education because it provides theoretical tools for understanding what technology is, analyzing important concepts, and deciding what technological content is most important to learn (Dakers et al., 2019). Just like technology education, philosophy of technology is a young discipline. Its primary concern is also technological artefacts, which really distinguishes it from the broader field of philosophy in general and philosophy of science in particular. Artefacts are human-made objects and have been thought of as uninteresting and even non-existent in the philosophy of science, because they have been considered as mere applications of science (Dusek, 2006; Meijers, 2009). The philosophy of technology and engineering, on the other hand, concerns itself with philosophical questions related to the human-made world, notably technological artefacts and related knowledge-based practices primarily in engineering but also in other technical professions such as architecture (Meijers, 2009).

One of the most influential philosophers to develop the artefactual perspective is Carl Mitcham (Meijers, 2009). Due to the focus on artefacts in technology education, the philosophical works most cited in a technology education context have related strongly to the artefactual perspective. Mitcham's (1994) *Thinking through Technology* has therefore been particularly influential on technology education research, making it a foundation for many current technology education studies (two of several recent examples are Svenningsson, 2020; Xu, Williams, & Gu, 2021). Mitcham (1994) writes:

> Artifacts–material objects such as tools, machines, and consumer products–are what most readily come to mind when the word "technology" is mentioned. […] Technology as object is the most immediate, not to say the simplest, mode in which technology is found manifest, and it can include all humanly fabricated material artifacts whose function depends on a specific materiality as such. (Mitcham, 1994, p. 161)

Because philosophy of technology really started as a field of research after the broader diffusion of computers and information technology in society, and as "tools, machines, and consumer products" are increasingly becoming digital and computerized, our digitalized society has been the object of much philosophical analysis (e.g. Borgmann, 1999; Brey & Hartz Søraker, 2009; Feenberg, 2017). However, abstract conceptions of technology beyond the material artefacts have not yet found their way from philosophy to technology education to any greater extent, for example, in relation to computation and information technology; examples could be coding, information processing, computational thinking, artificial intelligence (AI), and more. One reason for this slow transition may be the preoccupation in technology education with the material aspects of the human-made world, which could make abstract aspects of technology invisible and even contested.

4 Post-Phenomenological Analysis of Digital Technology

There has indeed been a preoccupation with the ontology and epistemology of physical artefacts lately even in the philosophy of technology and engineering (see above and e.g. Hahn & Soentgen, 2011; Kroes & Verbeek, 2014). The philosophical concerns of the digital world – which has made attention to abstract aspects of technology particularly pertinent – have instead mostly been addressed in the philosophy of AI, mathematics, or computing, as well as computer ethics (Brey & Hartz Søraker, 2009). On the other hand, a central journal such as *Philosophy & Technology* is so broad in scope that it includes research from all of the above fields and thus in relation to both concrete and abstract aspects of technology (see, for example, No. 3, 2014; cf. Hansson et al., 2021).

The problem of coming up with a definition of technology that includes both analog, digital, abstract and concrete dimensions of objects and systems could be framed in post-phenomenological terms. Post-phenomenology deals with the human experience of technology and the different ways that this experience can be embodied, but it also offers ways of categorizing and defining technology through human–technology relations (Rosenberger & Verbeek, 2015). I will here consider *what* is being mediated in different such relations (Ihde, 2010), as a way of illustrating an integrated view of technology. If we look at post-phenomenology in Ihde's (1990) version, the relationship between humans and the world is seen as mediated through technology, often conceived of in the following simple model:

I → technology → world

Ihde exemplifies technology with a number of concrete instruments such as a telescope, a clock or a pair of eyeglasses, so the conception of technology is essentially physical or concrete, as is the human experience of technology as embodied (Ihde, 1990, p. 25; cf. Ihde, 1993). The corresponding post-phenomenological model for digital technology could be the same if one focuses on digital artefacts such as smartphones as concrete tools:

I → digital artefact (smartphone) → world

However, if we look at the abstract code or information as technology the model would have to be revised into something like this, in the context of programming:

I → interface → code

Or, when handling information and datasets:

I → algorithm → dataset (cf. Wellner & Rothman, 2020)

In both cases the material component of technology is missing, so it is a both abstract and digital form of technology that mediates the relation between I/human and world. Furthermore, the world is also abstract, so it is a virtual world. There is therefore also a corresponding problematic regarding what the technology mediates, the real or the virtual. Brey and Hartz Søraker (2009) write:

> The software constructs with which computer users interact, such as files, folders, and web pages, exist virtually rather than physically. Although they are realized and sustained by means of physical systems, they do not exist solely or primarily as physical entities. The existence of virtual objects, and correlated virtual spaces, actions and events raises questions regarding their ontological status: what is their mode of existence, and what is their place in philosophical ontology? Let us call nonphysical software-generated objects and spaces with which users interact *virtual entities*. Virtual entities are represented as part of the *user interface* of computer programs. They manifest themselves to the user through symbolic or graphical representations, and they interactively respond to actions of the user. Contemporary user interfaces are in most cases *graphical*, representing virtual objects as ordinary, manipulable physical objects. (pp. 1381–1382)

Thus, in many current digital settings and, for example, visual or graphic programming environments, the above post-phenomenological models could be conceived of as:

I → virtual object → virtual space

This is thus a far cry from Ihde's (1990) post-phenomenological conception of the relationship of humans with – concrete/physical – technology, and may be a reason that there is an unease about digital technology in technology education.

4.1 *Technologies as Technical Artefacts*

There have been attempts at resolving the duality between the concrete and the abstract, for example, to see software, programs and virtual worlds or objects as *de facto* technical artefacts. Turner (2014) goes some way towards explaining the relationship between the programming languages and the physical devices by referring to Kroes and Meijers' (2006) conception of the dual nature of technical artefacts, that is, their physical/structural and functional/intentional dimensions. In Turner's view, the abstract programming language can stand on its own as having a mathematical semantics, but once we include a specification as the functional dimension that Kroes and Meijers (2006) require for an artefact to be defined as technical, this changes the program into a technical artefact:

> Programming languages have a mathematical semantics and an implementation. It is in the mathematical object that structure is located. And the semantics can stand alone as a mathematical object, i.e., to be investigated as such. But when used as the specification, it is given governance over the implementation. The intentional aspect then comes into play: the agent gives the abstract semantics normative force over the physical implementation: they must be in extensional accord. Abstract structure and physical structure are linked by the intention to take the former as having normative governance over the latter. It is at this point that the piece of abstract mathematics takes on its functional guise. The agent's intention relates the abstract and the concrete. (Turner, 2014, pp. 393–394)

In other words, when viewing the program as a specification (function) of how a (physical) object should behave, we have merged the abstract and the physical components of digital artefacts and systems. A post-phenomenological model of technical artefacts would thus generally look like this:

I → technical artefact (function/structure) → world

And, more specifically, based on Turner (2014) a post-phenomenological model of a programming language as a technical artefact could be conceived of as such:

I → program specification/physical execution → world (virtual/real)

Many digital instruments such as digital telescopes, computerized tomography scanners and electronic microscopes mediate the world in this way via virtual entities – symbolic representation (Brey & Hartz Søraker, 2009) – and Ihde (2011) therefore calls this translational mediation. The world translates and even transforms through these technologies. There is thus less distinction between the virtual and the real if we acknowledge that the virtual is some form of representation of a real (or imagined) world.

4.2 *Technologies as Technologies of Representation*

There is thus from the perspective of the dual nature of technical artefacts a way of merging the concrete and the abstract into one conception of technology, which also translates into a reasonable post-phenomenological model. However, I want to suggest a further perspective that might also broaden the conception of technology. In a sense, virtual entities and objects are rather graphical-visual models of real, or imagined, phenomena, that is, types of representation (cf., Hallström & Schönborn, 2023). One way of merging the (conflicting?) perspectives on technology entailed by digitalization – epitomized by the dualisms concrete vs. abstract, and real vs. virtual – is to view digital technologies as *technologies of representation* (Denning & Tedre, 2019; Mindell, 2002). Representational technologies could include everything from simple control systems to computers to AI systems, and it would be possible to conceive of the concrete and abstract parts of these technologies as different components of their representational capacity. That is, a component could either be seen as representing (concrete, real) or represented (abstract, virtual), but part of the same representational system that makes up the technology (cf., Hallström & Kaijser, 2022). Furthermore, it is possible to see representational technologies as *semiotic* technologies, in which case even the abstract, representing components such as a piece of code are, in the semiotic sense, material. (Latour's (1999, 2005) actor-network approach is semiotic in the sense that the so-called *actants* could be construed as non-human, concrete or abstract, "actors".) The post-phenomenological model would thus look like this:

I → technology (representing/represented) → world (real/virtual)

James Watt's centrifugal flyball governor in his late 18th century modified steam engine was the first simple yet effective modern feedback mechanism to be widely used. Modern, more complex technological systems include a multitude of feedback loops, particularly for negative feedback, to make the systems operate in a stable and robust way, and to assist operators and users. Today such control systems use both analog and digital sensors to communicate information back to a regulator or control unit, which is often a computer program of some sort (Mindell, 2002). Whether we talk about the flyball governor or a digital sensor, they are both part of the representing components that help control the system by providing the (represented) information. In the case of the digital sensor, small currents represent and provide information whereas in the case of the flyball governor a lever represented and transmitted information directly to a steam valve (Glad & Ljung, 2000).

5 Towards a Unified Conception of Technology: Concluding Discussion and Implications

Technology could be conceived of both as *technical artefacts with a dual nature* and *technologies of representation*. The dual nature of technical artefacts, that is, their functional/intentional and physical dimensions, can be mirrored in the abstract programming language that on its own has a mathematical semantics but once we include a specification/intention the program becomes a *de facto* technical artefact since it must include a physical configuration to operate. Representational technologies could include everything from simple control systems to computers to AI systems. It would be possible to conceive of the concrete and abstract dimensions of these technologies as different components of their representational capacity; a component could either be seen as representing (concrete, real) or represented (abstract, virtual), but part of the same representational system that makes up the technology. In both these "dual" perspectives on technology, artefacts and systems could be viewed from a common point of view and may consist of abstract and concrete components that together make up the technology, regardless of whether it is analog or digital. See Figure 5.2 for a visualization of how these two perspectives play out in relation to the digital, analog, abstract, and concrete dimensions of technology.

Both these dual conceptions make sense in a technology education context. Denning and Tedre (2019) argue that "internally the computer does not process numbers and symbols. Computer circuits deal only with voltages, currents, switches, and malleable materials. The patterns of zeroes and ones are abstractions invented by the designers to describe what their circuits do"

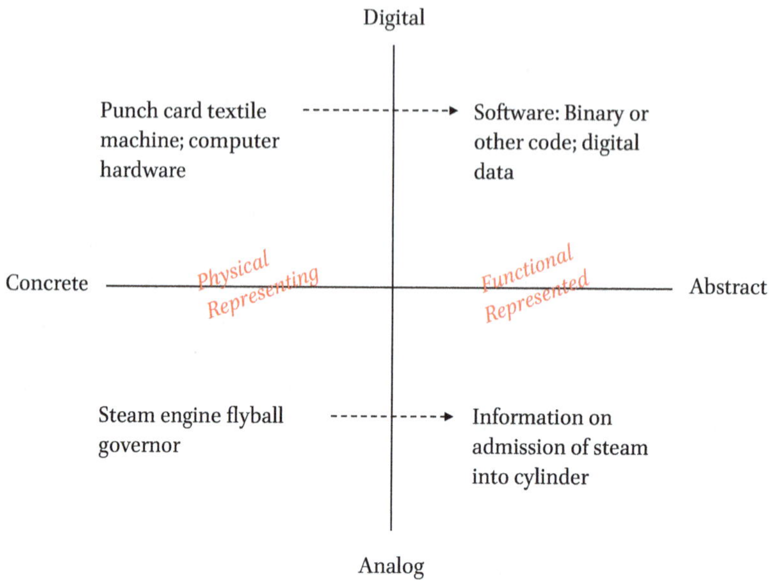

FIGURE 5.2 Schematic outline of digital, analog, abstract, and concrete dimensions of technology, in relation to the *dual nature of technical artefacts* and *technologies of representation*

(pp. 55–56). From the computer scientists' point of view, according to Turner (2014), they build:

> Programs, data types, type inference systems, etc. [that] seem to have an abstract guise that enables us to reflect and reason about them independently of any physical manifestation. [...] One can reason about these in a mathematical way that is independent of any physical representation. Much the same applies to programs, types, compilers, virtual machines, interpreters, etc. All of these notions seem to have an abstract guise that is independent of their physical realization or implementation. On the other hand, these objects must have a physical implementation that enables them to be used as artifacts in the physical world. For instance, a program that has no physical realization is of little use as a practical device for performing humanly intractable computations. (pp. 394–395)

In technology education, therefore, it would be limiting for students if the teaching merely focused on the programming in itself, because there needs to be a connection to a physical realization in the form of a design, product or technical solution. On the other hand, teachers in a study by Vinnervik

(2021) deemphasized procedural programming knowledge with the argument that the technology students were not going to become programmers. With regard to the two dual perspectives on technology discussed in this chapter, however, it is clear that technology teachers who deal with the digital world in general and programming in particular should see them as an integrated part of the designed, human-built world that is at the core of technology education. Knowledge about analog and digital technology cannot be separated, but it is, like technology, dual.

Thus, when teaching about, for example, programmed technological solutions (PTS) in a technology classroom, teachers need to include the nature of the abstract code as well as the concrete technical solutions that are to be programmed and controlled. As Cederqvist (2020) writes: "The ability to understand and handle the programming material encompasses both procedural and conceptual knowledge that refers not only to the material itself but also to a general understanding of programming concepts that are necessary for fulfilling the PTS" (p. 24). Such PTS projects should appeal to technology teachers since they could restore some of the design aspects of teaching programming, in the sense that the design encompasses both analog and digital, concrete and abstract dimensions and coding could be conceived of as a craft (Bratteteig, 2010; Lingel & Regan, 2014).

Note

1 The analysis on which this chapter is based also provided material for Hallström (2022).

References

Bjursten, E. L., Nilsson, T., & Gumaelius, L. (2022). Computer programming in primary schools: Swedish technology teachers' pedagogical strategies. *International Journal of Technology and Design Education*. Online first. https://doi.org/10.1007/s10798-022-09786-7

Borgmann, A. (1999). *Holding on to reality: The nature of information at the turn of the millennium*. The University of Chicago Press.

Bratteteig, T. (2010). A matter of digital materiality. In I. Wagner, T. Bratteteig, & D. Stuedahl (Eds.), *Exploring digital design: Multi-disciplinary design practices* (pp. 147–169). Springer.

Brey, P., & Hartz Søraker, J. (2009). Philosophy of computing and information technology. In A. Meijers (Ed.), *Philosophy of technology and engineering sciences* (Vol. 9). Elsevier.

Buckley, J., Seery, N., Gumaelius, L., Canty, D., Doyle, A., & Pears, A. (2021). Framing the constructive alignment of design within technology subjects in general education. *International Journal of Technology and Design Education, 31*(5), 867–883.

Cederqvist, A.-M. (2020). An exploratory study of technological knowledge when pupils are designing a programmed technological solution using BBC Micro:bit. *International Journal of Technology and Design Education*, 1–27. https://doi.org/10.1007/s10798-020-09618-6

Dakers, J. R., Hallström, J., & de Vries, M. J. (2019). Introduction. In J. R. Dakers, J. Hallström, & M. J. de Vries (Eds.), *Reflections on technology for educational practitioners: Philosophers of technology inspiring technology education.* Brill Sense.

Denning, P. J., & Tedre, M. (2019). *Computational thinking.* The MIT Press Essential Knowledge Series. The MIT Press.

de Vries, M. J. (2016). *Teaching about technology: An introduction to the philosophy of technology for non-philosophers.* Springer.

Dusek, V. (2006). *Philosophy of technology: An introduction.* Blackwell.

Feenberg, A. (2017). *Technosystem: The social life of reason.* Harvard University Press.

Glad, T., & Ljung, L. (2000). *Control theory: Multivariable and nonlinear methods.* Taylor & Francis.

Hahn, H. P., & Soentgen, J. (2011). Acknowledging substances: Looking at the hidden side of the material world. *Philosophy & Technology, 24*, 19–33. https://doi.org/10.1007/s13347-010-0001-8

Hallström, J. (2022). The dual nature(s) of technology: Towards a unified conception of digital and analogue technology. In D. Gill, J. Tuff, T. Kennedy, S. Pendergast, & S. Jamil (Eds.), *PATT 39: PATT on the edge technology, innovation and education, St. John's, Newfoundland and Labrador, Canada, June 21st–24th, 2022.* Memorial University of Newfoundland.

Hallström, J., Hultén, M., & Lövheim, D. (2014). The study of technology as a field of knowledge in general education: Historical insights and methodological considerations from a Swedish case study, 1842–2010. *International Journal of Technology and Design Education, 24*(2), 121–139.

Hallström, J., & Kaijser, A. (2022). Socially constructed and society shaping: Investigating characteristics of technological systems for technology education. In J. Hallström & P. J. Williams (Eds.), *Teaching and learning about technological systems: Philosophical, curriculum and classroom perspectives.* Springer Nature.

Hallström, J., & Schönborn, K. J. (2023). Models and modeling in STEM education: Nature, roles, and implementation. In R. J. Tierney, F. Rizvi, & K. Erkican (Eds.), *International encyclopedia of education* (4th ed., pp. 112–116). Elsevier.

Hansson, S. O., Belin, M.-Å., & Lundgren, B. (2021). Self-driving vehicles—An ethical overview. *Philosophy & Technology*, 1–26. https://doi.org/10.1007/s13347-021-00464-5

Ihde, D. (1990). *Technology and the lifeworld: From garden to earth*. Indiana University Press.

Ihde, D. (1993). *Philosophy of technology: An introduction*. Paragon House.

Ihde, D. (2010). *Embodied technics*. Automatic Press.

Ihde, D. (2011). Stretching the in-between: Embodiment and beyond. *Foundations of Science, 16*(2), 109–118.

International Technology and Engineering Educators Association. (2020). *Standards for technological and engineering literacy: The role of technology and engineering in STEM education*. www.iteea.org

Jones, A., Buntting, C., & de Vries, M. J. (2013). The developing field of technology education: A review to look forward. *International Journal of Technology and Design Education, 23*(2), 191–212.

Kroes, P., & Meijers, A. (2006). The dual nature of technical artefacts. *Studies in History and Philosophy of Science, 37*, 1–4.

Kroes, P., & Verbeek, P.-P. (Eds.). (2014). *The moral status of technical artefacts*. Springer.

Latour, B. (1999). *Pandora's hope: Essays on the reality of science studies*. Harvard University Press.

Latour, B. (2005). *Reassembling the social: An introduction to actor-network theory*. Oxford University Press.

Lingel, J., & Regan, T. (2014). "It's in your spinal cord, it's in your fingertips": Practices of craft in building software. In *Proceedings of the 17th ACM conference on computer supported cooperative work & social computing* (pp. 295–304). https://doi.org/10.1145/2531602.2531614

Meijers, A. (2009). General introduction. In A. Meijers (Ed.), *Philosophy of technology and engineering sciences* (Vol. 9). Elsevier.

Mindell, D. A. (2002). *Between human and machine: Feedback, control, and computing before cybernetics*. The Johns Hopkins University Press.

Mitcham, C. (1994). *Thinking through technology: The path between engineering and philosophy*. The University of Chicago Press.

Rosenberger, R., & Verbeek, P.-P. (2015). A field guide to postphenomenology. In R. Rosenberger & P.-P. Verbeek (Eds.), *Postphenomenological investigations: Essays on human-technology relations*. Lexington Books.

Rossouw, A., Hacker, M., & de Vries, M. J. (2011). Concepts and contexts in engineering and technology education: An international and interdisciplinary Delphi study. *International Journal of Technology and Design Education, 21*, 409–424. https://doi.org/10.1007/s10798-010-9129-1

Solomonidou, C., & Tassios, A. (2007). A phenomenographic study of Greek primary school students' representations concerning technology in daily life. *International Journal of Technology and Design Education, 17*(2), 113–133.

Svenningsson, J. (2020). The Mitcham score: Quantifying students' descriptions of technology. *International Journal of Technology and Design Education, 30*, 995–1014. https://doi.org/10.1007/s10798-019-09530-8

Turner, R. (2014). Programming languages as technical artifacts. *Philosophy & Technology, 27*(3), 377–397.

Vinnervik, P. (2021). *När lärare formar ett nytt ämnesinnehåll: Intentioner, förutsättningar och utmaningar med att införa programmering i skolan* (Diss). Umeå University.

Wellner, G., & Rothman, T. (2020). Feminist AI: Can we expect our AI systems to become feminist? *Philosophy & Technology, 33*(2), 191–205.

Williams, P. J. (2018). Technology education: History of research. In M. J. de Vries (Ed.), *Handbook of technology education*. Springer.

Xu, M., Williams, P. J., & Gu, J. (2021). Developing an instrument for assessing technology teachers' understandings of the nature of technology. *International Journal of Technology and Design Education*. https://doi.org/10.1007/s10798-021-09698-y

CHAPTER 6

Framing Computational Thinking and Digital Competence in Technology Education

Helena Isaksson Persson and Arnold Pears

Abstract

Computational Thinking (CT) has developed as a worldwide priority area for compulsory school education since it was proposed as a central 21st century skill by Wing in 2006. CT and Digital Competence are highly visible parts of the European Union and Swedish discourse on strategic workforce development and the Swedish innovation agenda. This chapter explores the emerging role of CT and digital competence in the Swedish compulsory school curriculum. We focus on the subject of technology, its historical role in equipping generations of young Swedes for a career in the technology and engineering domains, and the development of the subject during the 20th century as a result of shifts in government policy, and under the influence of curricula reforms. We conclude that "Technology" is a natural home for CT and related skills such as systems design and development of programmed technical solutions in the current educational system, and propose that the relevance and identity of the technology subject be strengthened through systematic integration of a higher degree of CT into classroom practice.

Keywords

computational thinking – technology – compulsory school curriculum – programming – digital competence

1 Introduction

The impact of digital transformation on our lives and global policy in the last two decades cannot be understated. Equipping future generations to understand and create responsible technological solutions is a complex endeavour, and clearly requires significant curricula reform (Pears, 2019). In this regard it is vital that the political discourse surrounding Computational Thinking (CT)

© HELENA ISAKSSON PERSSON AND ARNOLD PEARS, 2024 | DOI:10.1163/9789004687912_006

be both explored, and well understood. The predominant discourse claims CT to be a central component of giving citizens agency in the society resulting from the transformations of the third industrial revolution (Wing, 2006, 2008), however, this discourse is problematic for a number of reasons (Tedre, 2021; Pears et al., 2021).

The technology subject in Swedish compulsory school has its roots in three major knowledge traditions, the craftsmanship tradition, the engineering tradition, and humanities and social sciences tradition (Nordlöf et al., 2022). The goal of this type of education has been, and remains, centred on the societal need for a well educated, intellectually and practically equipped workforce. These traditions have been central to the emergence of Sweden as a powerful technological nation, despite a comparatively small population (Hartvik, 2013, p. 22).

Besides these traditions, technology education in Sweden has its roots in vocational education for children and young adults; further education for craft apprentices in the mid-19th century is an example. Ideas about the competencies needed to equip the labor market characterise Swedish society's view of technology education (Hallström, 2013; Konstfack, 2010; Statistiska centralbyrån [SCB] 1977, 1984). Such arguments, relating to the utility of technology education, are still current and are directly connected to the introduction of digital technologies in Swedish compulsory schooling. In the 1980s the students learned to master computers through programming; in the 1990s, information technology was the primary focus. The 2018 curriculum, however, placed primary emphasis on digitization and programming (Riis, 1991, 2019; Skolverket, 2018a, 2018b, 2018c). A result of these discourses is that an enduring theme emerges, where knowledge and competences related to digital technologies are emphasized as important for the future labor market.

This chapter discusses: (1) how Swedish society values technology, and how this affects the development of technology education in Sweden; (2), the introduction of digital technology into Swedish schools, and the effect this has on policy in terms of teaching and learning. These two perspectives form our foundation as we discuss the framing of computational thinking in Swedish education (K-12) in general and in the technology subject in particular.

2 Technology Emerges as a Subject in the Swedish Curriculum

In order to position technology as a subject in the Swedish school curriculum it is useful to explore the historical origins of the subject, and its evolution and changing status in the national curriculum. From its humble origins as a vocational training programme, with strong links to supplying Swedish industry with a well equipped workforce, technology has evolved into a subject intended to provide an holistic approach to integrating science and engineering.

2.1 *From Vocational Education to a Technology School Subject*

Views on technical knowledge have changed over time. Before 1918, lower technical vocational training in Sweden is called technical teaching and learning, "[...] teknisk undervisning" (Statistiska centralbyrån [SCB], 1984, p. 15). An example of early technical teaching is the Swedish Association of Handicrafts School (Svenska Slöjdföreningens skola) which provided education from the middle of the 19th century (Konstfack 2010; Statistiska centralbyrån [SCB], 1984). In 1850 the youngest students admitted were 10 years of age, though subsequently the minimum age for admission was raised to 14 (Statistiska centralbyrån [SCB], 1984). This is a clear reminder of a different Swedish society where most children needed to work for a living. Another form of technological education from this time is technical secondary schools *tekniska elementarskolor*, despite the name, at first these were more general than technically oriented but later evolved to provide students with basic technical knowledge aimed at industrial activities such as mechanics, chemistry and building technology (Statistiska centralbyrån [SCB], 1977). Over time, lower level technical education was considered ill adapted to the needs of modern industry. This resulted in the government bill 1918: 96, which proposed a reorganization of lower technical education into technical continuation schools (tekniska fackskolor) and technically general secondary schools (tekniska gymnasier) (Statistiska centralbyrån [SCB], 1977). The directives for educating younger children were also changed, and the resulting 1919 elementary school curriculum contained more technical content than the previous curricula (Hallström, 2009). These curriculum changes affected some school subjects more than others, for example sloyd (slöjd). The sloyd subject, previously general education for learning domestic and local craftsmanship traditions, became more oriented towards technology and design. This new focus can be derived from (1) new directives in the curriculum. In addition, pressure from the government and teachers arguing that the subject form a foundation for vocational training in (2) technology; and (3) arts and crafts (Hallström, 2009).

In addition to the technical education curriculum, many school systems were operating in parallel in Sweden during the first half of the 20th century. The older vocational training system was considered to be outdated, as it covered too few occupational areas, and was regarded as too specialised. In 1948 the school commission recommended merging the parallel school system into a nine-year elementary school for the general population. As most young people were supposed to start working after year nine, the commission proposed a final career-oriented year in the new elementary school. The alternative was to commence upper secondary studies or choose a general theoretical final year (SOU, 1948, p. 27). The first curriculum in the compulsory school system was launched in 1962 (Riksarkivet, n.d.). An elective subject, "technology orientation", was made available to all students in year 8. Students interested in technology could in the ninth and final year, choose technology in their upper secondary studies or general technology vocational training (Skolöverstyrelsen, 1962). A revision in 1969 eliminated these specializations, but the elective subject remained, now called Technology (Skolöverstyrelsen, 1969a, 1969b). In the curriculum of 1980, Lgr80, through its close affiliation with the science subjects, Technology became a compulsory subject (Skolöverstyrelsen, 1980b). With reference to theory and application it was emphasized that science and technology should be treated as a whole, thus establishing a unique subject (Skolöverstyrelsen, 1980a).

2.2 *Computing in Schools, a Twenty-Year Perspective*

In Sweden during the late 1980s and early 1990s management of primary and secondary school transferred from the state to the municipalities. This change and technological development in communication came to create investments in *Information and Communication Technology* (ICT) during the 1990s (Skolverket, 2000). ICT should promote the student's ability to independently seek knowledge.

Compared to the 1980s investments on hardware, such as computers in primary and secondary school, the 1990s were characterized by government investments in software. Development of teaching strategies was not in focus; teachers introduced adapted or developed their own software.

The 1994 compulsory school curriculum (Utbildningsdepartementet, 1994a) includes the computer as an important tool, used for information searching, creating images and communication in subjects such as Language, Arts and Crafts. In Mathematics new mathematical methods and models led to new applications with the use of computers (Utbildningsdepartementet, 1994a, 1994b). The Technology subject syllabus mentions problem solving as one of the characteristics associated with technology development. Computers are not explicitly mentioned, but, electronic information processing is suggested as a technological solution for students to explore in order to become familiar with technology (Utbildningsdepartementet, 1994a).

Towards the end of the 1990s, and increasingly prevalent after the millennium, computers and communication systems such as the Internet have established their positions in education for children and young people. Competence development for teachers, however, was often seen to be about being able to use computer programs rather than discussing computers and the Internet from an educational perspective (Skolverket, 2000, 2013).

In the curriculum for the compulsory school from 2011 the computer is referred to as a tool in the subjects Swedish and Art, and a new term, *digital*, appears. Digital is combined with the words tools, technologies, and media in several subjects. In Mathematics, for example, digital technologies are used to investigate problems. Within Technology, problem solving is still emphasized and digital sketches and models are introduced as tools with which to document and represent constructions. The teaching scope grows to include communication, and information systems such as the Internet, computers, and mobile phones (Skolverket, 2011).

Although schools were investing in IT equipment, the investments were not accompanied by deployment of appropriate teaching and learning strategies (Skolverket, 2013). The Swedish National Agency for Education conclude in a report from 2013 that 90% of students in grades 7–9 (students of approximately 13–16 years of age) are not using IT when studying mathematics. Also in the Science, Technology and English subjects computer use is low. The report defines areas of use and skills related to computers as software, interactive whiteboards, the Internet, intranets, digital cameras, projectors and email as IT. Students in grades 7–9 and upper secondary school should use IT when searching for and critically reviewing information from the Internet. They should also gain experience in the use of software for word processing, calculations, presentations and when editing images, movies and music (Skolverket, 2013).

2.3 *Digitalization and the Programming Era*

In Sweden a number of successful IT companies such as Klarna,[1] Skype[2] and Spotify[3] were founded after the turn of the millennium. A 2015 article published in a major Swedish newspaper, describes Stockholm as a hub for ICT innovation and entrepreneurship. The authors are the municipal commissioners for finance of Stockholm, the CEO of Stockholm Chamber of Commerce and the founder and chairman of the well-known music streaming provider Spotify. Programmer is, in the article, claimed to be one of the most common occupations in the region. The authors describe coding as a new language and argue for the introduction of coding in schools. By teaching coding and programming, they argue, the school will be brought into the future and this effort would also increase the opportunities for more people to become programmers (Wanngård et al., 2015). This argument echoes the government bill of 1918 (Statistiska centralbyrån [SCB], 1977). It seems apparent that in Sweden, commercialisation of new inventions often leads to industry and political stakeholders raising their voices in campaigns to adapt the education system to the new technology.

Following in the wake of these debates we see a government decision in 2017 announcing the introduction of a digitization strategy for the education system (Ministry of Education and Research, 2017). The goal of the strategy is described as being about democracy; all children should get the knowledge they need. Creating and understanding digital solutions and tools is valuable knowledge in a future labor market, learning to program is a way to develop such competences (Ministry of Education and Research, 2017). Responding to this shift in policy a revised curriculum for compulsory school was published in 2018 (Skolverket, 2018a, 2018b) with the aim of strengthening students' digital competence (Ministry of Education and Research, 2017). The Swedish National Agency for Education also published a research overview with the goal of sharing knowledge about how digitization affects education. Programming is described as a digital competence, and computational thinking as an important prerequisite for understanding programming (Skolverket, 2018c).

The revision of the curriculum affects all school subjects. The Swedish National Agency for Education justifies this comprehensive revision with the argument that digitization has a great impact on the individual and society and thus needs to permeate the school system (Lingman et al., n.d.). The conclusion is that everyone, students, teachers, and others who are active within the school, need digital competence. Four aspects that together provide adequate digital competence are presented on the Swedish National Agency for Education's website (Skolverket, n.d.):

- Understanding the impact of digitization on society, e.g. development of future professions and careers; the impact of digital services on everyday life; the individual's impact on societal development.
- Using and understanding digital tools and media. Examples of tools are physical devices such as computers and smartphones, software, and internet-based services. Examples of media are platforms for communication and creation like e-books, social media, and other internet-based services.
- Having a critical and responsible approach e.g., in the verification of sources and information; in the individual development of knowledge; in communication with others.
- Solving problems and turning ideas into action. By use of, e.g. curiosity, creativity, and entrepreneurship; tools and combination of tools. Through cooperation with others. With programming.

A representative from the Swedish National Agency for Education describes computational thinking as the logic needed in problem solving and thus central to programming (Häger et al., n.d.).

3 Computational Thinking Emerges as an Educational Priority

The relevance of computational thinking as a skill vital to empowering all citizens has become prevalent worldwide after the initiatives taken by Wing (Wing, 2011) to introduce her definition of the concept into the Northern American educational research agenda (Saqr et al., 2021). Swedish policy makers could not be seen to ignore this important international trend, and discussion ensues in terms of the integration of computational thinking into Swedish compulsory education.

3.1 *The Roots of Computational Thinking*

Computational thinking, however, is neither easy to understand or introduce, and has a complex and troublesome history. With roots in the early days of manual computation, computational thinking is often associated with understanding the design procedures required to craft highly efficient computational solutions to what often were considered intractable problems. It should be noted that this type of computational challenge is not new. The notion of using "computers" to perform routine operations in order to solve large-scale numerical problems pre-dates the notion of automating the process with machinery. Early examples of "computers" can be found in the history of the navigation tables of Great Britain, and in the collections of logarithms, sines and cosines and other mathematical reference tables long used as an aid in lengthy calculations requiring numerical solutions. To compile such tables was an herculean task, requiring concentrated planning and considerable effort, not to mention an enormous number of "computers" and "comparators" who would perform the computations and compare the results prior to inclusion in the final book of tables (Craik, 2003).

These "computers" were often clergy or other mathematically educated and reliable persons, and their collaboration around the systematic compilation of navigational and other types of mathematical table cannot be underestimated (Robson & Stedall, 2008). However, the process was fraught with error. Individuals made mistakes, and it was not uncommon for a comparator to either receive two erroneous computations to compare, or to make a mistake in either the comparison or transcription. Thus emerged an urgent priority, to mechanise (automate) these computations to eliminate the element of human fallibility.

As computation moved beyond specialised areas of application to form the basis for automation of many processes in modern society, the potential for a second technological revolution emerged (Tedre, 2014). Automation is on the cusp of reaching such a level of sophistication that many activities hitherto considered the province of human agency will almost inevitably be automated in the near future. Examples include e-judges in courts of law and self-driving vehicles, to name but two. In this context computational thinking has been advanced as a necessary skill. It is necessary that everyone should receive a grounding in relation to these technologies.

3.2 *What Is Computational Thinking?*

The current focus of CT activity in classrooms are programmed solutions that primarily relate to developing programming abilities. It is not clear that those skills are those one will require to operate effectively as a citizen in the increasingly technological environment typical in developed societies worldwide. Quite different skills are required to assess whether it is feasible that the world will make a transition to self-driving electric vehicles within the next 20 years, or to reason about the implications of inference algorithms applied to large data sets that lay the foundation for Google Translate, or social media feed recommender systems.

Pears and Tedre (Pears et al., 2021; Denning & Tedre, 2019) argue that the focus of current curricula on programming and development of imperative language, sequential programming skills and competencies is not sufficient. However, the question of defining CT remains. Denning and Tedre (Denning & Tedre, 2021) discuss the origins and potential definitions of CT, but fail to reach a concrete definition of the exact skills, competencies and values that are associated with the concept. However, they do propose a high level definition:

> Computational thinking is the mental skills and practices for designing computations that get computers to do jobs for us, and for explaining and interpreting the world in terms of information processes. (Denning & Tedre, 2021)

A systematic review of literature in the area for the period 2016–2021 with a focus on CT in school education, proposes that CT definitions can be grouped into three categories (Bocconi et al., 2022); those that are generic; those that are operational; and those that are expressed in connection to educational and curricula frameworks. Generic definitions provide no explicit guidance to teachers, and require substantial interpretation. For instance, the definition proposed by Denning and Tedre above. Operational definitions provide concrete guidance in terms of competencies and practices with a clear connection to the disciplinary ideas of computing. Finally, definitions connected to educational systems and curricula are based on problem solving methods that are derived from computer science tasks or activities that are directly applicable in computer science.

A major problem with the generic definitions is that they relate to computing as a discipline and rely on the reader having a strong connection to computer science as a research discipline. However, these definitions are very vague, and ignore much of advanced computing knowledge, they typically involve concepts such as abstraction and data representation, which are important, but do not enable learners to reason about concurrent execution, race conditions or mutual exclusion. For those without an extensive disciplinary background the type of abstract guidance they provide runs a substantial risk of perpetuating the "Computer Science = programming" misconception that has plagued computing since the 1960s. As Tedre puts it we must "abandon the belief that rule-based 'traditional' programming is a central aspect and building block in developing next generation computational thinking" (Tedre, 2021).

Operational and educational and curricula models are also problematic as they combine concepts at both high and low levels of abstraction, e.g. debugging and abstraction are included together with data representation, looping and sequences as discussed by Dagienė and colleagues (Dagienė et al., 2022). There seems to be scant coverage in these types of CT definition for concepts dealing with models of execution and computing behaviours at a more fundamental level, e.g. control-flow based, lambda calculus, or functional and dataflow execution. In addition, vital concepts such as concurrency, and the statistical and inferential solutions inherent in the large data models which underly and enable recent AI (artificial intelligence) and ML (machine learning) solutions are ignored in most CT definitions.

Focussing on operational definitions can provide some practical guidance to teachers, however. For example, operational definition is used in the study (Jocius et al., 2020) proposing an educational framework of CT based on solving problems, designing systems, and understanding human interactions with systems drawing on fundamental computing concepts. This definition emphasises the significance of recognising and utilising problem-solving skills, including abstraction, decomposition, and pattern recognition. Another widely recognised operational level definition is that of Weintrop et al. (2016) who divide CT into data practices, modeling and simulation practices, computational problem-solving practices, and systems thinking practices.

However, from the practical point of view, definitions of CT themselves often mean nothing to teachers. Zhang and colleagues (Zhang 2020) emphasize that student learning of CT is likely to be compromised, at least in the short term, as many teachers have very limited knowledge of CT, and furthermore are incapable of translating the multiple fuzzy definitions proposed in the research literature into concrete classroom activities.

What is needed in these definitions is guidance as to how CT equips future generations with the sophisticated awareness, and technological insight, required to comprehend, interpret and take intellectual control of the computing systems that permeate our world, and whose effect is predicted to grow substantially? What "habits of mind" should be painstakingly inculcated during compulsory education to equip future generations with personal agency and insight with regard to the systems all around them?

John Sanford and colleagues propose an approach deeply founded in mathematical problem solving, with an associated focus on mathematical modelling, and ability to test alternatives that would be precluded by a manual approach. Here, automation of computation and the ability to compute many alternatives provide extra power in relation to manual calculation (Sanford & Naidu, 2016). This type of approach may prove useful as a means of thinking about CT in the context of mathematics and physics, however, the focus of CT in technology must be less mathematical, and more focussed on design, regulation and automation of technical systems, and the construction and analysis of programmable/programmed technical artefacts and systems.

In this context CT can be seen as providing mechanisms for unpacking and inspecting computational designs and systems. Computational thinking and "unplugged" activities are designed to make the computational elements and process which are normally invisible tangible. It is just this process of unpacking and reasoning about computation that provides pupils with the ability to reason about advanced programmed computational systems and processes. Using this definition also provides a stronger link to key elements of the Technology subject curriculum and learning outcomes. Providing concrete insight into how to understand and analyse computing solutions and link this to sustainable and ethical problem solving practices.

4 Computational Thinking and Digital Competences in Sweden

During the period that Sweden introduces computers and software in schools, Seymour Papert is developing his new and progressive ideas on learning in interaction with computers. During the 1980s, Papert develops ideas which he terms "constructionism" and designs learning spaces for children for learning mathematics and physics in interaction with "things" (computational artefacts). It is in the interaction and communication with these "things" that learning takes place, and the child is in control (Papert, 1980/2020). However, these ideas and the Logo language that accompanied them, while gaining international attention did not establish a solid foothold in the Swedish school.

It was not until Jeannette Wing revived the debate on the role of computing as a central way of thinking for the 21st century in the article Computational Thinking (Wing, 2006) that attention in school education once again focussed on the role of computers in problem solving. Wing claimed that society as a whole could, and should, learn from computer science when solving everyday problems.

Initiatives in Sweden to introduce programming and computational thinking in schools prior to 2018 were at different levels in society and on different scales. These initiatives ranged from individual projects at local schools to larger projects conducted as joint ventures involving researchers and municipalities (Heintz et al., 2015). Most research on computational thinking in school settings is based on computer science and focuses on possibilities and challenges with the implementation of programming in, primarily, compulsory school. The legacy of Papert and Wing is clear. School subjects involved in these early initiatives are crafts, languages, mathematics, civics and technology (Chibas et al., 2018; Heintz et al., 2015, 2016, 2018; Mozelius et al., 2017; Palmer, 2017; Pears et al., 2017; Sjöberg et al., 2018; Tyrén et al., 2018; Vrieler, 2017).

Since 2018 the curriculum promotes programming and digital competencies. However, it is important to understand under which conditions the curriculum allows CT to emerge as prioritised subject matter. Two analyses form the basis of our conclusions in this regard. To more fully understand the situation we have performed a qualitative content analysis of the curriculum and a rapid literature review of research conducted in Sweden in which CT forms the central component (Isaksson Persson, 2022).

4.1 *Mapping the Curriculum*

In our analysis (Table 6.1), the Swedish words digital (digital), programmering (programming), and datalogi/sk/t (computational) are used as references. Combinations or variations of all these words were included in our initial search, and each of the word combinations is counted as one reference term. For example, datalogiskt tänkande (computational thinking) is a combination of datalogi/sk/t (computational) and tänkande (thinking) and considered as one reference term or string.

Reference counting is a powerful tool for revealing patterns and revealing the context in which a text is used (Prior, 2023). We use reference counting to show how government directives affect the opportunities offered for teaching and learning in the target areas. Table 6.1 shows the presence of references in syllabi or other parts of the curriculum. These references are categorised into, *Digital*; *Programming*; and *Computational thinking*. The word Yes indicates the presence and the word No the absence of a keyword (Table 6.1).

TABLE 6.1 Content analysis showing the presence of keywords in the Swedish curriculum *Läroplan för grundskolan, förskoleklassen och fritidshemmet 2011* (revised 2018)

Curriculum syllabi (rows 1–21)	**Reference count**		
	Digital	**Programming**	**Computational thinking**
1: Art	Yes	No	No
2: Biology	Yes	No	No
3: English	No	No	No
4: Physics	Yes	No	No
5: Geography	Yes	No	No
6: Home and consumer studies	No	No	No
7: History	Yes	No	No
8: Physical education and health	Yes	No	No
9: Chemistry	Yes	No	No
10: Mathematics	Yes	Yes	No
11: Modern languages (optional)	Yes	No	No
12: Mother tongue tuition (optional)	Yes	No	No
13: Music	Yes	No	No
14: Religion	Yes	No	No
15: Civics	Yes	Yes	No
16: Sami (optional)	Yes	No	No
17: Crafts	Yes	No	No
18: Swedish	Yes	No	No
19: Swedish as a second language (optional)	Yes	No	No
20: Sign language for the hearing (optional)	No	No	No
21: Technology	Yes	Yes	No
Other parts of the curriculum			
Fundamental values and tasks of the school	Yes	No	No
Overall goals and guidelines	Yes	No	No
Preschool class	Yes	No	No
School-age educare	Yes	No	No

TABLE 6.2 The presence of references to the keyword programming in the Swedish curriculum *Läroplan för grundskolan, förskoleklassen och fritidshemmet 2011* (revised 2018)

Keyword: Programming		
References		**Reference count**
1.	Programmering (Programming)	11
2.	Programmeringsmiljöer (Programming environments)	2
3.	Programmeras (Programmed)	1
Total		14

There are three syllabi which reference the keyword *Programming*, Civics, Mathematics and Technology. In 18 out of 21 syllabi there are references to the keyword *Digital* but no reference at all to computational thinking (Table 6.1).

Tables 6.2 and 6.3 show the references the keywords are based on. The variation of references to the keyword *Digital* (Table 6.3) is greater in comparison to references to the keyword *Programming* (Table 6.2). Furthermore, the difference in the number of references to the keyword *Digital* (178 references, Table 6.3) is markedly higher than to the keyword *Programming* (14 references, Table 6.2).

The most common terms identified in our analysis are *Digital tools* followed by *Digital media* and *Digital aid/s* (Table 6.3). The term *Digital tools* can be found in almost all syllabi and parts of the curriculum. *Digital media* is found in about a third of syllabi and in the parts of the curriculum that describe overall goals and guidelines, and guidelines for the preschool class. *Digital aid/s* can only be found in the optional courses *Modern languages* and *Mother tongue tuition*. Several language curricula provide repetitive search results for these terms in the context of language teaching.

The analysis shows that the curriculum is dominated by references to digitization with emphasis on tools and to some extent media. Words such as communication, tool, and media feature prominently in the Swedish National Agency for Education's four aspects of digital competence as previously stated (see Section 2.3). *Solving problems and translating ideas into action*, is one of the four aspects, and programming is discussed within the scope of this aspect. Computational thinking is referred to as the logic used in problem solving with programming and is therefore considered important but not described in detail (Häger et al., n.d.).

TABLE 6.3 The presence of references to the keyword digital in the Swedish curriculum *Läroplan för grundskolan, förskoleklassen och fritidshemmet 2011* (revised 2018)

Keyword: Digital		
References		**Reference count**
1.	Digital/a kompetens (Digital competence)	3
2.	Digitalisering/en/s (Digitization)	7
3.	Digital/a teknik/er (Digital technology/techniques)	8
4.	Digital bildbehandling (Digital image processing)	1
5.	Digital bearbetning (Digital processing)	1
6.	Digital kommunikation (Digital communication)	1
7.	Digital form (Digital form)	2
8.	Digitala hjälpmedel (Digital aid/s)	23
9.	Digitala medier (Digital media)	34
10.	Digitala miljöer (Digital environments)	13
11.	Digitala modeller (Digital models)	10
12.	Digitala sammanhang (Digital contexts)	2
13.	Digitala skisser och ritningar (Digital sketches and drawings)	1
14.	Digitala utvecklingen (Digital development)	1
15.	Digitala verktyg (Digital tools)	71
Total		178

We note in this analysis that governing documents omit computational thinking; does this mean that this area is of no relevance to Swedish teaching and learning? To investigate that issue further, we now turn to an analysis of prior research.

4.2 *Rapid Literature Review*

Although CT is not specifically mentioned in the Swedish curriculum, our rapid literature review suggests that it is a vital emerging research field.

Our investigation follows the standard process for a rapid review in the sense that the selection process is systematic, has a narrow focus, and is completed within a limited time (approximately 12 months, not full-time) (Denscombe, 2016; Folkhälsomyndigheten, 2017; Ganann et al., 2010). Therefore, we do not claim to cover all research, but only that which emerged through the search procedure.

The aim of the overview is to examine whether computational thinking is of interest for research and researchers with a focus on Swedish education. Follow-up questions to this overall purpose are related to definitions and perspectives on computational thinking that are covered by the research. In order to find relevant research we used the databases SCOPUS, ERIC, and Web of Science. The purpose of the search was to identify research concerning the Swedish school system which includes results and perspectives connected to CT. Different combinations of keywords were tested. The best results are given by combinations of the following terms, *Computational Thinking*; *school*; *education*; *Sweden*; and *programming*. Four combinations were ultimately chosen, these keyword combinations and examples of search strings from the databases are shown in Table 6.4. The term *programming* is added to broaden the search, while the term *school* is added to exclude studies related to higher education. Each keyword combination is given a code, the first letter of each database to track the origin of the results (Table 6.4).

Based on four keyword combinations (Table 6.4), the result is four parallel search and selection processes in relation to each database. The following selection process was used:

Procedure
- STEP 1: Identification, screening titles.
- STEP 2: Identification, reading abstracts.
- STEP 3: Removing duplicates.

TABLE 6.4 Keywords used to find relevant research in SCOPUS, ERIC, and Web of Science

	Keyword combinations	Code
	education-Computational Thinking-Sweden	eCTS
	school-Computational Thinking-Sweden	sCTS
	education-programming-Sweden	epS
	education-school-programming-Sweden	espS
Database	Examples of search strings	Database + Code
ERIC	Years 2002–2021 (All years). References ERIC-education AND Computational Thinking AND Sweden	EeCTS
SCOPUS	Years 2015–2021 (All years). References (school AND computational PRE/0 thinking AND Sweden)	SsCTS
Web of Science	ALL FIELDS: (education programming Sweden) Timespan: All years.	WepS

- STEP 4: Read full texts.
- STEP 5: Inclusion, selections of texts.

The selection process resulted in a collection of 24 research articles or papers which are attributed to 43 authors. Of these publications 21 are found in Web of Science, 11 in Scopus and 2 in ERIC. Only 3 texts are not in Web of Science but in Scopus. Consequently, the majority of the selected texts can be found in Web of Science.

The earliest study meeting the criteria is published in 2015 and the latest in 2021, the same year the search for publications ended. Of the publications 10 are published before, and 14 after, the revision of the Swedish curriculum in 2018. In terms of research focus, 3 publications concern research in preschools, 20 concern primary school or grades K-12 and 1 upper secondary school. The publications included in the analysis are listed in Appendix A.

Thematic analysis (Braun & Clarke, 2006; Ryan & Bernard, 2003) was used to scrutinize the 24 publications in relation to two research questions:

- (a) Which frameworks are used to define CT?
- (b) Which perspectives on CT are found in the publications?

The text of the publications is analysed both as a complete body of text, corpus (Braun & Clarke, 2006), and as individual texts. This means that the focus of the reading alternates between approaching the text as a whole or as individual publications. Another approach used in the analysis is semantic and identifies themes on an explicit level (Braun & Clarke, 2006). Explicit in this context means being close to the text and identifying descriptions that relate to the research questions. The readings alternate between the corpus and the individual publications with the aim of finding patterns in the texts. The patterns are then coded and the codes are compiled into themes.

4.2.1 Perspectives on Computational Thinking

The following text focuses on research question (b). However, the following short comment needs to be made in relation to question (a), which concerns the frameworks used to define CT. CT is seen as a universal competence used in problem solving and closely related to programming. The analysis of the publications shows that CT definitions and strategies for assessment of CT skills are still evolving (Mannila et al., 2020; Zhang et al., 2020). Despite an increased diversity in CT definitions, the legacy of Papert and Wing is clear. For further reading about the frameworks used in the publications, see Isaksson Persson (Isaksson Persson, 2022).

TABLE 6.5 Themes representing perspectives on computational thinking

The outcome of the thematic analysis. Themes representing perspectives on computational thinking (CT)
A. Definitions of CT
B. CT as problem solving
C. CT and programming
D. CT as a universal competence
E. CT and educational perspectives
F. CT in school subjects

The second research question, (b) concerns perspectives on CT found in the publications. The results are presented as six themes in Table 6.5. Themes (A)–(D) are in line with the findings of the research question (a). Themes (E) and (F) focus on educational issues. Examples of sub-themes for (E) *CT and educational perspectives* are curriculum (considerable focus is placed on the curricula revision of 2018), assessment of CT skills, progression of CT skills, and CT for young children. The theme (F) *CT in school subjects*, maps CT to school subjects. Theme (F) is further developed in Table 6.6.

Table 6.6 shows how CT is contextualised in the Swedish school subjects. Civics, Mathematics, and Technology are identified as subjects with programming objectives. CT is treated in the contexts of problem solving in Mathematics and automatic control in Technology. Little is documented concerning CT and Civics but one publication describes a cross-disciplinary project with Mathematics and Civics (Table 6.6). However, CT and programming activities are also found in other subjects. In Physics the students work with simulations, and in Art with visual interfaces. CT is also associated with Craft and Languages (Table 6.6).

Other examples of classroom activities including CT include game design, animation, visualisation, simulations, control, and automation. A common block-based programming resource in these activities is the software Scratch. But other resources besides Scratch are also visible, programming languages e.g. Python; robots e.g. BlueBots and Lego EV3; embedded systems e.g. Arduino and Micro:bits (Chibas et al., 2018; Sjöberg et al., 2019; Svensson et al., 2020; Vinnervik, 2022; Tyrén et al., 2018). Thus, activities and resources differ from subject to subject (Table 6.6).

TABLE 6.6 Themes representing computational thinking related to Swedish school subjects

Themes	Computational thinking (CT) related to school subjects. Summaries of the themes
Mathematics	CT in Mathematics is discussed in relation to problem solving.
Technology	CT in Technology is linked to the control of objects, for example robots. Simulations using micro:bits are also described.
Art	CT in Art. Students redesign a game, including both coding and aesthetic features (Bowden, 2019; 2020).
Civics	CT in Civics. An interdisciplinary activity (Civics and Mathematics) contextualises mathematics when the students analyse voting results with programming to learn about the Swedish electoral system (Sjöberg et al., 2019).
Craft	CT in Craft. Ahmed et al. (2019) study students' perceptions of programming. Students associate programming with Craft. This association is explained by the fact that construction is a prominent element in programming and computational thinking.
Language	CT in Language. There are several languages in the curriculum. Compulsory languages are Swedish/Swedish as a second language and English. Students associate programming with language as there are similarities in structures and syntax. They also associate with language when using Swedish and English in programming (Ahmed et al., 2019). Heintz et al. (2015) see possibilities to introduce computational thinking concepts during language lessons.
Physics	CT in Physics. Students analyse physics problems by programming simulations that visualise phenomena through animations. Svensson et al. 2020, claim that simulations and animations help students understand concepts not only in Physics but also in other fields of science.

5 Conclusion

Technology as a subject in the Swedish compulsory school system has a long history, and a central role in positioning Sweden as a technological nation. One of the dominant discourses surrounding the subject is the need to equip future citizens, both in terms of agency and understanding of the technologies around them, and importantly as highly skilled contributors to Sweden's continued technological excellence.

A strong emphasis on digital competencies in the Swedish national curriculum (Table 6.1) aligns well with this historical narrative. Programming is seen as a digital competence to solve problems and transform ideas into action. Computational thinking is claimed to be the logic needed to enable this (Skolverket, 2018c). CT is thus broadly conceived as being necessary in developing a desired competence profile in Swedish education. Our content analysis shows that this relationship is not reflected in the overall curriculum or in the learning outcomes of individual subjects (Table 6.1). We conclude, on the basis of our analysis, that the interest of Swedish authorities to promoting teaching and learning in CT appears to be weak.

On the other hand, the research overview underlying the thematic analysis shows that CT is already integrated into classroom practice through a variety of activities even in subjects without programming objectives (Table 6.6). We can assume that the curriculum's emphasis on developing digital competencies is the underlying explanation for this situation.

According to Bocconi et al. (2022), both basic computer science concepts and digital competence are necessary components for developing CT skills. The Swedish technology subject is an example of an educational context where both of these components are present and thus conditions for the development of CT skills are very promising.

In consequence, we conclude that the Swedish technology subject faces significant challenges if it is to remain relevant as the world enters the digital industrial revolution, and digital systems and technologies transform our way of life and the world around us. While overcoming climate and other sustainability challenges might make this revolution an important part of securing our future, this does not diminish the responsibility of the educational system to provide every citizen with the basic knowledge and skills needed to navigate this world, have agency, and sustain our central democratic principles.

If programming and computational thinking are as important as they appear to be then action is required, as our analysis shows that these skills and competencies are insufficiently represented in the current school curriculum. The technology subject is the ideal place in which to strengthen the programming and CT content of the Swedish curriculum. The subject is intended to prepare citizens in terms of technologies and competencies with technologies that are important to Swedish industry and society. What more appropriate place could be found for a greater emphasis on regulation and control of complex systems, systems thinking, and design of computationally intensive technological solutions than the technology curriculum.

Increasing the emphasis on CT and programming in the technology subject would also enhance the relevance of the content of the technology curriculum and give additional substance to a curriculum that is increasingly antiquated in its content. This would benefit all stakeholders, Swedish citizens, society and the public sector.

Notes

1 https://www.klarna.com/international/about-us/
2 https://www.skype.com/
3 https://newsroom.spotify.com/company-info/

References

Ahmed, G., Nouri, J., Norén, E., & Zhang, L. (2019, October 23–25,). Students' perceptions of programming in primary school. *WiPSCE'19*, Glasgow, Scotland UK, ACM International Conference Proceeding Series, https://doi.org/10.1145/3361721.3362112

Bocconi, S., Chioccariello, A., Kampylis, P., Dagienė, V., Wastiau, P., Engelhardt, K., Earp, J., Horvath, M. A., Jasutė, E., Malagoli, C., Masiulionytė-Dagienė, V., & Stupurienė, G. (2022). Reviewing computational thinking in compulsory education. In A. Inamorato dos Santos, R. Cachia, N. Giannoutsou, & Y. Punie (Eds.), *Publications office of the European Union, Luxembourg*. https://doi.org/10.2760/126955,JRC128347

Bowden, H. M. (2019). Problem-solving in collaborative game design practices: Epistemic stance, affect, and engagement. *Learning, Media and Technology, 44*(2), 124–143. https://doi.org/10.1080/17439884.2018.1563106

Bowden, H. M., & Aarsand, P. (2020). Designing and assessing digital games in a classroom: An emerging culture of critique. *Learning, Media and Technology, 45*(4), 376–394. https://doi.org/10.1080/17439884.2020.1727500

Braun, V., & Clarke, V. (2006). Using thematic analysis in psychology. *Qualitative Research in Psychology, 3*(2), 77–101.

Chibas, A., Nouri, J., Norén, E., & Zhang, L. (2018). Didactical strategies and challenges when teaching programming in pre-school. In *Edulearn18: 10th International Conference on Education and New Learning Technologies* (pp. 3345–3350). IATED. http://dx.doi.org/10.21125/edulearn.2018

Craik, A. D. D. (2003). The logarithmic tables of Edward Sang and his daughters. *Historia Mathematica, 30*(1), 47–84. https://doi.org/10.1016/S0315-0860(02)00018-6

Dagienė, V., Jevsikova T., Stupurienė, G., & Juškevičienė, A. (2022). Teaching computational thinking in primary schools: Worldwide trends and teachers' attitudes. *Computer Science and Information Systems, 19*(1), 1–24.

Denning, P. J., & Tedre, M. (2019). *Computational thinking*. The MIT Press.

Denning, P. J., & Tedre, M. (2021). Computational thinking: A disciplinary perspective. *Informatics in Education, 20*(1), 361–390. https://doi.org/10.15388/infedu.2021.21

Denscombe, M. (2016). *Forskningshandboken: för småskaliga forskningsprojekt inom samhällsvetenskaperna* (3rd ed.). Studentlitteratur.

Folkhälsomyndigheten. (2017). *Handledning för litteraturöversikter: förutsättningar och metodsteg för kunskapsframtagande baserat på forskningslitteratur vid Folkhälsomyndigheten* (2nd ed.). Folkhälsomyndigheten.

Ganann, R., Donna C., & Thomas, H., (2010). Expediting systematic reviews: methods and implications of rapid reviews *Implementation Science, 5*(56). https://doi.org/10.1186/1748-5908-5-56

Häger, J., Ramqvist, L., & Roole, G., (n.d.). *Att lösa problem och omsätta idéer i handling* [Audio podcast]. Skolverket. https://www.skolverket.se/om-oss/var-verksamhet/skolverkets-prioriterade-omraden/digitalisering/fyra-aspekter-av-digital-kompetens

Hallström, J. (2009). Technical knowledge in a technical society: Elementary school technology education in Sweden, 1919–1928. *History of Education, 38*, 455–474. https://doi.org/10.1080/00467600801995427

Hallström, J. (2013). "Teknisk bildning för hvar och en": framväxten av ett tekniskt kunskapsinnehåll i folkskolan, 1900–1930. In J. Hallström, M. Hultén, & D. Lövheim (Eds.), *Teknik som kunskapsinnehåll i svensk skola 1842–2010* (pp. 103–146). Skrifter från Forum för ämnesdidaktik, Linköpings universitet, No. 4. Möklinta: Gidlunds.

Hartvik, J. (2013). *Det planlagda och det som visar sig: klasslärarstuderandes syn på undervisning i teknisk slöjd*. Åbo akademi.

Heintz, F., & Mannila, L. (2018). Computational thinking for all: An experience report on scaling up teaching computational thinking to all students in a major city in Sweden. *ACM Technical Symposium on Computer Science Education (SIGCSE)*, 9(2), 65–71. ACM Inroads. https://doi.org/10.1145/3159450.3159586

Heintz, F., Mannila, L., & Färnqvist, T. (2016). A review of models for introducing computational thinking, computer science and computing in K-12 education. *2016 IEEE Frontiers in Education Conference (FIE)*, Eire, PA, USA (pp. 1–9). https://doi.org/10.1109/FIE.2016.7757410

Heintz, F., Mannila, L., Nygårds, K., Parnes, P., & Regnell, B. (2015). Computing at school in Sweden – Experiences from introducing computer science within existing subjects. In A. Brodnik & J. Vahrenhold (Eds.), *Informatics in schools. Curricula, competences, and competitions. ISSEP 2015. Lecture notes in computer science* (Vol. 9378, pp. 118–130). Springer. https://doi.org/10.1007/978-3-319-25396-1_11

Isaksson Persson, H. (2022, June 21–24). Has computational thinking a place in the curriculum? In D. Gill, J. Tuff, T. Kennedy, S. Pendergast, & S. Jamil (Eds.), *PATT 39: PATT on the Edge Technology, Innovation and Education, St. John's, Newfoundland and Labrador, Canada.* Memorial University of Newfoundland.

Jocius, R., Albert, J., Andrews, A., Joshi, D., & Robinson, R. (2020). Infusing computational thinking into disciplinary teaching: From professional development to classroom practice. In D. Schmidt-Crawford (Ed.), *Proceedings of society for information technology & teacher education international conference 2020* (pp. 1771–1775). Association for the Advancement of Computing in Education (AACE). https://www.learntechlib.org/p/215952

Konstfack (2010, October 21). *History.* University of Arts, Crafts and Design. https://www.kth.se/en/eecs

Lingman, L., Ramqvist, L., & Skoglöf, M. (n.d.). *Om skolans digitalisering* [Audio podcast]. Skolverket. https://www.skolverket.se/om-oss/var-verksamhet/skolverkets-prioriterade-omraden/digitalisering/fyra-aspekter-av-digital-kompetens

Mannila, L., Heintz, F., Kjällander, S., & Åkerfeldt, A. (2020, October 28–30). Programming in primary education: Towards a research based assessment framework. *Workshop in Primary and Secondary Computing Education* (*WiPSCE '20*). Virtual Event, Germany. https://doi.org/10.1145/3421590.3421598

Ministry of Education and Research. (2017, March 9). *Regeringen beslutar om nationell digitaliseringsstrategi för skolväsendet.* Regeringskansliet. https://www.regeringen.se/informationsmaterial/2017/10/regeringen-beslutar-om-nationell-digitaliseringsstrategi-for-skolvasendet/

Mozelius, P., & Öberg, L. M. (2017). Play-based learning for programming education in primary school: The Ostersund model. *Proceedings of the 16th European Conference on E-Learning,* (*Ecel 2017*), 375–383.

Nordlöf, C., Norström, P., Höst, G. & Hallström, J. (2022). Towards a three-part heuristic framework for technology education. *International Journal of Technology and Design Education, 32*, 1583–1604. 10.1007/s10798-021-09664-8

Palmér, H. (2017). Programming in preschool – with a focus on learning mathematics. *International Research in Early Childhood Education, 8*(1), 75–87. https://doi.org/10.1007/s10643-019-00976-y

Papert, S. (1980/2020). *Mindstorms: Children, computers, and powerful ideas.* Basic Books.

Pears, A., Barendsen, E., Dagienė, V., Dolgopolovas, V., & Jasutė, E. (2019). Holistic STEAM education through computational thinking: A perspective on training future teachers. In S. N. Pozdniakov & V. Dagienė (Eds.), *Informatics in schools: Focus on learning programming* (pp. 41–52). https://doi.org/10.1007/978-3-030-33759-9_4

Pears, A., Dagiene, V., & Jasute, E. (2017). Baltic and Nordic K-12 teacher perspectives on computational thinking and computing. In V. Dagiene & A. Hellas (Eds.), *Informatics in schools: Focus on learning programming* (pp. 141–152). https://doi.org/10.1007/978-3-319-71483-7_12

Pears, A., Tedre, M., Valtonen, T., & Vartiainen, H. (2021). What makes computational thinking so troublesome? *2021 IEEE Frontiers in Education Conference* (*FIE*), *2021*, 7, https://doi.org/10.1109/FIE49875.2021.9637416

Prior, L. (2003). *Using documents in social research.* SAGE.

Riis, U. (1991). *Skolan och datorn: satsningen Datorn som pedagogiskt hjälpmedel 1988–1991*. Univ., Tema Teknik och social förändring.

Riis, U. (2019). Realiserad skolpolitik i en föränderlig värld – tidig datalära i grundskolan. *Vägval i skolans historia, 2019*(1). https://undervisningshistoria.se/realiserad-skolpolitik-i-en-foranderlig-varld-tidig-datalara-i-grundskolan/

Robson, E., & Stedall, J. (2008). *The Oxford handbook of the history of mathematics*. OUP Oxford.

Ryan, G., & Bernard, G. (2003). Techniques to identify themes. *Field Methods, 15*(1), 85–109. https://doi.org/10.1177/1525822X02239569

Riksarkivet. (n.d.). *Grundskolan införs*. https://riksarkivet.se/grundskolan-infors

Sanford, J. F., & Naidu, J. T. (2016). Computational thinking concepts for grade school. *Contemporary Issues in Education Research* (CIER), *9*(1), 23. https://doi.org/10.19030/cier.v9i1.9547

Saqr, M., Ng, K., Oyelere, S. S., & Tedre, M. (2021). People, ideas, milestones: A scientometric study of computational thinking. *ACM Transactions on Computing Education, 21*(3). https://doi.org/10.1145/3445984

Sjöberg, C., Nouri, J., Sjöberg, R., Norén, E., & Zhang, L. (2018). Teaching and learning mathematics in primary school through scratch. In *Edulearn18: 10th international conference on education and new learning technologies* (pp. 5625–5632).

Sjöberg, C., Risberg, T., Nouri, J., Norén, E., & Zhang, L. (2019). A lesson study on programming as an instrument to learn mathematics and social science in primary school. *13th International Technology, Education and Development Conference (Inted2019)*, 2230–2235. https://doi.org/10.21125/inted.2019.0622

Statistiska centralbyrån (SCB). (1977). Promemoria från informationsenheten. *Elever i icke-obligatoriska skolor 1864–1970. Pupils in secondary schools in Sweden 1864–1970*. Nr: 1977:11. Statistiska centralbyrån. http://share.scb.se/OV9993/Data/Historisk%20statistik/Historisk%20statistik%20f%C3%B6r%20Sverige%201700-1900-tal/Elever-i-icke-obligatoriska-skolor-1864-1970.pdf

Statistiska centralbyrån (SCB). (1984). Elever i skolor för yrkesutbildning 1844–1970. *Pupils in schools of vocational education in Sweden 1844–1970*. Statistiska centralbyrån. http://share.scb.se/OV9993/Data/Historisk%20statistik/Historisk%20statistik%20f%C3%B6r%20Sverige%201700-1900-tal/Elever-i-skolor-for-yrkesutbildning-1844-1970.pdf

Skolverket. (2000). *IT i skolan mellan vision och praktik: en forskningsöversikt*. Stockholm: Skolverket.

Skolverket. (2011). *Läroplan för grundskolan, förskoleklassen och fritidshemmet 2011*. Skolverket.

Skolverket. (2013). *It-användning och it-kompetens i skolan*. Skolverket. http://www.skolverket.se/publikationer?id=3005

Skolverket. (2018a). *Curriculum for the compulsory school, preschool class and school-age educare 2011: Revised 2018*. Skolverket.

Skolverket. (2018b). *Digitaliseringen i skolan - möjligheter och utmaningar* (Forskning för skolan). https://www.skolverket.se/publikationer?id=3971

Skolverket. (2018c). *Läroplan för grundskolan, förskoleklassen och fritidshemmet 2011: reviderad 2018* (Femte upplagan). Skolverket.

Skolverket. (n.d.). *Fyra aspekter av digital kompetens*. https://www.skolverket.se/om-oss/var-verksamhet/skolverkets-prioriterade-omraden/digitalisering/fyra-aspekter-av-digital-kompetens

Skolöverstyrelsen. (1962). *Läroplan för grundskolan*. Stockholm: Kihlström.

Skolöverstyrelsen. (1969a). *Läroplan för grundskolan 1 Allmän del*. Utbildningsförl. http://hdl.handle.net/2077/30902

Skolöverstyrelsen. (1969b). *Läroplan för grundskolan. 2, Supplement: kompletterande anvisningar och kommentarer. Teknik*. Utbildningsförl. http://hdl.handle.net/2077/31037

Skolöverstyrelsen. (1980a). *Datorn i skolan: DIS: SÖ:s handlingsprogram och slutrapport SÖ-projekt 628*. Skolöverstyr.

Skolöverstyrelsen. (1980b). *Läroplan för grundskolan Allmän del: mål och riktlinjer, kursplaner, timplaner*. Liber Läromedel/ Utbildningsförl. http://hdl.handle.net/2077/31016

SOU. (1948). *1946 års skolkommissions betänkande med förslag till riktlinjer för det svenska skolväsendets utveckling*. https://weburn.kb.se/metadata/718/SOU_704718.htm/

Svensson, K., Eriksson, U., & Pendrill, A. M. (2020). Programming and its affordances for physics education: A social semiotic and variation theory approach to learning physics. *Physical Review, Physics Education Research, 16*(1). https://doi.org/10.1103/PhysRevPhysEducRes.16.010127

Tedre, M. (2014). *The science of computing: Shaping a discipline*. Taylor & Francis. https://books.google.se/books?id=I2tYBQAAQBAJ

Tyrén, M., Carlborg, N., Heath, C., & Eriksson, E. (2018). Considerations and technical pitfalls for teaching computational thinking with BBC Micro:bit. *Paper presented at the ACM International Conference Proceeding Series*, Part F137702 (pp. 81–86). https://doi.org/10.1145/3213818.3213829

Utbildningsdepartementet. (1994a). *Kursplaner för grundskolan* [Elektronisk resurs]. Utbildningsdep. http://hdl.handle.net/2077/30959

Utbildningsdepartementet. (1994b). *Läroplaner för det obligatoriska skolväsendet och de frivilliga skolformerna Lpo 94 : Lpf 94*. [Elektronisk resurs]. Utbildningsdep.

Vinnervik, P. (2022). Implementing programming in school mathematics and technology: Teachers' intrinsic and extrinsic challenges. *International Journal of Technology and Design Education, 32*, 213–242. https://doi.org/10.1007/s10798-020-09602-0

Vrieler, T. (2017). Digital capital - A platform for developing computational thinking. *17th Koli Calling International Conference on Computing Education Research, (Koli Calling 2017)*, 205–206. https://doi.org/10.1145/3141880.3143782

Wanngård, K., Rankka, M., & Lorentzon, M. (2015, February 10). Stockholms stad inför dataslöjd på skolschemat. *Dagens Nyheter*. https://www.dn.se/debatt/stockholms-stad-infor-dataslojd-pa-skolschemat/

Weintrop, D., Beheshti, E., Horn, M., Orton, K., Jona, K., Trouille, L., & Wilensky, U. (2016). Defining computational thinking for mathematics and science classrooms. *Journal of Science Education and Technology, 25*(1), 127–147.

Wing, J. M. (2006). Computational thinking. *Communications of the ACM, 49*(3), 33–35. https://doi.org/10.1145/1118178.1118215

Wing, J. M. (2011). Research notebook: Computational thinking—What and why? *The Link magazine, 6*, 20–23.

Zhang, L. C., Nouri, J., & Rolandsson, L. (2020). Progression of computational thinking skills in swedish compulsory schools with block-based programming. *ACE 2020 - Proceedings of the 22nd Australasian Computing Education Conference*, 66–75. https://doi.org/10.1145/3373165.3373173

Appendix A

All publications used in the research review.

1. Ahmed, G., Nouri, J., Norén, E., & Zhang, L. (2019, October 23–25). Students' perceptions of programming in primary school. *WiPSCE'19*, Glasgow, Scotland UK, ACM International Conference Proceeding Series. https://doi.org/10.1145/3361721.3362112
2. Bowden, H. M. (2019). Problem-solving in collaborative game design practices: Epistemic stance, affect, and engagement. *Learning, Media and Technology, 44*(2), 124–143. https://doi.org/10.1080/17439884.2018.1563106
3. Bowden, H. M., & Aarsand, P. (2020). Designing and assessing digital games in a classroom: An emerging culture of critique. *Learning Media and Technology, 45*(4), 376–394. https://doi.org/10.1080/17439884.2020.1727500
4. Bråting, K., & Kilhamn, C. (2020). Exploring the intersection of algebraic and computational thinking. *Mathematical Thinking and Learning, 23*(2), 170–185. https://doi.org/10.1080/10986065.2020.1779012
5. Chibas, A., Nouri, J., Norén, E., & Zhang, L. (2018). Didactical strategies and challenges when teaching programming in pre-school. In *Edulearn18: 10th International Conference on Education and New Learning Technologies* (pp. 3345–3350). IATED. http://dx.doi.org/10.21125/edulearn.2018

6. Heintz, F., Mannila, L., Nygårds, K., Parnes, P., & Regnell, B. (2015). Computing at school in Sweden - Experiences from introducing computer science within existing subjects. In A. Brodnik & J. Vahrenhold (Eds.), *Informatics in schools. Curricula, competences, and competitions /Lecture Notes in Computer Science and General Issues* (Vol. 9378, pp. 118–130). Springer. https://doi.org/10.1007/978-3-319-25396-1_11
7. Heintz, F., Mannila, L., & Färnqvist, T. (2016). A review of models for introducing computational thinking, computer science and computing in K-12 education. In *2016 IEEE Frontiers in Education Conference (FIE)* (pp. 1–9). Eire, PA, USA. https://doi.org/10.1109/FIE.2016.7757410
8. Heintz, F., & Mannila, L. (2018). Computational thinking for all: An experience report on scaling up teaching computational thinking to all students in a major city in Sweden. *ACM Technical Symposium on Computer Science Education (SIGCSE), 9*(2), 65–71. ACM Inroads. https://doi.org/10.1145/3159450.3159586
9. Humble, N., Mozelius, P., & Sällvin, L. (2019). On the role of unplugged programming in K-12 education. *Proceedings of the European Conference on e-Learning, ECEL, 2019*, 224–230. https://doi.org/10.34190/EEL.19.049
10. Humble, N., Mozelius, P., & Sällvin, L. (2020), Remaking and reinforcing mathematics and technology with programming – teacher perceptions of challenges, opportunities and tools in K-12 settings, *International Journal of Information and Learning Technology, 37*(5), 309–321. https://doi.org/10.1108/IJILT-02-2020-0021
11. Kjällander, S., Mannila, L., Åkerfeldt, A., & Heintz, F. (2021). Elementary students' first approach to computational thinking and programming. *Education Sciences, 11*(2). https://doi.org/10.3390/educsci11020080
12. Mannila, L., Heintz, F., Kjällander, S., & Åkerfeldt, A. (2020, October 28–30). *Programming in primary education: Towards a research based assessment framework.* Workshop in Primary and Secondary Computing Education (WiPSCE '20). Virtual Event, Germany. https://doi.org/10.1145/3421590.3421598
13. Mozelius, P., & Öberg, L. M. (2017). Play-based learning for programming education in primary school: The Ostersund model. In *Proceedings of the 16th European Conference on E-Learning (Ecel 2017)* (pp. 375–383).
14. Otterborn, A., Schönborn, K. J., & Hultén, M. (2020). Investigating preschool educators' implementation of computer programming in their teaching practice. *Early Childhood Education Journal, 48*(3), 253–262. https://doi.org/10.1007/s10643-019-00976-y
15. Palmér, H. (2017). Programming in preschool – with a focus on learning mathematics. *International Research in Early Childhood Education, 8*(1), 75–87. https://doi.org/10.4225/03/5a6d25a6bfb03
16. Pears, A., Dagiene, V., & Jasute, E. (2017). Baltic and Nordic K-12 teacher perspectives on computational thinking and computing. In V. Dagiene & A. Hellas

(Eds.), *Informatics in schools: Focus on learning programming* (pp. 141–152). https://doi.org/10.1007/978-3-319-71483-7_12

17. Sjöberg, C., Nouri, J., Sjöberg, R., Norén, E., & Zhang, L. (2018). Teaching and learning mathematics in primary school through scratch. In *Edulearn18: 10th International Conference on Education and New Learning Technologies* (pp. 5625–5632).
18. Sjöberg, C., Risberg, T., Nouri, J., Norén, E., & Zhang, L. (2019). A lesson study on programming as an instrument to learn mathematics and social science in primary school. *13th International Technology, Education and Development Conference (Inted2019)* (pp. 2230–2235). https://doi.org/10.21125/inted.2019.0622
19. Stigberg, H., & Stigberg, S. (2020). Teaching programming and mathematics in practice: A case study from a Swedish primary school. *Policy Futures in Education, 18*(4), 483–496. https://doi.org/10.1177/1478210319894785
20. Svensson, K., Eriksson, U., & Pendrill, A. M. (2020). Programming and its affordances for physics education: A social semiotic and variation theory approach to learning physics. *Physical Review Physics Education Research, 16*(1). https://doi.org/10.1103/PhysRevPhysEducRes.16.010127
21. Tyrén, M., Carlborg, N., Heath, C., & Eriksson, E. (2018). Considerations and technical pitfalls for teaching computational thinking with BBC Micro:bit. *Paper presented at the ACM International Conference Proceeding Series,* Part F137702 (pp. 81–86). https://doi.org/10.1145/3213818.3213829
22. Vinnervik, P. (2022). Implementing programming in school mathematics and technology: Teachers' intrinsic and extrinsic challenges. *International Journal of Technology and Design Education, 32,* 213–242. https://doi.org/10.1007/s10798-020-09602-0
23. Vrieler, T. (2017). Digital capital - A platform for developing computational thinking. *17th Koli Calling International Conference on Computing Education Research, (Koli Calling 2017)*, 205–206. https://doi.org/10.1145/3141880.3143782
24. Zhang, L. C., Nouri, J., & Rolandsson, L. (2020). Progression of computational thinking skills in Swedish compulsory schools with block-based programming. *ACE 2020 - Proceedings of the 22nd Australasian Computing Education Conference,* 66–75.

CHAPTER 7

Visual Programming as a Tool for Developing Knowledge in STEM Subjects

A Literature Review

Karin Stolpe and Jonas Hallström

Abstract

There is currently a trend toward introducing computational thinking in schools, and one popular tool to carry this out is through visual programming. A literature review focusing on visual programming and its effects on learning is still lacking, however; especially in STEM (science, technology, engineering, and mathematics) subjects. The aim of this literature review is to investigate and synthesise the findings of research studies on what students potentially learn in STEM subjects from visual programming. Database searches resulted in 25 studies that were analysed qualitatively. Results showed that by engaging in visual programming students do learn to code, but several basic programming concepts are still challenging. Visual programming for learning in STEM subjects is limited concerning science education, whereas it seems as if visual programming could work as a methodical glue for other STEM subjects. By introducing visual programming, our findings indicate that mathematics, technological, and engineering knowledge and skills, as well as basic programming knowledge and skills, are practised and reinforced. However, teachers need to be aware of which competences are in play in the specific teaching and learning modules, in order to support students' learning.

Keywords

visual programming – STEM subjects – learning – literature review

1 Introduction

Computing has a strong connection to STEM (science, technology, engineering, and mathematics) because it has been conceived of variously as a science, a branch of mathematics, and/or a branch of technology and engineering

© KARIN STOLPE AND JONAS HALLSTRÖM, 2024 | DOI:10.1163/9789004687912_007 (e.g. De Mol & Primiero, 2014; Hallström, 2022). The current trend of introducing computational thinking (CT) and programming into school curricula worldwide (Denning & Tedre, 2019a; Wing, 2006) is paralleled by an increasing focus on STEM education; either as individual subjects or as interdisciplinary cooperation between STEM subjects (Hallström & Schönborn, 2019; Tang & Williams, 2019). Visual programming is, for example, used in STEM education to promote subject-specific skills, although little is known about the effects of visual programming on the development of such skills, particularly in technology and engineering.

Visual programming languages are descendants of Seymour Papert's LOGO, which was introduced in the late 1960s with the ambition of promoting CT to children: developing students' cognitive and generic skills such as metacognition, problem solving capability, and creativity (Papert, 1980, 1996). However, research about the effects of learning to program with LOGO on students' cognitive skills or skills in other subjects—the so-called "transfer hypothesis"—was inconclusive (Björklund, 2016; Denning & Tedre, 2019b; Pea, 1987; Scherer et al., 2018; Voogt et al., 2015; Xu et al., 2019). Even though many of today's visual programming languages largely stem from LOGO, there are many differences concerning not only hardware and software but also the technological and educational contexts in which these languages are being implemented.

1.1 *Aim and Research Question*

The aim of this literature review is to investigate and synthesise the findings of research studies on what students potentially learn in STEM subjects from visual programming, in primary and lower secondary education from 2011 throughout 2021. The research question to be answered is: What do students potentially learn in terms of subject specific knowledge and skills in STEM subjects when working with visual programming, according to the reviewed studies?

2 Background and Earlier Research

The term "computational thinking" (CT) was originally coined by Papert (see e.g. Papert, 1996) to denote the metacognitive, procedural thinking said to result from programming. Jeannette Wing recently reintroduced the concept and claimed that "[c]omputational thinking is the *thought process* involved in formulating a problem and expressing its solution(s) in such a way that a computer—human or machine—can effectively carry it out" (Wing, 2017, n.p.). Denning and Tedre (2019a) show, however, that computational thinking is a complex set of mental skills and practices that humans have developed for hundreds of years, and that it involves both an engineering design (software and hardware) and a scientific explanation. For the purposes of this chapter, suffice it to say that computational thinking is of empirical interest in the sense that most of the studied works on visual programming also claim to be about CT.

A major tool of education in computational thinking has therefore become visual programming. In schools, programming is often connected to or merged with one or more STEM subjects (e.g. Tucker-Raymond et al., 2019). In some European countries, programming was either seen as part of STEM in the curriculum, or there were plans to integrate it more fully with STEM subjects; in Malta and Belgium, for example (European Schoolnet, 2015). Following a long curriculum debate on computational thinking, Sweden updated the syllabi in technology and mathematics education for compulsory schools to include programming components in 2018 (Digitaliseringskommissionen, 2016; Skolverket, 2017). The new curriculum of 2022 still includes programming—for example, visual programming—in mathematics education and technology education (Skolverket, 2022).

This review will involve an inductive exploration of the character of the findings of research studies on visual programming as a tool for learning in STEM subjects. We define visual programming software as various kinds of graphical or visual environments in which coding is carried out by means of manipulating blocks, symbols, or pictures (e.g., Scratch), but also similar programming software for robotics, such as Lego Mindstorms.

There are some previous reviews focusing on programming and its relation to learning in STEM subjects. Benitti et al. reviewed the educational potential of programming robots in schools. The authors were able to show some gains in student learning in STEM subjects such as mathematics, physics, and programming, although there were also several studies indicating no improvement in learning (Benitti, 2012). Sullivan and Heffernan conducted a similar review focusing on how robotic construction kits function as computational manipulatives in K–12 STEM education. The results, although limited, indicated that robots may help students develop scientific literacy, improve their conceptual understanding of concepts in physics related to forces and motion, and help in modelling biological systems (Sullivan & Heffernan, 2016).

Furthermore, Scherer et al. (2018) concluded that there was a positive transfer of programming skills to situations that required creative thinking, mathematical skills, and metacognition, as well as those requiring spatial skills and reasoning. Literacy and school achievement in subjects other than mathematics, on the other hand, benefited the least from learning to program.

Although the above results are illuminating, there has not yet been a literature review focusing on *visual* programming and its effects on subject specific knowledge and skills in STEM subjects. Earlier studies have explored older software and/or only partially focused on visual programming (cf. Scherer et al., 2018).

3 Research Methodology

3.1 *Data Collection: Literature Searches and Selection*

The procedures for performing a literature review employed in this study were inspired by systematic reviews and partly based upon Kitchenham (2004), and Grant and Booth (2009), but some adjustments and refinements have been made to adapt the procedures to the sample

The literature searches were primarily conducted in the electronic database Educational Resources Information Centre (ERIC). ERIC was selected because it covers the most prominent journals in education research. In addition, a complementary search was conducted in Google Scholar, during the initial search stage. Google Scholar was selected because of its very broad scope, in order to find relevant articles that were not, for some reason, indexed in ERIC. Thus, the literature searches in both databases covered the period from 2011 to 2016, but for the 2017–2021 period only ERIC was used because of the abundance of search hits.

The searches in ERIC were carried out in three rounds: a primary and a secondary round, the former being the most extensive and thorough and the latter being complementary, for the years 2011–2016, as well as a third round for the years 2017–2021. In the primary round, the following search terms were used in a full-text search: "Computational thinking" AND programming OR "programming languages" OR "visual programming" AND "elementary education"/"primary education"/"secondary education". This search resulted in 60 hits, after delimiting it to only peer-reviewed articles in journals with a good reputation; that is, included in Web of Science. The primary search in ERIC was also complemented with a search in Google Scholar, using the following search terms: "Computational thinking" AND programming OR "programming languages" OR "visual programming" AND "elementary education". The number of hits was large (141) because it was not possible to automatically apply the criterion of only peer-reviewed journal articles. The sorting of the 141 hits to find only peer-reviewed articles indexed in Web of Science was therefore performed manually, after which 52 articles remained, 27 of which did not appear in the ERIC search. The final outcome of the primary search round was therefore 60 plus 27 articles, a total of 87 peer-reviewed articles.

In the second round, keywords were used as search terms, this time in a full-text search. The terms used in the secondary ERIC search were as follows: "programming" OR "programming languages" AND "elementary education" OR different grades (e.g., DE "Grade 1" OR DE "Grade 2", etc., up to "Grade 9"). This search resulted in 32 articles, after delimiting the search to peer-reviewed articles in Web of Science journals in the second round. Due to the (so far) open nature of the searches, they identified a total of 119 individual peer-reviewed articles, so an additional search in Google Scholar was deemed unnecessary. For the years 2017–2021 in the third round, we only searched ERIC with the search terms ((DE "Programming" OR DE "Coding" OR DE "Computer Literacy") AND (DE "STEM Education")) AND ((DE "Secondary Education") OR (DE "Primary Education") OR (DE "Elementary Education")) which resulted in 55 articles.

After the three search rounds, further delimitations using several inclusion criteria were applied manually to the 174 articles (cf. Kitchenham, 2004). First, only articles on general education—elementary/primary or secondary education but not areas like special needs education—and only grades 1 through 9, were included. Secondly, only articles that included students' visual programming were included, so articles only dealing with other programming software or teachers' development of programming skills were excluded. Third, only articles focusing on STEM subjects (science, technology, engineering, mathematics) or visual programming seen as part of STEM education were included, so studies relating to all other school subjects were excluded. Fourth, the articles included in the final selection were only those that discussed completed empirical studies, so review articles, pilot studies or semi-completed studies, conference papers, and discussion/position papers were excluded.

The inclusion criteria drastically reduced the number of articles. Thus, based on the above inclusion criteria, our review included a total number of 25 unique scientific studies selected for analysis. Studies using qualitative, quantitative, and mixed-methods designs were represented among these 25 articles. For a step-by-step illustration of the whole selection process, see Figure 7.1.

3.2 *Data Analysis*

In this study, we considered it important to synthesise the actual findings of the included studies, to be able to answer our research questions based on data that was as "raw" as possible. Therefore, we were inspired by a method of data analysis called *interpretive synthesis* (Evans, 2002). The first step consists of tabulation; that is, presentation of the main characteristics and major findings of each included study in a table (see Table 7.1). Based on this table, we also formed a narrative summary of the major findings, as presented in the Findings section (Evans, 2002).

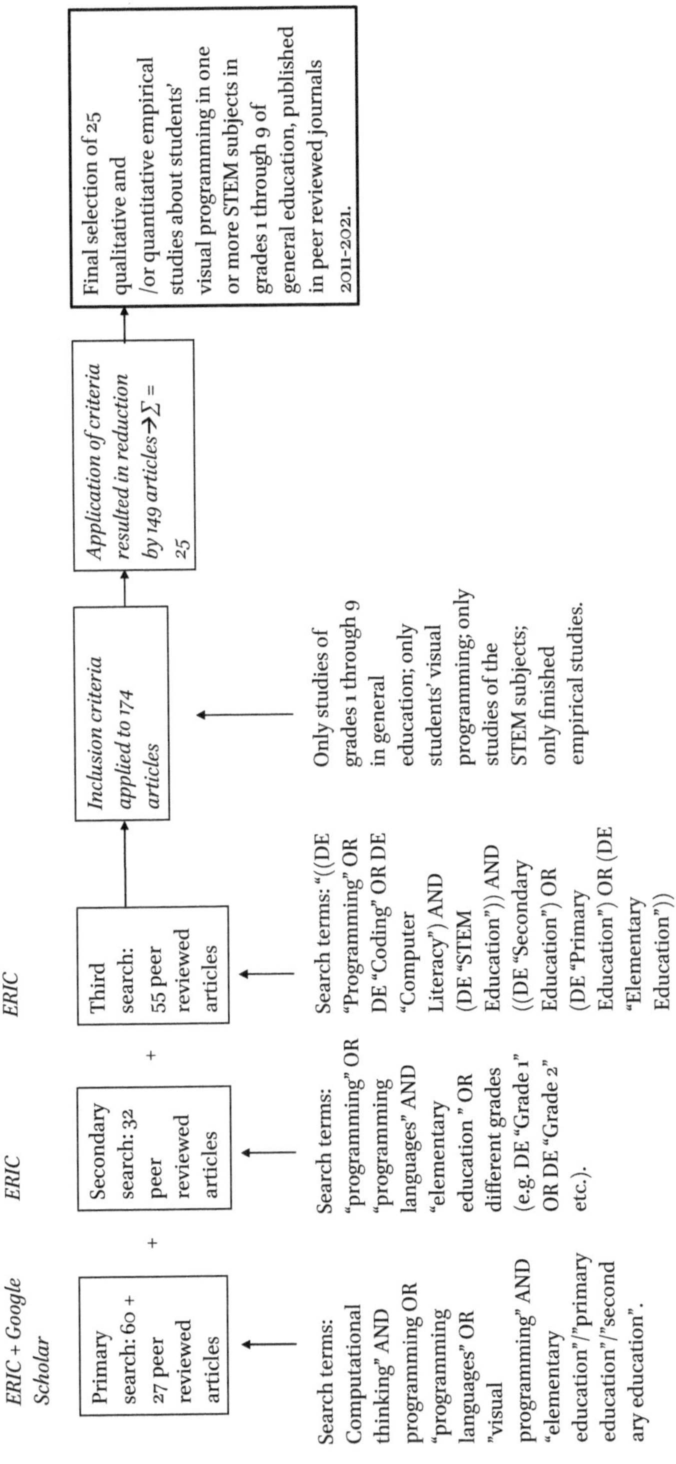

FIGURE 7.1 The process of selecting the 25 articles for the literature review

TABLE 7.1 Summary of the twenty-five included articles in terms of aim/topic, sample, programming language/software, study design, data collection, and major findings

Article	Aim/topic	Sample	Programming language/software	Study design and data collection	Major findings
Boyd Harlow and Emerson Leak (2014)	How can collective idea and skill development be mapped and what classroom interactions lead to skill development and innovation?	20 third-grade students, 8–9 years old, and their teacher	TurtleArt—tool for teaching geometry	Qualitative. Data was collected over one school year. Participant observers, video recordings, two video cameras. Observable ideas (memes) were used as the tracer unit in the analysis.	Programming activities made students share ideas with each other concerning programming and mathematics. Ideas on how to program something more effectively were most often prompted by questions from the teacher. Ideas that led to using commands to obtain new outputs were prompted by questions or comments from students.
Castledine and Chalmers (2011)	Are LEGO robotics an effective problem-solving tool? Investigate problem-solving strategies students are engaged in.	23 grade 6 students	LEGO Mindstorms software program	Qualitative. Two weeks of daily one-hour lessons. Two problems for the students to solve—The Race and The Maze. Observation of student problem-solving discussions, and collection of student questionnaire and software programs.	Results indicate that student efficacy seems to play an important role when it comes to their willingness to seek their own solutions in programming issues. As efficacy increased, the students searched more actively for solutions to their problems. However, students were not able to discuss the actual problem-solving skills they used during the activities.

(*cont.*)

TABLE 7.1 Summary of the twenty-five included articles in terms of aim/topic, sample, programming language/software, study design, data collection, and major findings (*cont.*)

Article	Aim/topic	Sample	Programming language/ software	Study design and data collection	Major findings
Dickes, Voss Farris, and Sengupta (2020)	How classroom teachers' actions and interactions with an agent-based programming tool impacted how computing was adopted and integrated into existing classroom practice. RQs: 1. How did the teacher's emphasis on mathematising and measurement support the development of sociomathematical norms around model 'goodness', and how were those norms adopted by the students? 2.Did these norms shape in any way the development of students' computational models and computational thinking?	15 third grade students, of which eight were female and seven male. The classroom teacher, Emma, was a first-year teacher in the school district.	ViMAP; an agent-based, visual programming language that uses the NetLogo modelling platform as its simulation engine	A design-based, microgenetic study in which an agent-based programming and computational modelling platform—ViMAP—was integrated with existing elementary science and math curricula through lessons co-designed and taught by the classroom teacher across a period of seven months. A dialogical re-positioning of coding is utilised, where disciplinarily grounded meanings of code emerge through the construction of computational utterances—i.e., computer models and complementary conversations and physical models that serve as mathematical and scientific explanations—through the use of socio-mathematical norms.	One finding that emerged is that curricular integration of computing in K–12 STEM contexts is a complex and challenging endeavour for both teachers and students that involves the introduction and adoption of new literacies (e.g., programming) alongside disciplinary ideas that students already find challenging to understand. The classroom teacher's emphasis on mathematising and measurement is suggested to be key in learning activities that help to meaningfully integrate programming as an integral component of STEM work in the classroom in a manner that also helped her connect two new literacies—computational thinking and modelling—with her regular mathematics and science curricular needs. The classroom teacher, in partnership with researchers, found and created opportunities to integrate agent-based programming with her regular science and mathematics curricula by iteratively developing sociomathematical norms.

(*cont.*)

TABLE 7.1 Summary of the twenty-five included articles in terms of aim/topic, sample, programming language/software, study design, data collection, and major findings (*cont.*)

Article	Aim/topic	Sample	Programming language/ software	Study design and data collection	Major findings
Emara, Hutchins, Grover, Snyder, and Biswas (2021)	RQ1: what is the nature of collaborative problem solving (CPS) regulation activated by students when they work in groups on three different types of physics modelling tasks? RQ2: How do students' self- and shared regulatory activities correlate with the performance? RQ3: How do action patterns and problem-solving strategies derived using natural language processing (NLP) emerge across collaborative computational modelling tasks?	14 9th grade students, mean age 15 years, attending a selective program designed to immerse high school students in advanced academic experiences in a university setting.	NetsBlox which is an extension of the Snap! With custom domain-specific blocks that help learners to focus on physics concepts.	Students' visual and verbal behaviour were recorded using a screen-capture software that recorded activities on the screen, video, and audio. Qualitative data analysis—multimodal analytics that combined student discourse and activity data from the computer-based model building environment.	Student CPS strategies evolved over time; e.g., use of more systematic model-building action sequences. Students use more persuasive and explanatory words such as "because" when they engage in more open-ended challenge tasks. However, the study does not contribute on students' learning of CT or STEM in detail.

(*cont.*)

TABLE 7.1 Summary of the twenty-five included articles in terms of aim/topic, sample, programming language/software, study design, data collection, and major findings (*cont.*)

Article	Aim/topic	Sample	Programming language/ software	Study design and data collection	Major findings
Gülmez and Özdener (2015)	The purpose was to compare narrative tools with flowchart model tools, in terms of student success and motivation. Determine the effect of academic success on success in developing algorithms.	84 students who had taken programming education in elementary school. Participants ranging in age from 12 to 14. The students were novice programmers.	Scratch (Exp group 2) and Fcpro tool (Exp group 1). Fcpro is a flow chart model tool ensuring the creation of an algorithm by bringing the codes through the drag-and-drop method.	Quantitative. One of the groups (Exp group 1) used the flowchart model tool and the other (Exp group 2) used the narrative tool. The student groups were exposed to the tools for eight weeks. A pre-test was conducted to determine whether the groups were equal. At the end of applications, two post tests were used to compare the effects of the two visualisation tools on algorithm development success. A motivation test was conducted before and after the intervention.	There was no significant difference in students using flowchart model tools compared with students using narrative tools (Scratch) in terms of their success on the use of variables. There was a significant difference in favour of students using narrative tools on the use of both conditions and loop expressions. There was no difference in terms of the two groups' motivation. There was a positive significant correlation between the post-test averages and academic achievements in English course, and there was a positive significant correlation at a medium level in Turkish, maths and information technologies courses. Achievements in information technologies, maths, Turkish, and English, have a positive significant correlation at a medium level with the algorithm development achievement.

(*cont.*)

TABLE 7.1 Summary of the twenty-five included articles in terms of aim/topic, sample, programming language/software, study design, data collection, and major findings (*cont.*)

Article	Aim/topic	Sample	Programming language/ software	Study design and data collection	Major findings
Hébert and Jenson (2020)	Examining student learning through a making project, focused on the construction of e-textiles—in this case, wearable hats—run as part of a series of after school workshops and as an in-class unit in a school in Ontario, Canada. We asked: through participation in this making unit, what can we report that students learned about coding and circuitry? What elements of making did students struggle with and find the most accessible, and what were their key take-aways from participating in the project?	18 students from grades 6, 7, and 8 took part in the after-school workshops, 10 who identified as male and 8 as female. For the in-class portion of the project, participants consisted of 18 grade 6 students, 10 who identified as male, 8 as female.	LilyPad Arduino	This study examined student learning through a making project and the construction of e-textiles—wearable hats—in a unit delivered in both a series of after-school workshops and as in-class lessons in a school in Ontario, Canada. The workshops took place after school, over 12 sessions, each two hours in length. Groups were divided by gender. There were 10 in-school sessions. A pre-test on programming and electronics was taken by students prior to the making intervention, and a post-test was conducted afterwards. Students were also interviewed at the end of the project about their past experience making wearables, coding, and sewing, and what they learned through the process of participating in the unit. Altogether, 15 students were interviewed.	Findings demonstrated that students in two distinct learning environments showed measurable increases in their knowledge of coding and circuitry after making an e-textile wearable. Students' self-reported areas of challenge and where they required the most assistance in making wearables also calls attention to gaps in knowledge around sewing and coding, while student responses that they were not likely to take up making outside of the classroom speak to the need to more explicitly incorporate making into the curriculum as a means of supporting twenty-first century competency development and learning in STEM.

(*cont.*)

findings (*cont.*)

Article	Aim/topic	Sample	Programming language/ software	Study design and data collection	Major findings
Kalelioğlu (2015)	Investigates, at a macro level, the effect of teaching code.org site on reflective thinking skills towards problem solving, and whether there is a gender difference.	32 primary students in grade 4 (10 years old) attending a computer course at a private school in Turkey. 17 females and 15 males.	code.org	Qualitative/quantitative. The course was taught for one hour per week over a period of five weeks. Reflective Thinking Skill Scale Towards Problem Solving: questioning, reasoning, and evaluation. Focus group structured interviews were performed after completion of the five weeks.	Generally, there were no differences between pre- and post-test results concerning students' reflective thinking skills towards problem solving. However, there was a slight increase in female students' such skills. Students performed higher (tried to complete more levels) than in their normal classroom activities, and females performed slightly better than males. The students enjoyed programming with code.org and found it to be easy, and many said that they wanted to learn more about programming.
Kalelioğlu and Gülbahar (2014)	This study tried to explore the effectiveness of using Scratch in the process of teaching children programming in terms of contribution to problem solving skills.	49 primary school students (5th grade) who were attending a computer course at a private school.	Scratch	Qualitative/quantitative. One hour per week over a period of five weeks, to write the programs at school (Hello World, Parrot, Aquarium programs and Maze project). Problem Solving Inventory—self-perception levels about their problem-solving skills. 24 items with three factors: 12 items for self-confidence in their problem-solving ability, 7 items for self-control, and 5 items for avoidance.	There was no difference between the pre- and post-test results regarding students' problem-solving skills. To solve their problems, the students asked the teacher or a peer. Students solved problems e.g., by adding new items to the programs or by doing what the teacher told them to do. The complexity of long blocks and the absence of special effects were mentioned as non-favourite aspects of the applications. All students stated that they liked programming, that they wanted to improve themselves in programming and that they wanted to create larger, more difficult 3D games.

(*cont.*)

TABLE 7.1 Summary of the twenty-five included articles in terms of aim/topic, sample, programming language/software, study design, data collection, and major findings (*cont.*)

Article	Aim/topic	Sample	Programming language/ software	Study design and data collection	Major findings
Kim, Park, and Tjoe (2021)	The study aims, through a set of three robotics-coding activities and by building on students' prior geometric knowledge of measures of single angles (specifically, acute, right, obtuse, and straight angles), to introduce elementary school students to the concepts of special angle pairs in geometry (namely, complementary and supplementary angles). To what extent would robotic coding activities interact with mathematical problem solving and critical thinking skills in	The study involved 24 elementary school students (four 4th graders and 20 5th graders, nine males and 15 females, ages 9 to 10).	Sphero SPRK+, Sphero Edu app	The 24 students who enrolled in this two-week STEM summer school program participated in the current study voluntarily. Students' participation was part of the two-week, three-hours-per-day STEM summer school program that was led by two mathematics education faculty (the first and second authors) and four preservice elementary teachers, and was geared towards the Hispanic community in a southern state of the U.S. Students in the current study had little to no prior exposure to any computer programming activities. Specifically, they were not familiar with Sphero SPRK+ or the Sphero Edu app prior to the current study. A pre-test assessment was administered to students. Included in the pre-test assessment were elementary items in geometry involving measures of single angles (e.g., acute, right, obtuse, and straight angles) and those of special angle pairs (e.g., complementary and	Students in the present study discovered the concepts of special angle pairs in geometry—namely, complementary, and supplementary angles—as they learned to navigate the immediate feedback from the robot Sphero SPRK+ into a trial-and-error mathematics problem-solving process. Students' experiences in these three coding activities revealed, to a certain extent, that engaging in reflective play could be shaped into meaningful teachable moments where students could participate in a doing with learning pedagogical method using educational robotics. These activities had transferability implications that might afford STEM learning access and opportunities for students to develop not only mathematical reasoning skills, but also problem solving and critical thinking skills operable to a coding environment.

(*cont.*)

TABLE 7.1 Summary of the twenty-five included articles in terms of aim/topic, sample, programming language/software, study design, data collection, and major findings (*cont.*)

Article	Aim/topic	Sample	Programming language/ software	Study design and data collection	Major findings
	the process of the development of new mathematical concepts in measures of complementary and supplementary angles at the elementary school level?			supplementary angles). It was expected that the 4th and 5th graders in the current study would not have been aware of the terms and concepts of complementary and supplementary angles, as those topics were part of the 7th grade common core state standards of mathematics. An assessment similar to the pre-test assessment was administered to students as the post-test assessment. Following the pre-test assessment and preceding the post-test assessment, students participated in three robotics-coding activities: driving, boomerang, and bowling.	

(*cont.*)

TABLE 7.1 Summary of the twenty-five included articles in terms of aim/topic, sample, programming language/software, study design, data collection, and major findings (*cont.*)

Article	Aim/topic	Sample	Programming language/ software	Study design and data collection	Major findings
Kwon, Kim, Shim, and Lee (2012)	Develop a programming tool through which students are able to experience procedural programming intuitively. It proposes a tool, Algorithmic Bricks	1st-grade elementary school students. 10 participants in an experimental group (A-Bricks), 14 in a control group (Scratch).	Algorithmics Bricks (A-bricks) and Scratch.	Quantitative. Five weeks of 10 sessions, each designed to last about 60 minutes. The experimental group (A-brick) used an actual robot to solve problems whereas the control group using Scratch solved problems by moving a virtual robot on the screen. Data 1) Pre survey of student logical thinking ability; 2) Post-test of convenience, understanding, and enjoyment; 3) Usability count of error occurrence; 4) Robot programming task.	The Scratch group asked more for help than the A-Bricks group and the A-bricks group rated understandability and usability higher than the Scratch group. Both groups enjoyed the programming activity. The Scratch group's problems appeared at level 3, while in the A-Bricks group's problems appeared at level 7.
Louca, Zacharia, Michael, and Constantinou (2011)	The development of a framework for analysing and evaluating student-constructed models of physical phenomena through the use of computer-based programming.	Two classes of sixth-graders, 20 in each	Microworlds Logo (MW) – textual programming tool (TPT) and Stagecast Creator (SC) – graphical programming tool (GPT)	Qualitative/quantitative. The students of each class met once a week for 80 minutes, for a total of 7 months and always worked in pairs. In class 1, students used SC programming tool for developing computer-based models of the phenomenon studied, whereas in class 2, students used MW programming.	On a weekly basis, all the models that the students from each class (SC and MW) developed were collected, in total 220 models all versions included. In this study, only initial and final versions, which accounted for 80 model versions per condition. The SC group (graphical programming tool) were able to make programs that were more complex.

(*cont.*)

TABLE 7.1 Summary of the twenty-five included articles in terms of aim/topic, sample, programming language/software, study design, data collection, and major findings (*cont.*)

Article	Aim/topic	Sample	Programming language/ software	Study design and data collection	Major findings
Miller (2019)	How mathematical patterns and structures can be promoted through engagement with coding.	132 students aged 7–8 years old; N =40 students in an experimental group; the rest constitutes the control group.	Scratch, coding robots (not defined)	Intervention-design. Pre- and post-tests, analysis of video data from the lessons.	The program focusing on coding had significant effect on students' capability to identify and generalise patterns and structures.
Moreno-León, Robles, and Román-González (2016)	Whether the use of programming with Scratch as an educational tool accelerates the learning curve, in terms of academic performance.	Three different schools, two 6th grade classes and one 2nd grade. In total three teachers and 129 students.	Scratch	Qualitative/quantitative. The participating teachers received a four-week training course in how to use Scratch as an educational tool. They then divided their classes into a control group and an experimental group, in a quasi-experimental design. Teachers designed a pre-test. The control group worked the unit without programming, while the experimental group included programming activities in Scratch. After at least four weeks of classes, a post test was taken.	The results from the three schools varied a lot. In two of the schools (grade 6), there were differences between pre- and post-test. The effect sizes varied between 1.19 and 0.06, the latter can be considered as a null effect. The computational thinking aspects varied between the different schools. However, students enjoyed working with Scratch, even though some of the schools' students found it harder than others.

(*cont.*)

TABLE 7.1 Summary of the twenty-five included articles in terms of aim/topic, sample, programming language/software, study design, data collection, and major findings (*cont.*)

Article	Aim/topic	Sample	Programming language/ software	Study design and data collection	Major findings
Mouza, Marzocchi, Pan, and Pollock (2016)	How can equitable CS pedagogical practices be applied in after-school programs for middle school students? Does participation impact middle school learning of CS concepts, computational practices and attitudes?	Four CS undergraduates in their junior or senior year, partnered with a middle school teacher, to implement an after-school program for 52 students, 30 boys and 22 girls.	Scratch	Qualitative/quantitative. 9-week after-school CS program. 9 sessions, 90 minutes per session. Two groups - Group A, which had no prior experience of Scratch, and group B which included several students who had some prior experience with Scratch through introductory lessons in a similar program. Data a) observations of after-school program meetings; b) CS undergraduates' entries on their field experience; c) pre/post-tests of middle school students' learning of CS concepts, and attitudes and motivation; d) Scratch programs developed by students.	Statistically significant improvement from the pre- to the post test in six out of 10 questions: Handling an event, loop (forever), conditionals (if), conditional test, data (modifying variable) and parallelism (broadcast). A majority used sprites that exist in the platform. 71% of the projects were interactive, e.g., question/answer or the sprite is controlled to perform a certain task. 29% were non-interactive, e.g., the project could be described as a story with no information needed from the user. Most students showed evidence of appropriate code organisation and documentation. Positive changes in students' attitudes towards programming; but the changes were not statistically significant. Boys felt more confident with their computer skills than the girls did.

(*cont.*)

TABLE 7.1 Summary of the twenty-five included articles in terms of aim/topic, sample, programming language/software, study design, data collection, and major findings (*cont.*)

Article	Aim/topic	Sample	Programming language/ software	Study design and data collection	Major findings
Nugent, Barker, Lester, Grandgenett, and Valentine (2019)	What is the impact of the WearTec intervention on students' (a) knowledge of circuitry, programming, and engineering design and (b) self-efficacy in making an e-textile product? Are there differential effects for minorities and women?	1426 upper elementary students (808 treatment and 618 control) from 29 schools within a Midwestern state.	LilyPad Arduino	Quasi-experimental design with WearTec intervention in formal/informal learning environments on e-textiles, sewing, circuitry, microcontrollers and programming. Identical pre and post instruments were administered to students in both the treatment and control groups at baseline and following completion of the WearTec activities. The instruments included an assessment of wearable technologies self-efficacy and knowledge (circuitry, engineering design, and computer programming).	Results indicate that wearable technology's integration of engineering, computing, and aesthetics promises to be an excellent interdisciplinary context to support students' STEM learning and attitudes at the upper elementary level. The wearable technology instruction was found to promote learning of both males and females and under-represented and majority students when compared with a control group. But while self-efficacy of both male and female students was significantly higher than the control groups, results showed that males' confidence exceeded that of females in the areas of programming and circuitry.

(*cont.*)

TABLE 7.1 Summary of the twenty-five included articles in terms of aim/topic, sample, programming language/software, study design, data collection, and major findings (*cont.*)

Article	Aim/topic	Sample	Programming language/ software	Study design and data collection	Major findings
Okita (2014)	Does initial knowledge building matter to the output carried out by an external source (i.e., robot or another human)?; (2) Does initial learning in specific environments continue to influence performance, after being exposed to alternative environments?; (3) Do benefits extend across different computing platforms?	41 students: 26 in fifth grade and 15 in sixth grade between the ages of 9 to 11. No prior experience or knowledge of robotics.	LEGO Mindstorms NXT 2.0 and NXT-G software and ROBOTC. Physical platform: picture cards from "You Robot" and Ladybug Robot.	Quantitative. Students in the high-transparency group learned to control robots using visual programming. Low-transparency group learned to control robots using syntactic programming. Learning phase 1: days 1–5, learning phase 2: days 6–10. 1 hour session for 10 consecutive days, during which students learned to program robot movements using abstract concepts of speed, distance, and direction. Mid test at the end of day 5, and post-test at the end of day 10. The mid test included problems that measured performance on familiar materials and transfer. Post-test measured understanding of basic concepts after students experienced both high- and low-transparency learning environments and new problems.	Students who initially learned programming in a low-transparency environment (text-based) seemed to be able to transfer their knowledge to visual and text-based programming problems. Students who initially learned programming in a high-transparency environment had problems applying their knowledge to text-based problems. Even after being exposed to the low-transparency environment, the group that initially learned programming in a high-transparency environment, performed lower.

(*cont.*)

TABLE 7.1 Summary of the twenty-five included articles in terms of aim/topic, sample, programming language/software, study design, data collection, and major findings (*cont.*)

Article	Aim/topic	Sample	Programming language/software	Study design and data collection	Major findings
Portelance, Strawhacker, and Bers (2016)	Which ScratchJr programming blocks do young children choose to use after a tailored programming curriculum?	62 students: 21 students in kindergarten, 17 1st graders, 24 2nd graders	ScratchJr	Quantitative. 12 lessons to each class, twice a week for 6 weeks for a period of 1 h. Researchers taught each lesson. Textual representation of 977 projects written by the students. Researchers recorded open-ended field notes.	The students were dedicated to moving the character around the screen and hence the students used the Motion blocks most frequently.
Sáez-López, Román-González, and Vázquez-Cano (2016)	The main objective of the study is to analyse the benefits and possibilities of coding with a Visual Programming Language through projects and activities in primary education.	107 primary school students from grade 6 in five schools in Spain. 60.7% girls, 39.3% boys. Control group with 32 students.	Scratch	Quantitative. Application took place over two academic years in 20 one-hour sessions integrated in sciences and arts. Students created their own material based on the project goals. Visual Blocks Creative Computing Test (VBCCT) assessed computational concepts, and use of different commands. A questionnaire that analysed learning processes and students' attitudes, as well as observations.	The implemented program improved students' ability to understand programming concepts and logic. The students were enthusiastic, motivated, and happy to work with the approach. Understanding of computational concepts, an active approach through project-based learning, usefulness, motivation, and commitment underscore the importance of implementing visual programming through active methodologies in primary education.

(*cont.*)

TABLE 7.1 Summary of the twenty-five included articles in terms of aim/topic, sample, programming language/software, study design, data collection, and major findings (*cont.*)

Article	Aim/topic	Sample	Programming language/ software	Study design and data collection	Major findings
Slangen, van Keulen, and Gravemeijer (2011)	What (key) concepts do pupils use to describe their knowledge, experiences and understanding of robotics? How does the pupils' conceptual understanding of robotics develop.	Six pairs of 11–12 years, last year of elementary. Three pairs were investigated during the whole trajectory.	LEGO Mindstorms NXT robots	Qualitative. Teaching experiment consisted of six video-taped 2-h lessons which were conducted during normal school hours but outside the classroom with pairs of pupils. The pupils, their mutual interactions, the construction and testing activities and the conversations with the researcher were focused. A webcam recorded the pupil's programming, while screen captures were made continuously with Camtasia 5.	At the beginning of the program, the students only used the 'move' blocks to create Reasoning-Acting programs. However, they developed their ability to see the systems as wholes as well as different subsystems along the experiment. Designing, constructing and programming a robot forced the students to relate the subsystems (e.g., the position of the sensor) to the whole construction. Most pupils quickly became familiar with the programming language. However, two or more synchronous actions appeared to be difficult, and they fell back to Reasoning-Acting loop thinking. Complex conditional reasoning (if… then…) were replaced with unconditional sequencing (first do this, then do that, do it for 10 seconds, etc.) of actions. The robot seemed to work as physical modelling which helped students to reason about how the robot should be programmed.

(*cont.*)

TABLE 7.1 Summary of the twenty-five included articles in terms of aim/topic, sample, programming language/software, study design, data collection, and major findings (*cont.*)

Article	Aim/topic	Sample	Programming language/software	Study design and data collection	Major findings
Su, Yang, Hwang, Huang, and Tern (2014)	Explore the effects of instructional tools (the ASP tool and the traditional tool) and programming instructional modes on Scratch programming performance.	135 sixth-graders (66 boys and 69 girls) mean age of 12 years old from four classes in Taiwan.	Scratch	Quasi-experimental: 6-week Scratch programming class. ASP-tool supported problem-solving instruction class, the ASP-tool-supported traditional instruction class, the traditional tool supported problem-solving instruction class, the traditional tool-supported traditional instruction class. Pre-test—prior knowledge test, and survey of prior computer backgrounds. Post-test—programming achievement test.	Students in the problem-solving-based teaching mode performed better than students in the traditional teaching mode did. In the traditional teaching mode, students' performance was not affected by the tool support. Students who received ASP tool support in conjunction with a problem-solving-based teaching approach performed significantly better than the other groups.
Sullivan and Bers (2016a)	Examination of young children's attitudes and ideas about technology and engineering products and performance on robotics and programming tasks.	18 K, 16 1st grade, 11 2nd grade	KIWI robotics	Quantitative. Participating children completed an eight-week robotics and programming curriculum in their classroom taught by research assistants from a university. Lessons were taught once a week and lasted ca. 1 hour. Pre- and post-test design.	Two tasks were carried out, "The Robot Parts Task" and the "Solve-Its tasks". In the simplest tasks, there was no statistically significant difference between grades, gender or the interaction between grades and gender. However, in the harder tasks, the boys' mean score was significantly higher than the girls' were.

(*cont.*)

TABLE 7.1 Summary of the twenty-five included articles in terms of aim/topic, sample, programming language/software, study design, data collection, and major findings (*cont.*)

Article	Aim/topic	Sample	Programming language/ software	Study design and data collection	Major findings
Sullivan and Bers (2016b)	Evaluating a curriculum on robotics and its effects on basic robotics and programming skills	15 pre-K, 18 K, 16 1st grade, 11 2nd grade	KIWI (Kids Invent With Imagination) robotics	Quantitative. Over the course of 8 weeks, four classrooms completed an introductory robotics and programming curriculum taught by research assistants. Lessons were taught once a week and lasted ca. 1 h.	All students performed better for Sequencing tasks than for Repeat Loop and Conditional Statements Tasks. The mistakes they made were of different types; some were syntactical (logical programming errors), and others were story related (program made syntactical sense but did not match the sequence of the story). There was a statistically significant difference between different grades for the hard sequencing tasks, where older students performed better.
Sullivan and Lin (2012)	Examine the relationship of middle school students' perceptions of the ideal science student to their problem-solving activity and conceptual understanding in robotics.	Twenty-six 11- and 12-year-olds (22 boys) attending summer camp for academically advanced students.	LEGO Mindstorms robotics kit & Robolab software.	Videotaped activities from a summer camp for academically advanced 11- and 12-years old students were analysed. The students were working on a problem in which they should complete individually the construction of a robot by adding sensors, wheels, and structural support. Then, they were asked to write a program that caused the robot to follow an angled black line on a paper track.	Five different domain specific problem-solving strategies used by the students were identified by the researchers, as well as four domain general strategies. The results also showed that students with a process-oriented view of being an ideal science student, for instance, being interested and motivated, were more likely to use domain specific strategies than students with a traits-based view of being an ideal science student; for example, being intelligent.

(*cont.*)

TABLE 7.1 Summary of the twenty-five included articles in terms of aim/topic, sample, programming language/software, study design, data collection, and major findings (*cont.*)

Article	Aim/topic	Sample	Programming language/ software	Study design and data collection	Major findings
Sung, Ahn, and Black (2017)	The study investigates the effects of embodied activity and computational perspective-taking practice integrated in mathematics learning.	66 kindergarten and first-grade students enrolled in an after-school coding program	Scratch Jr.	Intervention with controlled randomised groups, pre- and post- and delayed tests to measure learning outcome.	One conclusion from the study is the importance of the activities the students perform, rather than relying solely on the use of programming software. Designing a learning environment that promotes explicit thinking processes is the key in deciding how to introduce programming. It is more powerful to prepare students in appropriate learning activities that support computational perspectives and practices and then expose them to digital tools.
Tan, Samsudin, Ismail, and Ahmad, 2020	The main aim of the presented research was to examine the interaction between student's gender and achievements in learning concepts of electricity via STEAM integrated approach utilising Scratch.	59 lower-secondary school students in Malaysia; 30 females and 29 males.	Scratch	Quasi-experimental design engaging male and female group were used in this research with the intention of studying the gender differences in learning concepts of electricity achievement. For both groups, the lessons on the electricity were taught using the STEAM integrated approach by using Scratch. A pre-test (Electricity Achievement Test, EAT) was carried out before the intervention to both	The findings proved that the STEAM integrated approach achievable via Scratch could narrow the gap between male and female in learning concepts of electricity. The findings of ANCOVA indicated that the intervention had similar positive effects on male and female students' achievement in learning concepts of electricity. This research also provided a new method and an alternate connective framework

(*cont.*)

TABLE 7.1 Summary of the twenty-five included articles in terms of aim/topic, sample, programming language/software, study design, data collection, and major findings (*cont.*)

Article	Aim/topic	Sample	Programming language/ software	Study design and data collection	Major findings
				groups. After the pre-test, they were given the intervention for two months from October to November. The intervention was carried out by using two lessons over one (1) week in three (3) months. Each lesson lasted for 90 minutes. Both groups underwent the same intervention, and the differences were the gender of the groups and taught by two different science teachers with a similar background in terms of qualification and years of teaching experience. Both groups of students were instructed with the STEAM integrated approach using Scratch. After the lessons, the students were required to design the games and animated stories based on the electricity topic by using Scratch. Also, a post-test was carried out to both groups after the intervention. The objective of the post-test was to identify the effect of the teaching approach in increasing students' learning concepts of electricity achievement.	for learning concepts of electricity via art, and showed that both males and females were able to understand the topic of electricity, which reduced gender biases and disparity in the field of science.

The second step of data analysis occurred in conjunction with writing the aforementioned narrative. For this purpose, we utilised another qualitative and interpretive method in line with Braun and Clarke's (2006) description of thematic analysis in six phases. According to this method, we treated the findings of the 25 previous studies just as we would any kind of qualitative data or empirical material. The initial two phases were: (1) to familiarise ourselves with the data, which included repeated reading of the tabulation of the entire dataset (based on the original 25 research articles); and (2) to generate initial codes, which meant that all text sections of each article relating to the corresponding research questions were labelled with a descriptive code. This process was inductive. The next phase (3) was to generate themes, which meant sorting the codes into a hierarchical order. The subsequent phase, (4) reviewing themes, involved revising and refining themes to minimise any overlap between them. The next phase (5) consisted of defining and, finally, naming themes, to be presented as headings. The final phase (6) consisted of compiling exemplary data to exemplify the themes/headings, the results of which are presented in the Findings section (Braun & Clarke, 2006). In the Discussion, our thematic analysis and interpretive synthesis are summarised based on the findings, resulting in conclusions in response to the research question.

4 Findings

Results of the data analysis are presented in this section. This study specifically focused on visual programming software, and the most common software programs used in the included studies are Scratch and ScratchJr (11 out of 25), and different LEGO software (4 out of 25). The remaining studies utilised either lesser-known visual software, or software created specifically for these studies. The findings are presented under three headings, which constitute the main themes. The themes all relate to the research question. These themes are: (1) mathematical knowledge and skills, (2) technological knowledge and skills, and (3) programming knowledge and skills.

4.1 *Mathematical Knowledge and Skills*

Depending on the specific programming tasks, opportunities for learning mathematical calculation and arithmetic may arise. In one study, primary students aged 8 to 9 years used TurtleArt, which is a tool for teaching geometry. Through observations, the researchers identified that students shared mathematical ideas about geometry, such as how to calculate the degrees of an angle (Boyd Harlow & Emerson Leak, 2014).

In a study that involved sixth graders working with programming robots, students worked in groups and discussed the objectives of the robot's design, construction, and programming. Students were also able to articulate their mathematical strategies (Castledine & Chalmers, 2011). Furthermore, these students used a variety of calculation methods, including estimation, addition, subtraction, and division.

Even though neither of these studies presented evidence of students' development of mathematical skills, they both showed that students discussed and shared mathematical ideas that may lead to the development of their mathematical content knowledge (Boyd Harlow & Emerson Leak, 2014; Castledine & Chalmers, 2011). This seems to be evident even if the task is not explicitly based in mathematics.

Similarly, Kalelioğlu (2015) did not test skills development in mathematics *per se*, but students were asked what skills they thought they had improved through the programming intervention. Several stated that they felt they had improved their mathematical and geometrical knowledge, although the study did not show evidence of the nature of such knowledge, and this was only what the students themselves expressed. However, the students had worked on Code.org, which contains, for example, tasks for calculating angles, so calculation was indeed part of the programming curriculum.

In a study of 84 students aged 12–14 years, Gülmez and Özdener (2015) identified a correlation between students' overall academic success and their success in algorithm development. Their results also showed that there was a significant positive correlation between average programming post-test results and achievement in mathematics.

Dickes et al. (2020) claim that curricular integration of computing in K–12 STEM contexts is a challenging task for both teachers and students because it involves the introduction and adoption of new literacies (e.g., programming) alongside disciplinary ideas that students already find challenging to understand, such as mathematics. The classroom teacher's emphasis on mathematising and measurement as key learning activities helped to meaningfully integrate programming as an integral component of STEM work in the classroom in a manner that also helped them connect computational thinking and modelling with regular mathematics curricular needs (Dickes et al., 2020).

Kim et al. (2021) wrote that students in their study discovered the concepts of special angle pairs in geometry—namely, complementary and supplementary angles—as they learned to navigate the immediate feedback from the robot Sphero SPRK+ into a trial-and-error mathematics problem-solving process. Students' experiences in the coding activities revealed that engaging in reflective play could be shaped into meaningful teachable moments where students could participate in learning through pedagogical methods using educational robotics. These activities had transferability implications that might afford STEM learning access and opportunities for students to develop not only mathematical reasoning skills, but also problem solving and critical thinking skills operable in a coding environment.

However, Miller (2019) discusses that the teachers are not always aware of the mathematics in coding lessons. He concludes that there is a need to acknowledge mathematics, as it is not always apparent for the teachers. Miller (2019) shows that year 2 primary students improved their skills in recognising and deducing patterns when they had participated in a coding intervention. Each of the six lessons in the intervention focused on teaching mathematical concepts (e.g., drawing a square or moving a robot through a particular path) using Scratch and three coding robots. The in-depth analysis of the classroom activities shows that the students demonstrated higher levels of mathematical thinking in three different areas: "(1) converting their knowledge of quarter turns to work with metric measures of 90° turns, (2) orientation and perspective taking, and, (3) deducing a repeating pattern to provide a generalised code for making a square" (Miller, 2019, p. 923).

In summary, the studies identified that students shared mathematical ideas regarding geometry (Boyd Harlow & Emerson Leak, 2014) and articulated their mathematical strategies (Castledine & Chalmers, 2011). A correlation between students' overall academic success and their success in algorithm development was identified by Gülmez and Özdener (2015), as was a significant positive correlation between average programming post-test results and mathematics achievement. There is thus some evidence that programming activities could inform the development of mathematical skills, especially if teachers are made aware of the mathematics content in coding lessons (Miller, 2019).

4.2 *Technological Knowledge and Skills*

Some studies identified that when students programmed robots, they were also able to understand the software program, the different parts of the robot, and the sensors as a technical system. For example, at the beginning of a programming course, a systems approach occurred only occasionally among 11- to 12-year-old students when working on designing and programming robots (Slangen et al., 2011). The students were required to compare a well-functioning robot with one that had three different kinds of bugs. One bug was a gear that was not attached properly, another was an incorrectly plugged cable, and the third was a software bug. The researchers concluded:

> This analytic approach seemed to work well, since all pairs detected and repaired the bugs. In this problem solving context, pupils successfully identified subsystems or parts and their proper function. (Slangen et al., 2011, p. 459)

Hence, this kind of debugging task fostered an approach in which the students practiced separating different sub-systems. In the same study, students were asked to program a robot from scratch to perform a pre-defined task. They had to analyse the task, make a plan for the work, identify which functions the robot needed, and then construct, program and test the robot. Going through all these steps required students to identify the whole system as well as its subsystems and how different parts were adapted to each other. Even though Slangen et al. (2011) stated that the systems thinking the students developed was largely implicit, they also concluded that there was a development in how the students talked about construction as a whole:

> The ability to 'see' properly is no longer attributed to the robot as such, but is related to the position of the sensor in the whole construction. (Slangen et al., 2011, p. 460)

Slangen et al. (2011) claimed that working with light sensors and how they function helped students to reason in terms of input/process/output. From there, the students were better prepared to reason about how the program should be written.

A study in which sixth-grade students worked with LEGO Robotics revealed similar results. The students were required to both construct their own robots and program the robots to solve a task (Castledine & Chalmers, 2011). The students were active in their learning, as they discussed the robot design and construction as well as the software programming, and progressively, as their self-efficacy increased, they began to link the programming to the various sensors and to the robot's performance.

Students working with robotics use both domain specific problem-solving strategies and domain general strategies (Sullivan & Lin, 2012). The students in this example (11–12 years old) were working on a problem where they were instructed to construct robots by adding, for example, sensors and wheels, so that they could program the robots to follow an angled line on a paper.

Tan et al. (2020) writes that their findings proved that a STEAM (science, technology, engineering, arts, mathematics) integrated approach achieved via Scratch could narrow the gap between male and female technology learning. The findings indicated that the intervention had positive effects on male and female students' achievement in learning concepts of electricity. This research also provided a new method and an alternate connective framework for learning concepts of electricity via art, and showed that both males and females were able to understand the topic of electricity, which reduced gender biases.

The findings of Hébert and Jenson (2020) demonstrated that students in two distinct learning environments showed measurable increases in their knowledge of coding and circuitry after making an e-textile wearable. Students' self-reported areas of challenge and where they required the most assistance in making wearables also calls attention to gaps in knowledge around sewing and coding, while student responses that they were not likely to take up making outside of the classroom speak to the need to more explicitly incorporate making into the curriculum as a means of supporting twenty-first century competency development and learning in STEM.

Similarly, Nugent et al. (2019) used a quasi-experimental, pre–post design with two groups (treatment and control) to measure the impact of a wearable technology intervention on students' knowledge of circuitry, programming, and engineering design in making a wearable e-textile product. Results indicate that wearable technology integration of engineering, computing, and aesthetics promises to be a good interdisciplinary context for supporting students' STEM learning and attitudes at the upper elementary level. The wearable technology instruction was found to promote learning of both males and females and under-represented and majority students when compared with a control group. But while self-efficacy of both male and female students was significantly higher than that of the control groups, results showed that males' confidence exceeded that of females in the areas of programming and circuitry.

In summary, depending on the assignment, it seems as though working with visual programming in the context of robotics can enhance students' technological systems skills. Programming robots may foster a systems approach, in which the software, the sensors, and the robot have to be seen as components of a larger technological system. Furthermore, through coding in Scratch and wearable e-textiles, understanding of electricity and circuitry could be enhanced both for boys and girls; although it seems that male students gain the most in confidence (e.g. Slangen et al., 2011; Nugent et al., 2019).

4.3 *Programming Knowledge and Skills*

Most of the studies included in this literature review showed that students developed their skills in programming through various activities. One of the studies reported inconclusive results, where differences between the participating schools varied largely (Moreno-León et al., 2016). Researchers found that some teachers found it harder to implement the programming activities than others. A few studies reported on differences between different teaching and learning approaches, and the following section will communicate these findings. This sub-theme focuses on how students learn to program and how they learn programming concepts. The results also report on which programming blocks students use in their programming activities, and how well they perform in the implementation.

ScratchJr is a programming environment inspired by Scratch and adapted for younger students, aged 5–7 years. In a study, Portelance et al. (2016) investigated which blocks children in kindergarten, first grade, and second grade used in their projects after being taught programming for twelve lessons. The blocks in ScratchJr are organised into different categories, depending on their function. The results show that the motion block, with "move right", was the most frequently used block across all ages. The trigger block, "start on green flag", was the second most commonly used block and the least commonly chosen block was the control flow block "stop". Some differences between the age groups were reported. However, the main conclusion was that children were able to use the blocks to create their own projects.

However, Sung et al. (2017) showed that the learning environment is crucial for young learners to develop their understanding of programming concepts. The researchers introduced ScratchJr to kindergarten and first-grade students. However, they concluded that it is most important to expose the students to appropriate learning activities that support computational perspectives and practices before exposing them to digital tools. Moreover, studies indicate that problem-solving–based learning, both individual (Su et al., 2014), and collaborative (Emara et al., 2021), improves students' programming skills.

Older students were able to make more complex programs when they used graphical programming tools (Louca et al., 2011). However, students who initially learned programming in a text-based programming environment seemed to be able to transfer their knowledge to visual and text-based programming problems (Okita, 2014). Students who initially learned programming in a graphic environment had to work harder to apply their skills in text-based environments, even when they had been trained in text-based programming.

Mouza et al. (2016) investigated the effects of a nine-week after-school program in Scratch for middle-school students. Data was collected using a pre- and post-test design, in which students were asked to match a Scratch script to a computing concept such as loop, conditional, variable, or parallelism. They were also asked to determine the result of executing a script. The findings indicated that there was, overall, significant improvement in students' knowledge of computing concepts after the course. However, no significant improvement was found in two out of the three questions dealing with the concept of loop. This was explained by the fact that students were not familiar with the language used in one of the questions. When analysing students' Scratch projects, Mouza et al. (2016) found that the most common block associated with a computing concept was the loop; loops were found in almost 75% of projects. Conditionals were the second most common concept (58%). In more than 85% of the cases where loops and conditionals were used, the code also worked as intended. Moreover, most of the students showed evidence of appropriate code organisation and documentation. The results indicate that the students both understood and were able to implement computational concepts such as loops, conditionals, data, and parallelism.

Kalelioğlu and Gülbahar (2014) showed that students did not improve their problem-solving skills after a five-week program using Scratch. The researchers suggest that the students added new blocks randomly using a trial-and-error approach, or because the teacher told them to do so.

Moreover, the results of Kwon et al. (2012) showed that the students found it easier to both understand and use programming activities that involve an actual robot. The control group used Scratch and solved problems by moving a virtual robot on the screen.

In their study, Sullivan and Bers (2016b) investigated children's programming knowledge after attending an eight-week robotics course. The children were of different ages, ranging from pre-kindergarten to second grade. The children in kindergarten, and the first and second graders all performed equally well on all the tasks. However, the children demonstrated a variety of mistakes as they were solving the tasks. Some mistakes were syntactical in character; which is to say that the children had made logical programming errors that rendered the programs dysfunctional for a real robot. Other mistakes were story-related, which means that the sequence did not match the intended actions of the robot. Overall, however, the children performed better on sequencing tasks than repeat loop or conditional statement tasks.

In another study conducted by the same researchers, similar results were found for the same programming curriculum (Sullivan & Bers, 2016a). In this study, the authors also reported on the absence of any statistical gender differences when it came to the simplest tasks. However, in the more advanced programming tasks, such as using repeated loops with numbers or sensors, boys performed significantly better than girls.

In a quasi-experimental study, Sáez-López et al. (2016) investigated the effect of an intervention using design-based research. The study was conducted with sixth graders in five different schools and the sample consisted of 107 students. The results showed a significant improvement from pre- to post-test in students' ability to understand programming concepts, after an active pedagogical intervention using a visual programming language. The test assessed students' understanding of different computational concepts, and the use of different commands and visual blocks. Important concepts were sequences, loops, conditional statements, parallel execution, coordination, event handling, and keyboard input (Sáez-López et al., 2016).

In summary, the reviewed studies show that students learn to operationalise computing concepts with visual software through different interventions (e.g. Sáez-López et al., 2016). However, students, at least when it comes to younger students, found it easier to build sequences than to use more complex functions such as loops and conditionals, although the latter were also commonly employed (Sullivan & Bers, 2016b). Sung et al. (2017) also pointed out the importance of the learning environment as a precondition for students in learning programming. Moreover, several studies showed improvements from interventions, but programming concepts such as sequences, loops, conditional statements, parallel execution, etc. were still challenging for students to learn, especially conditionals (Mouza et al., 2016; Sáez-López et al., 2016; Sullivan & Bers, 2016b). Boys performed better at the more advanced programming tasks, although this was only reported by one study (Sullivan & Bers, 2016a).

5 Discussion

The research question in this study is: What do students potentially learn in terms of subject-specific knowledge and skills in STEM subjects when working with visual programming, according to the reviewed studies? Three main themes emerged from the review of the included studies: mathematical knowledge and skills, technological knowledge and skills, and programming knowledge and skills.

In line with Scherer et al. (2018), we could also see improvement on STEM-related mathematical skills. When it comes to the more subject-specific competencies, this review references several studies that suggest various ways in which mathematical competencies and self-efficacy might be enhanced by visual programming. However, in regard to chemistry, physics, and biology education, none of the included studies point to any benefits of visual programming on learning; even though STEM was one of our inclusion criteria. This contrasts with an earlier review on the effects of robotics on STEM learning. Sullivan and Heffernan's review focused on how robotic construction kits function as computational manipulatives. Their findings, although limited, indicated that robots may help students develop their scientific literacy, improve conceptual understanding of physics concepts related to forces and motion, and help in modelling biological systems (Sullivan & Heffernan, 2016).

There were a couple of studies (Castledine & Chalmers, 2011; Slangen et al., 2011) performed in a technology education context. One of these indicated a development in students' systems thinking in terms of seeing how a system is made up of components and subsystems that relate to each other; the students could also analyse the system in terms of model input–process–output (Slangen et al., 2011). Visual programming thus helped these students acquire advanced systems thinking skills for their age (cf. Hallström & Klasander, 2017).

Earlier reviews in STEM learning and programming have focused on specific skills, rather than on subject specific improvement. By breaking down computational thinking into several narrower and more well-defined skillsets, it may be possible to determine the effects of programming on CT. Our results show some similarities with Scherer et al. (2018), in that we saw an effect of learning to program on programming skills, although the far-transfer effects we saw were less clear, especially for low achievers. We could see some positive effects on problem-solving strategies and capabilities, although overall the results were inconclusive, and in terms of abstract and logical thinking there was no observable effect. The latter could be compared to what Scherer et al. (2018) labelled "reasoning", which had a less than moderate effect.

The findings of this review regarding students' learning of programming skills are similar to a British study by Straw et al. (2017). They evaluated so-called "Code Clubs" in the UK, and more precisely what impact attending such a club might have on children's computational thinking, programming skills, and attitudes toward computers and coding more generally. They concluded that attending Code Club for a year did not impact upon children's computational thinking—determined by having the children take a test about simple programming (e.g., giving proper instructions), problem solving and logical thinking—beyond changes that would have happened anyway; that is, compared to a control group. Mixed results like these were also seen in studies during the 1980s in the wake of LOGO. In fact, it seems that inconclusive results persist, regardless of the definition of CT, as long as CT is defined in rather general terms (e.g., Voogt et al., 2015).

However, if we focus on more narrowly-defined programming knowledge and skills, there were some reported effects in the reviewed studies. Students can learn to operationalise computing concepts with visual software through different interventions; especially simpler sequential operations as opposed to conditionals which were more difficult to learn, especially for girls (Sullivan & Bers, 2016a).

6 Conclusion

The introduction of computational thinking through visual programming in schools all over the world is a very significant development that will affect school systems for years to come. Because of this, it is imperative that this introduction is based on solid research, but so far there have been few studies and few clear-cut results. One reason for the inconclusiveness of prior research may be the fuzziness and multiple definitions of the concept of computational thinking, but the inconclusiveness seems to remain, regardless of definition. However, in this literature review, we focus on skills and knowledge in STEM subjects. We conclude this chapter by summarising our main findings:

Visual programming for learning in STEM subjects is limited to mathematical, technological, and programming knowledge and skills. This means that studies focusing on the scientific subjects such as biology, physics, and chemistry have not been found in this literature review.

Engaging in visual programming can nonetheless likely help develop skills in all the aforementioned subjects. For mathematics, it seems important that teachers are aware that programming activities include mathematic components (Dickes et al., 2020; Miller, 2019). Even if it could be challenging for students and teachers to introduce new programming literacies into a mathematical context (Dickes et al., 2020), several studies indicate that it does lead to more conversations about mathematics among students (Boyd Harlow & Emerson Leak, 2014).

Moreover, working with robotics and programming robots seem to reinforce important technological knowledge and skills; such as, for example, technological systems thinking (Slangen et al., 2011), and circuitry, programming, and engineering design, especially in the context of the relatively new area of e-textiles (Hébert & Jenson, 2020; Nugent et al., 2019).

To conclude, it seems as if visual programming could work as a methodical glue for STEM subjects. By introducing visual programming in elementary school, our results indicate that mathematics, technological and engineering knowledge and skills, and basic programming knowledge and skills, are practised and reinforced. However, teachers need to be aware of which competences are at play in specific teaching and learning modules, in order to best support students' learning.

7 Limitations

There are some potential limitations to our selection of research articles in our review. First, the search algorithms of ERIC and Google Scholar are not entirely transparent, so the search results would probably have been slightly different had we made small changes in the Boolean search strings, or in the keywords. As a matter of fact, we once received slightly different results with the same search string on a certain day, compared to the day before. Because of this slight discrepancy, we also validated the selected studies at a research seminar with experts from STEM, ICT, and programming education. The final 25 studies were deemed a reasonable outcome of our searches and the application of the inclusion criteria. Secondly, there are other selection methods that we did not utilise; for example, snowball selection from reading lists of references in previous studies. The selected previous studies for this review might have been different had we used one or other of these methods, but we believe that the difference would have been small, given the previously-mentioned external validation (e.g. Robson, 2002). Thirdly, there is also a limitation related to the results of the included studies; namely, that not all of them are entirely transparent in regard to research design and methodology (see Table 7.1). It is thus sometimes difficult to compare them due, for example, to low transparency regarding what kinds of questions were asked, or how certain interventions were carried out, in more detail.

References

Benitti, F. B. V. (2012). Exploring the educational potential of robotics in schools: A systematic review. *Computers & Education, 58*(3), 978–988. https://doi.org/10.1016/j.compedu.2011.10.006

Björklund, L. (2016, June 15). *Datalogiskt tänkande, ett förhoppningsbaserat, generellt sätt att lösa problem?* CETIS' Summer Seminar, Rockelstad Castle, Sweden.

Boyd Harlow, D., & Emerson Leak, A. (2014). Mapping students' ideas to understand learning in a collaborative programming environment. *Computer Science Education, 24*(2–3), 229–247.

Braun, V., & Clarke, V. (2006). Using thematic analysis in psychology. *Qualitative Research in Psychology, 3*(2), 77–101.

Castledine, A.-R., & Chalmers, C. (2011). LEGO robotics: An authentic problem solving tool? *Design and Technology Education: An International Journal, 16*(3), 19–27.

De Mol, L., & Primiero, G. (2014). Facing computing as technique: Towards a history and philosophy of computing. *Philosophy & Technology, 27*(3), 321–326.

Denning, P. J., & Tedre, M. (2019a). *Computational thinking*. MIT Press.

Denning, P. J., & Tedre, M. (2019b). Thinking like a computer. *American Scientist, 107*(3), 187–189.

Dickes, A. C., Farris, A. V., & Sengupta, P. (2020). Sociomathematical norms for integrating coding and modeling with elementary science: A dialogical approach. *Journal of Science Education and Technology, 29*(1), 35–52. https://doi.org/10.1007/s10956-019-09795-7

Digitaliseringskommissionen. (2016). *Digitaliseringens effekter på individ och samhälle – fyra temarapporter. Delbetänkande av Digitaliseringskommissionen.*

Emara, M., Hutchins, N. M., Grover, S., Snyder, C., & Biswas, G. (2021). Examining student regulation of collaborative, computational, problem-solving processes in open-ended learning environments. *Journal of Learning Analytics, 8*(1), 49–74.

European Schoolnet. (2015). *Computing our future, computer programming and coding: Priorities, school curricula and initiatives across Europe.* E. Schoolnet.

Evans, D. (2002). Systematic reviews of interpretive research: Interpretive data synthesis of processed data. *Australian Journal of Advanced Nursing, 20*(2), 22–26.

Grant, M. J., & Booth, A. (2009). A typology of reviews: An analysis of 14 review types and associated methodologies. *Health Information & Libraries Journal, 26*(2), 91–108.

Gülmez, I., & Özdener, N. (2015). Academic achievement in computer programming instruction and effects of the use of visualization tools; at the elementary school level. *British Journal of Education, Society & Behavioural Science, 11*(1), 1–18.

Hallström, J. (2022). The dual nature(s) of technology: Towards a unified conception of digital and analogue technology. In D. Gill, J. Tuff, T. Kennedy, S. Pendergast, & S. Jamil (Eds.), *PATT 39: PATT on the edge technology, innovation and education, St. John's, Newfoundland and Labrador, Canada, June 21st–24th, 2022.* Memorial University of Newfoundland.

Hallström, J., & Klasander, C. (2017). Visible parts, invisible whole: Swedish technology student teachers' conceptions about technological systems. *International Journal of Technology and Design Education, 27*(3), 387–405. https://doi.org/10.1007/s10798-016-9356-1

Hallström, J., & Schönborn, K. J. (2019). Models and modelling for authentic STEM education: Reinforcing the argument. *International Journal of STEM Education, 6*(1), 1–10.

Hébert, C., & Jenson, J. (2020). Making in schools: Student learning through an e-textiles curriculum. *Discourse: Studies in the Cultural Politics of Education, 41*(5), 740–761.

Kalelioğlu, F. (2015). A new way of teaching programming skills to K-12 students: Code.org. *Computers in Human Behavior, 52*, 200–210.

Kalelioğlu, F., & Gülbahar, Y. (2014). The effects of teaching programming via Scratch on problem solving skills: A discussion from learners' perspective. *Informatics in Education, 13*(1), 33–50.

Kim, Y. R., Park, M. S., & Tjoe, H. (2021). Discovering concepts of geometry through robotics coding activities. *International Journal of Education in Mathematics, Science, and Technology (IJEMST), 9*(3), 406–425.

Kitchenham, B. (2004). *Procedures for performing systematic reviews.* Keele University Technical Report TR/SE-0401.

Kwon, D.-Y., Kim, H.-S., Shim, J.-K., & Lee, W.-G. (2012). Algorithmic Brics: A tangible robot programming tool for elementary school students. *IEEE Transaction on Education, 55*(4), 474–479.

Louca, L. T., Zacharia, Z. C., Michael, M., & Constantinou, C. P. (2011). Objects, entities, behaviors, and interactions: A typology of student-constructed computer-based models of physical phenomena. *Journal of Educational Computing Research, 44*(2), 173–201.

Miller, J. (2019). STEM education in the primary years to support mathematical thinking: Using coding to identify mathematical structures and patterns. *ZDM, 51*(6), 915–927. https://doi.org/10.1007/s11858-019-01096-y

Moreno-León, J., Robles, G., & Román-González, M. (2016). Code to learn: Where does it belong in the K-12 curriculum? *Journal of Information Technology Education: Research, 15*, 283–303.

Mouza, C., Marzocchi, A., Pan, Y.-C., & Pollock, L. (2016). Development, implementation, and outcomes of an equitable computer science after-school program: Findings from middle-school students. *Journal of Research on Technology in Education, 48*(2), 84–104.

Nugent, G., Barker, B., Lester, H., Grandgenett, N., & Valentine, D. (2019). Wearable textiles to support student STEM learning and attitudes. *Journal of Science Education and Technology, 28*(5), 470–479. https://doi.org/10.1007/s10956-019-09779-7

Okita, S. Y. (2014). The relative merits of transparency: Investigating situations that support the use of robotics in developing student learning adaptability across virtual and physical computing platforms. *British Journal of Educational Technology, 45*(5), 844–862.

Papert, S. (1980). *Mindstorms: Children, computers, and powerful ideas.* Basic Books, Inc.

Papert, S. (1996). An exploration in the space of mathematics educations. *International Journal of Computers for Mathematical Learning, 1*(1), 95–123.

Pea, R. (1987). The aims of software criticism: Reply to professor Papert. *Educational Researcher, 16*(5), 4–8.

Portelance, D. J., Strawhacker, A. L., & Bers, M. U. (2016). Constructing the ScratchJr programming language in the early childhood classroom. *International Journal of Technology and Design Education, 26*(4), 489–504. https://doi.org/10.1007/s10798-015-9325-0

Robson, C. (2002). *Real world research: A resource for social scientists and practitioner-researchers* (2nd ed.). Blackwell.

Sáez-López, J.-M., Román-González, M., & Vázquez-Cano, E. (2016). Visual programming languages integrated across the curriculum in elementary school: A two year case study using "Scratch" in five schools. *Computers & Education, 97*, 129–141.

Scherer, R., Siddiq, F., & Sánchez Viveros, B. (2018). The cognitive benefits of learning computer programming: A meta-analysis of transfer effects. *Journal of Educational Psychology*, 1–29. http://doi.org/10.1037/edu0000314

Skolverket. (2017). *Förordning om ändring i förordningen (SKOLFS 2010:37) om läroplan för grundskolan, förskoleklassen och fritidshemmet.* Skolverket.

Skolverket. (2022). *Läroplan för grundskolan, förskoleklassen och fritidshemmet. Reviderad 2022.*

Slangen, L., van Keulen, H., & Gravemeijer, K. (2011). What pupils can learn from working with robotic direct manipulation environments. *International Journal of Technology and Design Education, 21*(4), 449–469.

Straw, S., Bamford, S., & Styles, B. (2017). *Randomised controlled trial and process evaluation of code clubs.* National Foundation for Educational Research and the Raspberry Pi Foundation.

Su, A. Y. S., Yang, S. J. H., Hwang, W.-Y., Huang, C. S. J., & Tern, M.-Y. (2014). Investigating the role of computer-supported annotation in problem-solving-based teaching: An empirical study of a Scratch programming pedagogy. *British Journal of Educational Technology, 45*(4), 647–665.

Sullivan, A., & Bers, M. U. (2016a). Girls, boys, and bots: Gender differences in young children's performance on robotics and programming tasks. *Journal of Information Technology Education: Innovations in Practice, 15*, 145–165.

Sullivan, A., & Bers, M. U. (2016b). Robotics in the early childhood classroom: Learning outcomes from an 8-week robotics curriculum in pre-kindergarten through second grade. *International Journal of Technology and Design Education, 26*(1), 3–20. https://doi.org/10.1007/s10798-015-9304-5

Sullivan, F. R., & Heffernan, J. (2016). Robotic construction kits as computational manipulatives for learning in the STEM disciplines. *Journal of Research on Technology in Education, 48*(2), 105–128.

Sullivan, F. R., & Lin, X. (2012). The ideal science student: Exploring the relationship of students' perceptions to their problem solving activity in a robotics context. *Journal of Interactive Learning Research, 23*(3), 273–308.

Sung, W., Ahn, J., & Black, J. B. (2017). Introducing computational thinking to young learners: Practicing computational perspectives through embodiment in mathematics education. *Technology, Knowledge and Learning, 22*(3), 443–463. https://doi.org/10.1007/s10758-017-9328-x

Tan, W.-L., Samsudin, M. A., Ismail, M. E., & Ahmad, N. J. (2020). Gender differences in students' achievements in learning concepts of electricity via STEAM integrated approach utilizing scratch. *Problems of Education in the 21st Century, 78*(3), 423.

Tang, K. S., & Williams, P. J. (2019). STEM literacy or literacies? Examining the empirical basis of these constructs. *Review of Education, 7*(3), 675–697.

Tucker-Raymond, E., Puttick, G., Cassidy, M., Harteveld, C., & Troiano, G. M. (2019). "I broke your game!" critique among middle schoolers designing computer games about climate change. *International Journal of STEM Education, 6*(1), 1–16.

Voogt, J., Fisser, P., Good, J., Mishra, P., & Yadav, A. (2015). Computational thinking in compulsory education: Towards an agenda for research and practice. *Education and Information Technologies, 20*, 715–728. https://doi.org/10.1007/s10639-015-9412-6

Wing, J. M. (2006). Computational thinking. *Communications of the ACM*, *49*(3), 33–35.

Wing, J. M. (2017, September 8). *Computational thinking*. Trippel Helix Conference on Computational Thinking and Digital Competences in Primary and Secondary Education.

Xu, Z., Ritzhaupt, A. D., Tian, F., & Umapathy, K. (2019). Block-based versus text-based programming environments on novice student learning outcomes: A meta-analysis study. *Computer Science Education*, *29*(2–3), 177–204. https://doi.org/10.1080/08993408.2019.1565233

PART 2

Curriculum and Teacher Perspectives on Computational Thinking and Programming in Technology Education

CHAPTER 8

Programming in School Technology Education

An Insight into Teachers' Efforts to Unpack and Shape Programming in Technology Education

Peter Vinnervik

Abstract

The 2018 Swedish school curriculum reform introduced programming as new subject content in mathematics and technology education. The aim of the study reported in this chapter was to explore how school technology teachers transform, that is, understand and contextualise programming in technology education. In addition, the study identifies factors that influence the teachers' transformation process. Teacher interviews provide insight into the process, and the interviews were analysed using content analysis.

The results show teachers' efforts to integrate programming in the context of technological systems and construction work rather than as an isolated phenomenon. This line of reasoning broadly resonates with the intent conveyed in the technology education syllabus but at the same time teachers express uncertainty about what domain specific knowledge children should have about programming in terms of practices and concepts. The teachers face several challenges during the process of transforming programming into teachable content. These challenges are, for example, related to their own professional knowledge and understanding of programming, the availability and quality of teaching materials and assessment procedures. The interviews also reveal a situation where the interdisciplinary cooperation between mathematics and technology, as envisaged in the school curriculum, is largely absent. The chapter concludes with a discussion of potential consequences for teachers' reform work and school children's understanding of programming.

Keywords

programming – school technology – curriculum enactment

© PETER VINNERVIK, 2024 | DOI:10.1163/9789004687912_008

1 Reboot and Reform

During the last decade, there has been a considerable international urgency to reboot and reform school policies to increase and improve the use of digital technologies in teaching and learning, and to increase the cross-curricular digital competence of school children (Balanskat & Engelhardt, 2015). The concept of digital competence captures abilities considered necessary for active participation in society and encompass, for example, digital and social awareness, critical thinking and problem solving skills. Oftentimes, programming and other aspects originating from the field of computer science (CS) have been added to school curricula, justified as a medium that could help children develop digital competences and computational thinking skills (Juškevičiene & Dagiene, 2018).

As outlined elsewhere in this book, a Swedish school curriculum reform was publicly announced in the first half of 2017 and formally enacted in mid-2018. The overarching goal of the reform was to improve the opportunities for schoolchildren to build their digital competence (National Agency for Education, 2017). As part of this push for digital competence, programming was included in mathematics and technology education. Programming is characterised in the curriculum as both an enabler and a component of digital competence, as well as a tool that can enhance children's understanding of mathematics and technology (National Agency for Education, 2017).

This reform sets the stage for this chapter, which is based on interviews with 10 technology teachers in grades 7–9 (13–15 years) and their efforts to bring programming into the technology classroom. The chapter builds on the premise that it is important to understand how new policies influence teachers' work (Cohen & Ball, 1990; Hargreaves, 2005; Sentance & Csizmadia, 2017; Yadav et al., 2016) and to compare policy ideas that travel across borders and contexts (Ozga & Jones, 2006). This is arguably all the more pertinent at a time when the debate over the impact of technology and digital transformation on individuals and society has been intensified by recent advances in artificial intelligence.

The aim of the study reported in this chapter was to explore how school technology teachers transform, that is, understand and contextualise programming in technology education. In addition, the study identifies challenging factors that influence the teachers' transformation process. This chapter thus provides insight into how the new subject content is being shaped. The following research questions were posed:

1. How do technology teachers transform programming into technology teaching?

2. What challenges do teachers face in the process of transforming programming into technology teaching?

The chapter concludes with a discussion of potential consequences for teachers' reform work and for school children's understanding of programming.

1.1 *Some Concerns*

There are several concerns surrounding the introduction of programming in Swedish schools that can affect the quality, speed, and impact of the reform. Some of these are presented below.

A significant concern is related to the teachers' professional knowledge and understanding of the new subject content. Few in-service school teachers have received formal training to teach with and about programming, and the teacher training institutions face a similar challenge (Swedish Higher Education Authority, 2020). To address this lack of knowledge among teachers, the Swedish National Agency for Education (NAE) procured five-week introductory courses in programming from Swedish universities as of autumn 2017. Despite this effort, many teachers have reported a lack of skills in using programming in their teaching in a way that promotes subject understanding (Lärarnas Riksförbund, 2020). Furthermore, research shows that teacher opportunities for professional development are limited (Parding et al., 2018).

Another concern is related to reform clarity and complexity, that is, the level of comprehensibility of the reform message. What is it that teachers are supposed to do new or different and is the expected change feasible? An analysis of the Swedish school curriculum introduced in 2017 suggests that the programming reform message lack in detail in terms of what programming knowledge entails, how it interconnects with the host subjects (mathematics and technology), and what the learning progression in between school years is (Vinnervik, 2022b).

Further, the introduction of programming in technology education may be influenced by the fact that the history of technology as a standalone subject is relatively short, receiving its first dedicated syllabus in 1994 (Hallström et al., 2014). In addition, technology has been partially neglected in Swedish teacher training programmes. Until 2005, five weeks of study were considered to provide aspiring technology teachers with enough subject specific knowledge, which was 1/3 or less of the time devoted to the natural sciences. Moreover, less than 50% of in-service compulsory school technology teachers are formally qualified (NAE, 2019), and several of them have obtained their qualification based only on five weeks of subject-specific study.

An early review of the uptake of the reform emphasised the need for teachers' professional development in programming and the expanded use of digital tools for teaching and learning (The Swedish Schools Inspectorate, 2019). According to the review, programming was not covered in a way that fully captured the intentions conveyed in the school curriculum. Amongst the 27 schools in the review, the reform efforts seemed to have had a slightly greater impact in mathematics than in technology education. It is important to note, however, that the review was conducted 7–10 months into the reform, and that reform work often is a lengthy process (Fullan, 2007; Ornstein & Hunkins, 2018).

2 Curriculum as a Process

The study reported in this chapter is rooted in curriculum studies, which is a research perspective aimed at understanding the processes, conditions and contexts that influence how educational objectives are designed, how knowledge content is selected and how teaching can be conducted to facilitate effective learning (Linde, 2012; Lundgren, 1989). In curriculum studies, there are models that represent policy decision making as a process (e.g., Billett, 2006; Goodlad, 1979). The study reported in this chapter draws on a curriculum model differentiated in three levels, or arenas (Linde, 2012):

1. The formulation arena
2. The transformation arena
3. The realisation arena

The *formulation arena* is in its fundamental sense represented through the formal curriculum documents issued by the school authorities. These documents express the underlying philosophy, goals, content and learning objectives of education (Wahlström & Sundberg, 2018). The arena can also be seen to represent the process of negotiation and interplay between school authorities and various stakeholders, that precedes the formalisation of intent manifested through the curriculum documents.

The *transformation arena* represents the process in which teachers exert their professional knowledge and understanding to interpret and recontextualise, or transform, intentions as they are perceived from the formal curriculum documents (Linde, 2012) and design teaching and learning activities that cater to the needs of the students. The *realisation arena* represents the space where the perceived intentions of the formal curriculum are put into practice.

The process of transformation can be thought of as a continuous, dynamic and evolving process, and the teacher must continually assess the intended curriculum in the light of developments in the realisation arena, typically the classroom (Linde, 2012). The study presented in this chapter departed from the didactic questions *what-why-how* to examine the choices teachers make throughout transformation (Priestley, 2019; Shulman, 1986; Wickman et al., 2018) and thus to answer the first research question.

Curriculum studies also examine preconditional factors which influence how teachers and schools interpret the curriculum intentions and hence the outcome of the reform. Such differentiating factors can lead to inequalities between schools and school districts and thus to inequalities in children's education (Jarl & Rönnberg, 2019; Linde, 2012; Lipsky, 2010). This chapter introduces the reader to a conceptual framework of challenging factors that influence teachers' reform work (Finger & Houguet, 2009). The framework was developed during the implementation of technology education in Australian primary schools in the early 2000s. In the framework, challenges are organised into two main categories, *intrinsic* and *extrinsic* challenges. Figure 8.1 presents a visual overview of the framework, and each challenge is further explained in the findings section of this chapter.

There are five intrinsic challenges that represent the personal challenges teachers face during reform work and are related to factors such as their professional

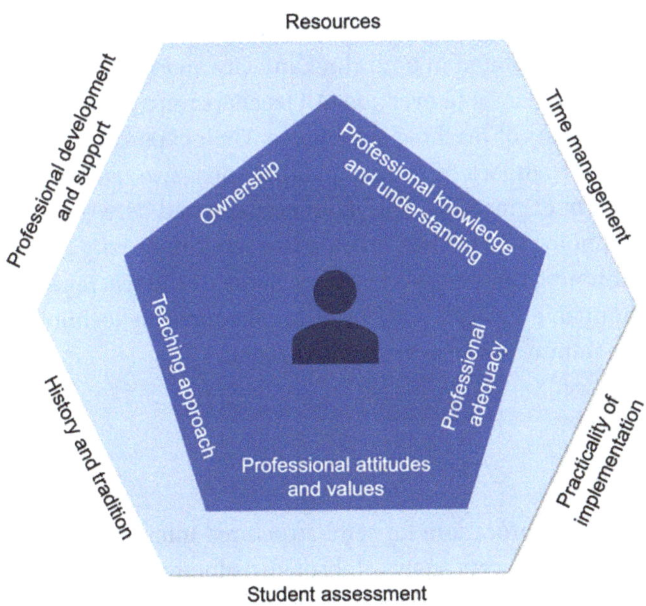

FIGURE 8.1 A visual representation of the conceptual framework of intrinsic and extrinsic challenges for implementing technology education (based on Finger & Houguet, 2009)

knowledge of and attitudes towards the intended innovation and educational change. Several of these are interconnected, for example, teachers' level of confidence (professional adequacy) grows with increased knowledge and understanding of the subject.

The extrinsic challenges are organised into six categories and capture environmental factors that may stand in the way for successful reform implementation. These challenges concern, for example, time management, the quality of which the reform message is communicated, which affects how teachers perceive the intended innovation and what they are expected to do new or differently. Other extrinsic challenges are related to access to resources (e.g., teaching materials), support, and assessment practises. The extrinsic challenges are also interrelated to some extent. The framework was used to identify challenging factors in the teacher interviews and is used in this chapter to organise the presentation of results for the second research question.

3 The Data Source

Ten grade 7–9 teachers, from nine public schools and one private, were interviewed 27 months after the reform was formally enacted in 2018, that is, 3.5 years after the intended curriculum change was first announced by the school authorities. The schools are located in both rural and city areas, and range in size from 15 children in grades 7–9 to over 300. All teachers were certified and had 10 to more than 20 years of teaching experience. Their expertise in technology education varied from lack of formal qualifications (two teachers) to two with a background in engineering. The six in between had between 5–25 weeks of study in technology. Three had taken a five-week university course in introductory Python programming for teachers. Some of the teachers had worked with programmable devices (e.g., Lego Mindstorms) in technology education prior to the reform, but to a very limited extent.

4 The Analysis

The interviews followed the procedure for semi-structured interviews (Kvale, 1997) and interview transcripts were analysed through deductive content analysis (Elo et al., 2014). To answer the first research question, the questions *what, why* and *how* (Priestley, 2019; Shulman, 1986; Wickman et al., 2018) were used as an analytical instrument to code the interview data. Table 8.1 shows the criteria used for coding for the first research question.

TABLE 8.1 Coding criteria, research question 1: Transformation

Classification category	Coding criteria
What	What knowledge of technology and programming teachers intend to address in their teaching
How	Organisational aspects of teaching in terms of teaching materials and support
Why	Rationale of choices made about content, learning goals and materials

Once the transcripts were fully coded, each transcript was reviewed a second time to make sure no key statements were left uncoded or coded incorrectly. The statements were then divided into sub themes, for example *how: teaching materials* and *how: support.*

The transcripts were subjected to a second round of analysis to look for factors that may have influenced the teachers' choices for how to integrate programming into teaching and learning activities. The first step was to group key statements into the primary categories of intrinsic and extrinsic challenges (Finger & Houguet, 2009). Subsequently, each statement classified under these primary categories was matched with the relevant challenge within the main category. When necessary, additional subthemes were created within each challenge. For example, the extrinsic challenge Resources was found to include three sub-themes: *monetary issues, teaching materials,* and *IT infrastructure.*

5 The Results

The results are presented separately for each research question. The results for the first research question are organised according to the classification categories of what, how and why. The results for the second research question are presented using the same structure as the conceptual framework of intrinsic and extrinsic challenges used to guide the analysis.

5.1 *Technology Teachers' Transformation of Programming*

5.1.1 Teachers' Choices of What Content to Cover

The first classification category, *what,* reveals the teachers' decision-making in terms of knowledge of technology and programming. Several teachers argued that structure, accuracy, and sequential thinking are significant qualities when programming. One teacher described in the following way:

> It is about computational thinking, thinking about stepwise instructions. To begin thinking in that way, you must first analyse what you want to happen and in what order. It is less important whether they can design scripts or write Python code.

Some teachers made a point about teaching some of the fundamental aspects of programming, such as step-by-step instructions (sequence of commands, or algorithms), conditional structures (if-else) and repetition. Knowing how to apply this knowledge in practice, that is, procedural knowledge (see McGill & Volet, 1997), was downplayed by most teachers. Rather, understanding programming in the context of how the broader notion of technological knowledge is presented in the curriculum was judged more important. The subject is interdisciplinary by nature, with a focus on analysing and solving technological problems, understanding how technology impacts society and individuals, and examining the forces that drive technological advancement. Specifically, teachers emphasised the importance of understanding technological systems as systems that are influenced by input and output, and that programmable circuits and devices only follow instructions. The children *"should understand that these systems are totally dependent on me programming them"*. Further, programming was closely connected to practical activity in terms of construction work: *"it [the technology syllabus] says you should make your own constructions"*. The teachers also touched upon the societal aspects of digital programmable technology.

Few teachers made an explicit connection between problem solving and programming. One teacher concluded that *"programming is really about solving problems"* but did not want to get entangled in programming technicalities and instead chose to concentrate on the broad 'technological problem', such as how various components of a technological construction fit together.

5.1.2 Teachers' Choice of Materials and Methods

The second classification category, *how*, reveals organisational aspects of how teaching is designed. This section describes the teaching support used to transform intentions into concrete teaching and learning activities.

Most teachers relied on tangible programmable robots and devices to introduce programming. Lego Mindstorms were used to explore the principles of control and regulation of technological systems, and how automation plays a significant role in people's lives. BBC Micro:bit microcontroller boards were popular because they were perceived as easy to get started with, robust, cheap, and well supported in terms of teaching materials and accessories that extend the functionality of the circuit, as *"you can connect a lot of entertaining stuff without adding too much complexity"*. A few teachers used Arduino boards, primarily because of its flexibility in situations where construction work involves electronic components (e.g., capacitors, diodes).

Block-based programming environments were also used to address computational concepts and the principles of control and regulation. The Blockly Games website were thought to offer a simple introduction to some core concepts and the significance of sequencing and accuracy:

> They learn how to stack things [commands] in a sequence. You must think things through; you cannot skip any steps in order to get the program where you want it to go.

Scratch is another block-based programming environment that was more frequently used, for both more deliberate goals, such as designing a functional virtual robot vacuum cleaner, and for free exploratory work, such as creating a game. The teachers seemed to favour block-based environments and believed they were relatively well served to comply with the perceived curriculum intentions.

Text-based programming environments (Replit IDE and the Arduino web editor) were used to a lesser extent. Neither of the text-based environments was the first choice; rather they were a result of the IT department's lockdown of the pupils' Chromebooks, which prevented any possibility of using appropriate software (see Section 5.2.2). The teacher would have preferred to use a dedicated Python interpreter on the Chromebooks.

Access to up-to-date technology textbooks for the current curriculum was limited. The few teachers who had access felt that it allowed them to align their classroom activities with the perceived objectives of the curriculum, since "the core content [of the syllabus] is represented there and it keeps me from falling off the rails".

In a few cases, external school resource centres were asked to assist with the teaching activities by providing teacher training, teaching materials (hardware) or assisting with actual teaching. These centres have various profiles, such as those in the arts, science, technology, or entrepreneurship. They can be part of a university's outreach program or run independently by municipalities.

5.1.3 Why Teachers Choose a Certain Content

The third classification category, *why,* addresses aspects influential for the choice of teaching content, learning goals and teaching materials. Three aspects emerged in the interview transcripts: The role of the formal curriculum, the role of external input, and the role of personal preference.

All teachers seemed to be aware of the context to which programming is linked in the curriculum. From a broader perspective, this awareness clearly guided the teachers during the transformation process. When it comes to the actual details, opinions differ. Although some considered themselves to have a clear idea of the intentions of the curriculum, others perceived a lack of clarity in the statements in the technology syllabus. This perceived lack of clarity was seen as potentially resulting in an undesirably wide variation in both the scope and direction of programming work in technology:

> It is rather unclear what terminology and [programming] concepts you should use. Some concepts are intuitive, but it can be at any level.

In the curriculum, programming is presented mainly as a tool, rather than as a knowledge area on its own merits. A teacher with engineering background found this somewhat unfortunate and argued:

> Programming, in my opinion, is valuable in and of itself. However, you are not expected to learn programming for the sake of learning it but rather to build technological constructions or solve mathematical problems.

This teacher believed that the lack of emphasis on programming as a skill and a practice hampers the transformation process.

Teachers' choices of what to bring into the technology classroom were also influenced by external input. While some chose to build on existing work, others had received tips and ideas from teachers they met elsewhere, external resource centres, the NAE or sales talk at school technology trade shows.

Some teachers carried a certain private interest for digital technologies, and this had influenced their decision making. A teacher at one of the schools were in control of a limited budget and had to spend spare time to evaluate programmable teaching materials that would be both appropriate and fit the budget.

5.2 *Challenges during the Transformation Process*

5.2.1 Intrinsic Challenges

All intrinsic challenges described in the Finger and Houguet (2009) framework appeared during the interviews. Examples of how these challenges emerged are presented below.

> Professional knowledge and understanding:
> You cannot just make things up

The challenge of *professional knowledge and understanding* of programming is a fundamental challenge for all teachers participating in the study presented in this chapter. "You cannot just make things up", as one of the teachers said emphatically. The teachers who had studied an introductory Python programming course were all convinced that they were standing on less shaky grounds after than before: "without the course, I would have fallen flat, definitely".

Understanding and using correct technological concepts and expressions is part of the key technological abilities presented in the technology syllabus. When new content and concepts are introduced into the curriculum, such as programming, without proper preparation teachers may be put in an awkward and challenging position, as shown by Nouri et al. (2020) and seen in the following quote:

> You can create a 'snurra' [spin] or a 'slinga' [loop] or whatever you want to name it by compiling a sequence of instructions [...] I am not a programming nerd, and many words have similar meanings, which can be perplexing for me as a teacher.

This is the voice of a devoted and well-educated technology teacher, with no formal programming experience.

> Professional adequacy:
> We cannot be the best in the room

The challenge of *professional adequacy* is related to the teacher's belief in her own capacity to teach a particular subject or topic.

One of the teachers, with over 20 years of experience as a teacher in both mathematics and technology, and a Python course to lean on, simply stated that "when it comes to programming, we cannot be the best in the room, just forget about it. We have too little training for that". This teacher believed that a pupil who spends "two weeks and three hours a night and finds it exciting, will know more than I do".

Another teacher, who had no interest in programming and technology education in general, said there is a "feeling of panic before each lesson" and saw no other choice but to inform the pupils that "here you have an illiterate when it comes to programming". To overcome the perceived lack of professional adequacy, some teachers made sure to script their teaching activities in detail, to avoid having to face potentially complicated and time-consuming debugging of code.

> Professional attitudes and values:
> I do not want to teach technology!

Professional attitudes and values are related to whether teachers perceive the intended change as meaningful and possible to implement. No teacher questioned the introduction of programming in technology education, quite the opposite. This did not mean that all teachers found it to be a walk in the park or were willing to teach programming. "I do not want to teach technology", said a teacher who was the only qualified technology teacher in the school, but felt more comfortable teaching mathematics and science. At the same time, at another school, there was a teacher with neither formal merits nor teaching experience in technology, that believed a general personal interest in contemporary technology made the teaching task a little easier, albeit the lack of formal training was acknowledged as troublesome and restricting.

> Teaching approach:
> Feedback so that I do not directly point out the problem

Teaching approach captures the willingness of teachers to modify their teaching approaches in accordance with what the intended change and children's learning needs call for. Implementing programming in the technology classroom presents teachers with new teaching challenges to which they must adapt:

> They can become frustrated because if a semi-colon is placed incorrectly, you are smoked… and we have numerous 7th graders that have difficulty distinguishing between uppercase and lowercase letters.

The teacher believed it was important to be able to guide the pupils towards a fundamental understanding of debugging, that is, how the errors could be identified and possibly fixed. At the same time, the teacher did not want to get stuck in syntax and miss broader programming principles. The children were therefore occasionally provided pre-made code snippets.

The challenge of debugging was also illustrated by another teacher who concluded that the coding process is "90% about debugging". The teacher was hesitant to spend time on learning from errors and said that the children lack patience: "there are many who do not enjoy this, you want things to happen immediately".

Debugging is also related to teachers' professional knowledge and understanding and can be a challenge for teachers, as illustrated by the following quote:

> It can be challenging to identify errors in the code when you are not used to doing so. It can also be challenging to understand which errors are frequent and to be able to provide feedback so that I do not instantly point out where the error is.

Ownership:
Comes at a cost

In terms of *ownership*, none of the teachers expressed that they felt completely comfortable or confident in the reform. Ownership has a 'cost', which includes professional development, teaching experience, sufficient and reliable resources and so on.

5.2.2 Extrinsic Challenges

Resources:
The economy is in free fall

The challenge *resources* capture matters related to teaching materials and the IT infrastructure of the schools.

Adequate equipment is necessary for technology education. However, most of the schools represented in the study experienced frequent problems with their equipment. One of the teachers bluntly stated that the municipality's finances are in "free fall" and that it is therefore difficult to make any major investments in the necessary equipment and materials. Two other schools had equipment (Micro:bit cards) only because they could borrow it from external resource centres. There were also a few schools who did not have any programmable physical devices at all due to financial constraints. A list of recommended equipment from the National Agency for Education was suggested, if only to assist in negotiating school expenses.

Most schools had textbooks from established publishers, but they were printed before the curriculum change and so did not include any material on programming. At two schools, digital textbooks that had been updated for the new curriculum were used. However, one of these teachers used a trial license, which eventually expired as the school could not afford the license costs. Instead of textbooks, the teachers relied on teaching materials provided by external actors (e.g., Lego, BBC). It was further described as challenging to find quality teaching materials that were perceived to align with the curriculum intentions. One teacher concluded that "there is not enough good equipment or teaching manuals available yet". In one of the schools, where technology textbooks were completely missing, the teacher had found textbooks on Python programming for mathematics and decided to use that book in the technology classroom to have something to build on.

One-to-one computing was implemented in all schools. The type of computers varied between iPads, Windows/Mac laptops, and Chromebooks. The Chromebook users encountered two types of problems. Either the required software was not available for the platform, or the municipal IT department had blocked users from installing additional software. As a result, teachers were forced to resort to time-consuming and inefficient solutions:

> We use Replit online. It is challenging to write and run code there. Sometimes the code disappears forcing the children to restart from scratch.

The school that relied on iPads ran into issues with apps that cost money which the school could not afford, as well as with devices that were too old and no longer supported by app providers. A few schools had their discarded Windows laptop carts around. On some occasions, when the iPads or Chromebooks failed to connect to the peripherals they were using in class, the teachers made the effort to bring the discarded computers back to life because the chances of making stable connections, and thereby meaningful lessons, increased significantly.

Time: The perennial challenge

Time management is treated as a separate challenge in the Finger and Houguet (2009) framework. Time, or the lack thereof, was mentioned repeatedly during the interviews. Teachers had to spend a significant amount of time searching for appropriate teaching materials and learning how to use them. Another time related issue became clear when the teachers who had completed a 200-hour Python programming course estimated that their respective schools compensated them for no more than half the time. The remaining time required to complete the course had to be taken from their personal time.

Furthermore, most teachers estimated that they would spend between 10–15 hours on programming related activities in technology education over the course of three school years. They did not anticipate the children's programming abilities progressing very far. However, there was some hope that the situation would improve over time if the coordination of mathematics and technology regarding programming could be improved.

> Practicality of implementation:
> Everything is open for interpretation

The challenge *practicality of implementation* encompasses aspects of how compatible the revised curriculum is with existing classroom practices and school culture. The clarity of the reform influences teachers' reform work (Fullan, 2007). In this particular case, several teachers perceived the programming statements in the curriculum as vague, which in turn affects the practicality of implementation:

> It is stated that programming should be used, but with little further guidance and that is a bit vague. On the other side, it may be a good thing because it opens up new possibilities. Yet, as a less experienced teacher, you require some assistance in getting started, a bit more concrete guidance.

The teacher in the above quote ponders the potential consequences of the perceived lack of clarity in the curriculum in terms of the nature of programming. Another teacher argued that the lack of clarity leads to a situation where "everything is open for interpretation", and that the reform message will be interpreted and transformed in significantly different ways.

It was suggested that the lack of reform clarity in the technology syllabus would be less of a problem if the core content statements in the mathematics syllabus had been clearer about which programming practises and concepts children should encounter. In such a situation, a technology teacher could potentially be in a better position to understand what programming knowledge to draw on in the technology classroom.

Another aspect that influenced the practicality of the reform work was the absence of guidance on programming languages. Some teachers believed that Python had become the default programming language in mathematics, while they did not see a similar consensus for technology education:

> It is rather unclear which programming language will become the main language that we should use. I don't think you should use different languages.

A challenge related to practicality has to do with learning progression throughout the school years. Several teachers taught in grade seven and had children from different schools in their classes. Most teachers had a fragmented knowledge about what the children had learnt about programming at their previous schools. The immediate solution to this problem was to assume that they had limited previous understanding and "start from the beginning".

They believed that their job would be "much easier if someone had made a learning progression that you could use as a template". 'Someone' in this case would be the National Agency for Education:

> It would have been useful to have instructional materials from the NAE, that would have made it easier for teachers. Programming is a very broad term that encompasses everything from giving instructions to a friend to writing complex routines in a programming environment. It would have been helpful to know which difficulty level is appropriate in the early school years and which in the later years.

Student assessment:
How to know what they know

Student assessment is a core challenge for teachers affected by the programming reform. The section in the technology syllabus that describes what technological knowledge pupils are expected to achieve (Knowledge requirements) contains no specific details related to the assessment of programming knowledge. That is, the perceived lack of clarity and direction in the curriculum also affects what programming knowledge teachers choose to assess:

> They should understand what a loop, a sequence, and an algorithm are, but the knowledge requirements section in the syllabus does not speak much about programming

The teachers concluded that the assessment of technological systems and construction work would also encompass assessment of programming knowledge, as one teacher put it: "It is not whether they can program. Programming is just a small part".

History and tradition:
Technology is continuously changing

The challenge of history and tradition covers aspects related to the novelty of the intended change and its context. While some teachers had a few years' experience of working with programming in the technology classroom, all of them were breaking new ground in some way. One of the teachers reflected on the history and tradition of technology education:

> I feel that technology is continuously changing and that we must adapt. The technology subject is still not settled, is it? I do not think so.

Professional development and support: Limited and unequal

The challenge *professional development and support* captures issues related to teachers' in-service training opportunities, collaboration, and access to technical/pedagogical support. Clearly, the programming reform requires extensive professional development. Three teachers had had the opportunity to study a five-week Python programming course. Even though they were pleased with the content knowledge the course had provided, they pointed out that they also need pedagogical content knowledge, PCK (Shulman, 1986), which is a perspective the course did not specifically address. The remaining seven teachers had no idea if and when they would have the opportunity to study programming. Furthermore, several teachers did not even know if they would be teaching technology next school year.

Swedish schools should have nearby technical and pedagogical support according to the national policy on how the school system can benefit from the digital transformation of society (Regeringsbeslut U2017/04119/S). In reality, this may be an elusive goal. Out of the schools in the study, only a few had technical-pedagogical support staff (ICT educators) in the organisation. The teachers at those schools felt that the support staff were not very useful, either due to a lack of subject knowledge or because they were difficult to get hold of. However, the resource centres on which two teachers relied were not only very appreciated, but also played a crucial role for these teachers. Without the resource centres, they would most likely have had no physical programmable devices to work with.

The curriculum conveys that programming is an interdisciplinary subject content and that it is in mathematics where children should learn the 'programming fundamentals' (Vinnervik, 2022b) that can be built upon in technology. In other words, cross-subject collaboration and support are expected. However, in this study, the collaboration between technology and mathematics was largely absent. When it occurred, it was simply because the teacher was teaching both subjects. With few colleagues to support them, some teachers had asked children who caught on fast for help to find teaching materials and peripherals for construction work. These children were also used as co-teachers in class.

6 Discussion

The study reported in this chapter examines the way technology education teachers incorporate programming into their teaching and challenges they encounter during this process. The research was conducted 27 months after the enactment of a reform that added programming to the Swedish school curriculum. The findings are relevant for gaining insight into how the intended curriculum is interpreted and transformed into teaching and learning activities and can inform future evaluations and curriculum revisions.

The intended curriculum and how it is perceived, as well as the context in which teachers make judgments about how to approach and transform new knowledge, all have an impact on how curriculum content, such as programming, is actually realised and ultimately perceived by school children (Adolfsson & Alvunger, 2018; Linde, 2012; Lundgren et al., 2004; Wahlström & Sundberg, 2015). The teachers we have heard from in this chapter perceive programming as a tool to help pupils get a more comprehensive understanding of technology, rather than as a goal in itself. Their teaching efforts focused mostly on construction work and technological systems, with programming serving as a tool to better understand the broader technological context. This strategy aligns with the objectives outlined in the written curriculum (Vinnervik, 2022b). The teachers did, however, believe that there are certain fundamental programming concepts that the children should learn, such as sequence of commands and conditional structures, although these ideas are not explicitly included in the curriculum documents. One of the teachers with an engineering background argued that programming is undervalued in the curriculum and that programming knowledge is also technological knowledge. Concerns have been raised that programming education may end up being more about basic procedures and routines than the broader technological and digital competencies outlined in the curriculum (Åkerfeldt et al., 2018), but the findings presented in this chapter seem to alleviate these concerns. It is also worth noting that few teachers referred to programming as an iterative problem solving process. This is a perspective on programming that is not clearly conveyed in the curriculum, although some links are made between programming and problem solving through the reinforced additions of digital competence in the curriculum (Vinnervik, 2022b).

A previous study (Vinnervik, 2022a) shows that teachers may face various intrinsic and extrinsic challenges in their efforts to comprehend and realise the programming reform message. Many of these challenges appear to be unresolved, as they resurface in the study reported in this chapter, which was conducted 2.5 years later. Regarding the intrinsic challenges, we have learnt that experienced technology teachers also feel that they lack necessary pedagogical content, or professional, knowledge, and that having to integrate programming into their teaching can even infuse the feeling of 'panic'. The lack of adequate professional knowledge is illustrated by the hesitancy some of the teachers expressed towards using debugging as a method of understanding the principles of programming. This is perhaps an opportunity not taken as debugging is seen as a vital part of programming and resonates with the ability to systematically test and retest which is a core competence of school technology. However, teachers must strike a good balance between paying attention to the small details of accuracy and the broader principles of programming and technological work. Another unresolved intrinsic challenge is that of ownership. It had been 3.5 years since the announcement of the reform (in March 2017) and the time of study presented here, but not a single one of these teachers expressed that they felt completely comfortable with the reform. That is, reform work takes time, but if the intrinsic and extrinsic challenges presented in this chapter are resolved, ownership can emerge.

The interviews also revealed several extrinsic challenges. One is related to how the reform message is communicated in the formal written curriculum, which influences how teachers transform programming into meaningful teaching and learning activities. The intended curriculum emphasises the use of programming as a multipurpose tool for learning mathematics and technology, but provides limited guidance on what specific programming-related content to be taught (Vinnervik, 2022b). As suggested by one of the teachers, the perceived lack of clarity in the technology syllabus regarding programming knowledge could have been remedied if the mathematics syllabus had provided more details (see Vinnervik, 2022b).

Furthermore, there are challenges related to the practicality of the reform implementation, for example, related to the lack of a clear learning progression through the school years. According to the Swedish Schools Inspectorate (2014) there is a risk that teachers may intentionally teach at a lower level due to uncertainty about the quality of technology education from previous years. This chapter provides evidence of some teachers who chose to start at the beginning, which may be valuable for some but less inspiring for others. Although the curriculum commentary materials address some aspects of learning progression, the teachers in this study paid little attention to them. It is also uncertain whether these texts provide any valuable information as the usefulness of the learning progression statements in them is questionable (Vinnervik, 2022b).

Related to learning progression is the challenge that assessment of programming knowledge presents teachers, and both issues are related to how the Swedish school curriculum is organised. While being structured around demarcated subject-based knowledge (Adolfsson, 2018), the knowledge requirements section of a Swedish subject syllabus does not target specific subject content to any detail. This can lead to greater or less control over the assessment process, depending in part on the teacher's professional knowledge of the subject content being taught (Rosenlund, 2019). The teachers we have heard from in this chapter believe that it leads to less control.

We have also learned that many of these teachers work independently, which can be challenging (Shulman, 1998; Yadav et al., 2016). The few instances where cross-curricular work occurred were when teachers taught both subjects. Research has demonstrated that collegial interplay is appreciated (Vinnervik, 2022a) and can foster professional development and improvement initiatives (Langelotz, 2014). The purposeful creation of networks or communities for workplace learning, according to Fullan (2007), will result in teachers willing to explore methods to improve their routines and how they bring new innovations into the classroom. In this case, more deliberate collaboration between mathematics and technology teachers could break the isolation, but unfortunately, as Fullan (2007, p. 75) concludes, it is rare for teachers to have working conditions where "sustained teacher innovation" thrives.

7 Conclusions

Creating lasting and sustainable change is a time-consuming and gradual process, especially when the reform involves a complex message, such as programming (Ahtiainen, 2017; Fullan, 2007; Hargreaves & Goodson, 2006). The results of this study confirm this assertion. In this chapter, we have met a group of teachers who believes that programming serves as a natural ingrediency in the technology subject particularly in the context of technological work. We have also learnt that more than 3.5 years after the reform was announced, teachers are struggling with various intrinsic and extrinsic challenges that need to be addressed with urgency to make the programming reform practical, effective, and sustainable. It is unfair to expect teachers to lead educational change of this character without proper training and resources to facilitate their efforts. Not to mention how pupils' rights to equal quality education is jeopardised.

References

Adolfsson, C. (2018). Upgraded curriculum? An analysis of knowledge boundaries in teaching under the Swedish subject-based curriculum. *The Curriculum Journal, 29*, 424–440. https://doi.org/10.1080/09585176.2018.1442231

Balanskat, A., & Engelhardt, K. (2015). *Computing our future: Computer programming and coding - Priorities, school curricula and initiatives across Europe. European Schoolnet (EUN Partnership AISBL).*

Billett, S. (2006). Constituting the workplace curriculum. *Journal of Curriculum Studies*, 1–15. https://doi.org/10.1080/00220270500153781

Cohen, D. K., & Ball, D. L. (1990). Policy and practice: An overview. *Educational Evaluation and Policy Analysis, 12*(3), 233–239.

Elo, S., Kääriäinen, M., Kanste, O., Pölkki, T., Utriainen, K., & Kyngäs, H. (2014). Qualitative content analysis. *SAGE Open, 4*(1), 215824401452263. https://doi.org/10.1177/2158244014522633

Finger, G., & Houguet, B. (2009). Insights into the intrinsic and extrinsic challenges for implementing technology education: Case studies of Queensland teachers. *International Journal of Technology and Design Education, 19*(3), 309–334.

Fullan, M. (2007). *The new meaning of educational change* (4th ed.). Teachers College Press.

Goodlad, J. I. (1979). *Curriculum inquiry: The study of curriculum practice*. McGraw-Hill.

Hallström, J., Hultén, M., & Lövheim, D. (2014). The study of technology as a field of knowledge in general education: Historical insights and methodological considerations from a Swedish case study, 1842–2010. *International Journal of Technology and Design Education, 24*(2), 121–139. https://doi.org/10.1007/s10798-013-9252-x

Hargreaves, A. (2005). Educational change takes ages: Life, career and generational factors in teachers' emotional responses to educational change. *Teaching and Teacher Education, 21*(8), 967–983. https://doi.org/10.1016/j.tate.2005.06.007

Jarl, M., & Rönnberg, L. (2019). *Skolpolitik: från riksdagshus till klassrum* (Tredje uppl). Liber.

Kvale, S. (1997). *The qualitative research interview*. Studentlitteratur.

Lärarnas Riksförbund. (2020). *Programmering—en skolreform utan program.*

Linde, G. (2012). *Det ska ni veta! En introduktion till läroplansteori*. Studentlitteratur.

Lipsky, M. (2010). *Street-level bureaucracy: Dilemmas of the individual in public services.* Russell Sage Foundation. http://www.jstor.org/stable/10.7758/9781610447713

Lundgren, U. P. (1989). *Att organisera omvärlden: en introduktion till läroplansteori.* Utbildningsförl.

National Agency for Education. (2017). *Få syn på digitaliseringen på grundskolenivå – Ett kommentarmaterial till läroplanerna för förskoleklass, fritidshem och grundskoleutbildning*. Skolverket.

National Agency for Education. (2019). *Skolverket statistical database (Siris).*

Nouri, J., Zhang, L., Mannila, L., & Norén, E. (2020). Development of computational thinking, digital competence and 21st century skills when learning programming in K-9. *Education Inquiry, 11*(1), 1–17. https://doi.org/10.1080/20004508.2019.1627844

Ornstein, A. C., & Hunkins, F. P. (2018). *Curriculum: Foundations, principles and issues* (7th ed.). Pearson.

Ozga, J., & Jones, R. (2006). Travelling and embedded policy: The case of knowledge transfer. *Journal of Education Policy, 21*(1), 1–17. https://doi.org/10.1080/02680930500391462

Parding, K., Sehlstedt, T., Johansson, A., Berg-Jansson, A., & Jakobsson, M. (2018). *Lärares arbetsvillkor i kontexten av marknadisering, privatisering, val och konkurrens*. http://urn.kb.se/resolve?urn=urn:nbn:se:ltu:diva-67117

Priestley, M. (2019). Curriculum: Concepts and approaches. Retrieved February 10, 2020, from https://mrpriestley.wordpress.com/2019/01/04/curriculum-concepts-and-approaches/

Sentance, S., & Csizmadia, A. (2017). Computing in the curriculum: Challenges and strategies from a teacher's perspective. *Education and Information Technologies*, 22(2), 469–495. https://doi.org/10.1007/s10639-016-9482-0

Shulman, L. S. (1986). Those who understand: Knowledge growth in teaching. *Educational researcher, 15*(2), 4–14.

Shulman, L. S. (1998). Theory, practice, and the education of professionals. *Elementary School Journal, 98*(5), 510–526. https://doi.org/10.1086/461912

Swedish Higher Education Authority. (2020). Utvärderingen av ämneslärarutbildningarna klar. Retrieved February 26, 2020, from https://www.uka.se/sitevision/proxy/kvalitet--examenstillstand/resultat-fran-granskningarna-hogskolekollen/hogskolekollen.html/svid12_794d1f901621e65b329c27f/427238778/files/1930738/2075414/Bedömargruppens_yttrande.pdf

The Swedish Schools Inspectorate. (2019). *Digitala verktyg i undervisningen*. https://larportalen.skolverket.se/#/modul/0-digitalisering/Grundskola/513-Digitala-verktyg-i-naturvetenskap

Vinnervik, P. (2022a). Implementing programming in school mathematics and technology: Teachers' intrinsic and extrinsic challenges. *International Journal of Technology and Design Education, 32*, 213–242. https://doi.org/10.1007/s10798-020-09602-0

Vinnervik, P. (2022b). An in-depth analysis of programming in the Swedish school curriculum—rationale, knowledge content and teacher guidance. *Journal of Computers in Education*. https://doi.org/10.1007/s40692-022-00230-2

Wahlström, N., & Sundberg, D. (2018). Transnational curriculum standards, curriculum reforms and classroom practices - an introduction. In N. Wahlström & D. Sundberg (Eds.), *Transnational Curriculum: Standards and classroom practices - The new meaning of teaching*. Routledge.

Wickman, P. O., Hamza, K., & Lundegård, I. (2018). Didaktik och didaktiska modeller för undervisning i naturvetenskapliga ämnen. *Nordic Studies in Science Education, 14*(3), 239–249. https://doi.org/10.5617/nordina.6148

Yadav, A., Gretter, S., Hambrusch, S., & Sands, P. (2016). Expanding computer science education in schools: Understanding teacher experiences and challenges. *Computer Science Education, 26*(4), 235–254. https://doi.org/10.1080/08993408.2016.1257418

CHAPTER 9

Discourses of Programming Teaching within Compulsory Education

Susanne Engström and Eva Björkholm

Abstract

Knowledge linked to programming has been extensively strengthened in curricula and syllabi in Swedish compulsory school, mainly regarding the subjects of mathematics and technology. The introduction of this content requires that technology teachers be trained in programming and how to teach it. In this chapter, we present an observation study of a professional development course for practicing teachers in compulsory education, focusing on introductory programming. The whole professional development course context, including the teachers' teaching practices and presentations of classroom projects, is explored as a discourse with the aim of finding governance steering strategies, normative values and knowledge content. The analysis shows that norms and values from the programming discourse within this professional development course become more important than the participating teachers′ professional view on how classroom teaching should be conducted. The code-content knowledge is taken for granted, but the artifacts tend to take away the focus from the specific code-content. A gender perspective is also highlighted, where coding is regarded as suitable for the stereotypical image of a female student. The overall aim seems to be that the students should find programming fun and interesting.

Keywords

technology education – programming – classroom discourse

1 Introduction

In a 2017 revision, knowledge linked to programming was strengthened in curricula and syllabi in Swedish compulsory school, mainly regarding the subjects of mathematics and technology (National Agency of Education, 2017). It was argued that this kind of knowledge is needed for students to manage

© SUSANNE ENGSTRÖM AND EVA BJÖRKHOLM, 2024 | DOI:10.1163/9789004687912_009 their digital world and deal with their digital tools. In addition, increasing interest in technology was highlighted as a way to get more people into technology training (Digitization Commission, 2014). When the curricula and syllabi were revised again in 2022, programming was emphasised even more, and some clarifications were made, but in all the same content was still expressed (National Agency of Education, 2022).

Such a vigorous introduction of new content involved substantial efforts in implementation processes, such as professional development activities for teachers, the purchase of equipment, etc., from 2017 and onwards. In different aspects of the implementation, several actors had great interest in striving for impact, such as university departments in computer science and teacher training; science centres; non-profit organisations; the business sector, as financing development projects; companies selling digital equipment; and publishers. In this context of forming a new topic in school, the values and norms mediated by the actors become crucial for what specific content is emphasised and how it is taught.

At the time the study presented in this chapter was conducted, previous studies had suggested that the programming content should involve decomposition of the problem, abstraction, logical thinking, algorithm creation, and debugging (Jaipal-Jamani & Angeli, 2017; Sullivan & Heffernan, 2016; Yadav et al., 2014). Research related to the use of robotics in K-12 education also proposed a learning progression that starts with sequencing abilities, advances to reasoning abilities such as casual and conditional reasoning, and results in greater systems understanding (Sullivan & Heffernan, 2016).

At the same time, some researchers argued that the teaching of introductory programming was problematic, since computing educators within the computing education community often teach with a narrow focus upon the technology tools, without explicitly taking students' views and needs into consideration (Berglund & Lister, 2010). The researchers claimed that "Computer scientists and academics /…/ are parts of, and carry, a certain culture, with its own values and norms. These values and norms need to be made explicit in the discourse of teaching." (ibid. p. 41). Such studies were, however, sparsely represented, but Barker & Garvin-Doxas, for example, observed in 2004 the communication climate in introductory computer science courses, concluding that the social environment could be characterised as impersonal, where it was easy to stay anonymous, socially distant, and very guarded since the students rarely disclosed personal information. Moreover, in a questionnaire-based study, secondary school computer science teachers were asked about the best ways to make the subject interesting (Black et al., 2013). A clear theme emerging from the teachers' comments was that students would be attracted by a feeling of in-the-moment enjoyment—"making computing fun".

In this chapter, we present a study of a professional development course for technology- and primary school teachers focusing on programming. The course was implemented in connection with the introduction of programming in the revised syllabus of 2017, during a period when an extensive number of Swedish teachers underwent in-service training in programming. The aim of this study is to examine what content related to programming emerges as important in the teaching throughout course implementation, and what implicit values and norms concerning this content are imbedded in the teaching practice. The following research question guides the analysis: What content and values concerning programming emerge as important in the teaching practice of the professional development course and the participating teachers' personal approach?

2 Theoretical Framework

As mentioned above, this chapter presents results from a study that generally focuses on programming within technology teaching, but the study object is a course for practising teachers. We are interested in investigating the discourse, i.e., emerging values and norms, and important content in the professional development course, while at the same time studying how that discourse has implications for technology teachers' teaching. Therefore, we broaden the discourse of the professional development course and involve teachers' technology teaching connected to the course.

Since we are interested in exploring the discourse, the theoretical frame of the study is inspired by the notion of "big D" (defined by Gee, 1990, as "Discourses" with a capital "D" is always more than just language). Gee explains how "Discourses are ways of being in the world, or forms of life which integrate words, acts, values, beliefs, attitudes, social identities, as well as gestures, glances, body positions and clothes" (p. 142). Since we, in this chapter, see programming as a subject area with connections to the school subject of technology (according to the curriculum), we suggest that the professional development course aims to give implications to a subject's teaching practice. We want to explore this specific practice seen as a discourse; we search for values and content that emerge as highly valued and what powerful strategies make values and content possible. Governmentality studies explore, by problematising or calling into question, who can govern, who is governed and how. Fimyar (2008) describes how the analytics of governmentality explore the practice of government in relation to ways in which "truth" is produced in social and cultural spheres. Therefore, such an analytic lens could make it possible to seek an open and critical relation to strategies for governing: their assumptions, their exclusions, their regimes of vision and spots of blindness (Fimyar, 2008). We take inspiration in Marie Öhman (2010) as well, and her arguments concerning how school subjects' practices have a relation to power, following Foucault (1982). Öhman (2010) expresses how "a school subject's practices, traditions and customs are often deeply rooted in the teaching practice, and often regard content as natural and obvious. With the aid of "[a Foucauldian power perspective], it becomes possible to study how the knowledge, norms and values included in an activity render certain ways of acting more reasonable and others less reasonable and thereby benefit certain ways of acting and being" (p. 406). When Öhman (2007, 2010) discusses governance strategies within the discourse of the teaching practice, she clarifies the notion based on Foucault's notion of governmentality. Thus Öhman (2010) places power in relation to a so-called transactional approach:

> Placing power in relation to the transactional perspective means not presupposing that power is something given, but something that is examined as it is manifested in a certain context. Thus, with the aid of Foucault's power perspective we can explain how students' actions are both facilitated and restricted in a teaching situation, and thereby determine what these processes look like and how they govern an individual's way of acting and being. (p. 397)

In this present study we want to investigate if there is a power relation between the teachers as computer scientists and the teachers participating in the course.

Here we lean back on the interpretation found in Öhman (2007) when she states: "governance that requires an individual freedom, in such a way that individuals are co-actors within their own governance. In other words, individuals become willing to act in line with the desirable order; an order including knowledge, beliefs and values, accepted within the society, "the normal"" (p. 129. translated by the authors). By linking governance and knowledge, it becomes possible to study how power operates in classroom actions and find such actions that are taken for granted, both in knowledge and approaches. It is possible to investigate and present a detailed analysis of how students act upon teacher actions and vice versa (Öhman, 2010).

In the study described in this chapter, the whole professional development course context, including teachers' own teaching and their presentations of their classroom projects, is explored as a discourse with the aim of finding governance steering strategies, normative values and knowledge content.

3 Method

3.1 *The Training Course*

The professional development course, also called the training course, was conducted by a science centre. The teaching was carried out by people skilled in programming with experience teaching both students and teachers but no formal teacher training. The course consisted of a total of six training sessions, each lasting a full day. The course would result in a small project on programming that the teachers implemented in their own classes, which was then presented and reported during the final session of the course (Figure 9.1).

Within the whole-day training sessions, the participants had the possibility to work hands-on with different programming tools such as Blue-Bot, Twine, Lego Mindstorms, and Scratch, as well as discuss different central concepts, gender aspects and assessments related to programming and technology.

3.2 *The Participants*

A total of 16 practicing teachers (teaching in grades 1–9) participated in the course, all of whom taught STEM-subjects (science, technology, engineering and mathematics). Three of the participating teachers, who taught in Grade 3, 4 and 7 (students aged 9, 10 and 13 years), respectively, were observed when implementing the project in their classes. These three teachers also reported, at the request of the whole group, their interest in being observed while teaching in their classes.

3.3 *The Data Collection*

Data were obtained by observing teaching in all six training sessions in the professional development course. Two researchers observed all training sessions at the science centre by sitting in each corner of the classroom for the entire course. Thereafter, how the classroom teaching connected to the projects was investigated. Two lessons, with durations of approximately one hour each and a few weeks apart, were observed for each teacher. Finally, in the sixth training session the teachers' presentations were observed, video filmed and transcribed. During all the observations, the two researchers, independent of each other, made field notes continuously. Immediately after each teaching or presentation session, the researchers met and discussed their individual observations and summarised their impressions.

FIGURE 9.1 The training course

3.4 *The Analysis*

All field notes and transcripts were read over and over again, searching for patterns of normative values, steering strategies and knowledge content. The results were then discussed by the two researchers and elaborated throughout the whole analysis process.

The analyses of the teaching within the training sessions, along with the teachers' classroom projects and presentations, were based on both identifying what knowledge content was highlighted in relation to programming and, to some extent, a discourse analytical perspective (Börjesson & Palmblad, 2015; Gee, 2014), which focused on the values and governance strategies within the two different teaching contexts (Öhman, 2007, 2010). We searched for specific words, emphases, body positions, gestures, and the ways with which persons are spoken to (related to Gee, 2014; Öhman, 2007, 2010 based on Foucault). One context is the training session combined with the project presentation, and the other is the classroom projects. Analyses were directed towards how power, in terms of possible paths of action, appears in an individual's everyday actions (Öhman, 2010). For example, within a teaching situation, the content that is offered to students contains ideas about which forms of learning and knowledge are important or significant and which are negligible (Östman, 1995).

4 Results

In the following section, we will present the results starting with the analysis of the observations of the teacher-training sessions. This is followed by the results of the classroom project observations. Finally, the results concerning the teachers' presentations of their projects within the teacher training course will be described.

4.1 *The Teacher Training Sessions—Normative Values Emerged*

4.1.1 The Obvious Knowledge Content

The specific knowledge content is seen as something obvious within the school context, and something implicitly obvious in itself. The knowledge content linked to programming consists of different aspects such as the formulation of well-defined thoughts, organisation of solutions and stepwise instructions, and structures of problem solving, i.e., coding. Moreover, the need for an exact language that cannot be misunderstood or interpreted in different ways is essential. This content is seen as something obvious and is not questioned and discussed:

Course leader: Of course, we think programming is important for many different reasons. Not the profession of programmer, but we need to inform about such a profession, such knowledge and skills. A lot of logic, problem solving comes into almost every subject. Give more students the chance to be good at something.

The intention of programming teaching seems to be the logic of programming and the essence of coding. The code must be seen separated from the artefact. It is important to learn the "craft" of coding, which is to test and make mistakes, to troubleshoot. Learning through extensive practising is seen as essential.

At the same time, in the teacher training session, the Blue-Bots and Lego robots seem to be the content since a great focus is placed on the robots and the controlling of the specific artefacts. The specific knowledge of coding is not made clearly visible. Also, when Scratch is used for programming at a computer, the teachers still focus on the artefacts (the aquaria) on the screen.

The intention of the progression of learning programming starts with using robots (Blue-Bots) and progresses to using the computer. The teaching does not highlight the code separate from the artefacts.

Moreover, teaching programming should be integrated in the ordinary teaching:

Course leader: The aim is that programming should become a natural part in the classroom. A common part of teaching, where you can include programming to where it fits your own teaching.

When discussing how programming can be integrated into ordinary teaching, the participant highlights practical, economic aspects, not educational aspects:

Teacher A: I must ask—what is the proposal for financial compensation for the schools? It's stressful with just two I-pads.

Course leader: Don't know, you can program very simply.

Teacher A: Oh yeah, but it's much easier if you teach the half class. /…/ I think it's great, it's not like that. But I want it to work in reality as well, that's my starting point.

The participants do not ask questions about the content and content-related activities. Their questions focus more on the economic aspect related to the specific equipment.

4.1.2 The Attitude of Narrowness, Worshipping the Nature of Programming and Its Possibilities

The attitude, both within the sessions and of the programmer teachers themselves, can be described in terms of no unnecessary talking, no discussions, just short and quick instructions. There is a "right way" to do things; learning sequences are fast; the aim is to inspire and be effective.

A great focus is on creative skills aiming to create something new without questioning or problematising and without any limits, as shown in the example below:

Course leader: Nothing is there if we don't put it there, it doesn't understand anything by itself. That is the charm of programming. You build your own world; you make it your own. With basic instructions, they can create amazing things.

A playful attitude can also be observed, showing an uncomplicated approach to related issues, and emphasising that programming is easy and without any limitations. In addition, there is an inviting attitude; it is important to share ideas and programming knowledge.

At the same time the importance of accuracy, patience and logic is highlighted in this context:

Course leader: Today it's often about fast entertainment, here it's boring and patient.

The discourse is characterised by both a playful and creative atmosphere and a well-structured, narrow, "boring" and "the right way to do things" approach.

4.2 *The Teacher Training Sessions—Steering Strategies Used by the Course Leaders*

4.2.1 The Strategy of Trivialising

Based on the course leaders' statements, compulsory schoolteachers are seen as programming incompetent. The teachers are spoken to with expressions like, "How do you feel?" "Do you have fun?" "Do not panic!" "Have fun!" "This is simple, as easy as writing a story." "Hello, we are going to play with robots!"

The teachers accept the course leaders' way of talking and adapting it to their expectations of them. They become "students"; they applaud, laugh, and trivialise themselves.

One of the course leaders tells about an 8-year-old student who managed to program a robot to dance after training for 3 hours during a summer course.

A teacher points to himself and says, "*Here is the stop*". The course leader confirms this. Several teachers say that they stop the students. On the other hand, this is not seen as a problem, as shown in the example below:

Course leader: a girl who will surely become an engineer, she sat quietly and repaired a robot. Be a little free, you want to walk around and ask: What have you done? You must withdraw. Not much is required of us [school teachers].

It seems thus unproblematic to disclaim the traditional teacher role of being a competent guide for the students.

4.2.2 The Strategy of Disjunction

The course leaders, well-educated computer scientists, distinguish between "WE" and "YOU", stating, for example, "what do WE mean by programming?" That is, WE, who are inside the programming context, unlike YOU, the participating teachers excluded from that context. The course leaders distinguish between WE, experts in programming, and YOU, novices in programming. In the following excerpt, the course leader talks about the leader team:

Course leader: When it comes to the high level that we use, then there is /../

The teachers are told "You will get some knowledge", and "Take some knowledge to your classes", "Something that make your students positive, interested and engaged", and "The students must believe that everything is possible".

The way the course leaders talk to the participating teachers, with a clear disjunction, stands in contrast to how they talk about themselves as highly qualified and the participants as unqualified.

4.2.3 The Injunction of Logic

The course leaders give orders to do everything right and keep to the programming logic, stating, for example, "*Follow the instructions*"! "*Troubleshoot*"! "*Be careful and accurate*"! In this discourse, well-structured thinking and systematic work is highly valued. The course leaders highlight how important it is to follow what they express as the "laws" and logic of programming.

4.3 *The Classrooms—Content and Values Highlighted*

In this section we present specific programming content, values and norms emerging in the three different observed classrooms. In each classroom some implicit steering strategies used by the teachers could be observed as well.

4.3.1 The Grade 3 Classroom

The knowledge content (the concept of loop, steering Blue-Bots in relevant contexts, and using a code language with consensus) is made clear in the teaching, and the students are given the possibility to learn. Small groups of students are told to program a Blue-Bot on a track and to line up cards with the corresponding code, showing how to steer the Blue-Bots (technology content). The tracks also relate to other subject contents, such as learning multiplication, the clock, and English words. However, after a while, the students first skip lining up the cards and instead program the robot directly. They then focus more on constructing the physical track, adding various obstacles made of Legos, wooden sticks, etc. As the students give up the coding activity, the progression of learning fails.

The values expressed by the teacher focus on the aims of the teaching sequence, and not giving up, as shown in the examples below:

Student E (to the teacher): Which card should we start with?
Teacher K: It's like reading books, start here.
Teacher K: You didn't give up, it will never be right at once, and I heard that you discussed very well.
Teacher K: I think you didn't talk to each other, everyone wanted to build different tracks, then it is not possible to program, you don't have to get your will through every time. Constructing was not the most important thing; it was the programming of the robot that was important.

Here we can see that it seems important to collaborate, as also shown in the following example:

Teacher K: Is it going well? You must somehow do it together, otherwise it will not work.

In this classroom the aims are in focus; the teacher gives feedback, and classroom discussion and collaboration are highlighted.

4.3.2 The Grade 4 Classroom

Here the knowledge content (decomposing a problem, describing an event, dividing the problem into constituents, different parts will be structured in a specific order—"that is programming!") is clarified. The students are told that they are going to "work with descriptive factual texts, and to program something that describes something that happens". They are asked to "write a story where you have several different choices", using the Twine program.

The learning progression dealing with the structure of programming is clear. However, the context of programming is about science and technology, more specifically atoms and molecules, which becomes problematic since the way of using programming related to this content is not logic for understanding science. The students create stories about water molecules acting like humans, and the specific science content is missing. The sequence has a focus on learning programming and the science is not relevant.

Values expressed by the teacher are the importance of a clear structure:

Teacher P: It is possible to follow, to know how it goes and that they [the cards with code] are connected to each other. Great.

Moreover, there seems to be a serious approach to programming (talking about useful knowledge, and not only to have fun), repeating the specific learning object, reminding the students what to do, and giving short and explicit guidance.

4.3.3 The Grade 7 Technology Classroom

In this technology classroom the students are using computers and supposed to program Lego robots to dance and walk into specific tracks by using sensors, etc. The knowledge content in focus is the artefacts, i.e., the Lego robots, while the programming content is unclear. Instructions are given to students about what to do in detail, without any feedback or discussions. The students are given tasks expressed in instructions, and the teacher helps with troubleshooting and pushing the students forward to solve problems and gain results without any collaboration and discussion:

Teacher W: I wonder if you lack wires and stuff. So. Now you see that it is lit. /…/ Test and see what happens! /…/ Click on a block and press the delete button!…/…/ Here you press download…then you can take it out…and then again, and then RUN!

The values expressed by the teacher are short instructions and efficiency, expressed as "Go on!" and "Have fun!" and "Yes! Now! Go! Download!", "Have you had fun with the robots?". The teacher seems to guide the students to fulfil the task and feel satisfaction rather than learning how to program.

4.4 *Teachers' Project Presentations—Normative Values Emerged*

The knowledge content highlighted in the teacher training course is reproduced and taken for granted without any critical attitude. The highly valued content involves abstract thinking, programming with loops, bugs, trouble shooting, and computational thinking. The teachers also express the importance of explaining central concepts related to programming as follows: *"understand the concepts of programming, code, loop, and bug".*

In addition, a two-sided approach to teaching is observed; on the one hand, programming must be fun, challenging, and raise the curiosity of students (students learn quickly and always want to do more), as seen in the example below:

Teacher, year 3: I let them loose, two and two, boys and girls. They thought it was so much fun. Great fun, they could have lasted for any length of time.

On the other hand, it conflicts with "usual teaching", since there is no explicit aim, no learning progression, and difficulties with subject integration:

Teacher, year 7: As an introductory project, it has worked very well, playful. It needs to be more "square-like" if it is to be included in the teaching.

Unexpressed views on students, related to the norm of a well-structured person, emerge and can be described as a clear gender coding in relation to programming. The girls are seen as stronger, more persistent, and precise, while the boys are careless and give up more easily. Some of the teachers have identified programming skills, and they see how girls seem to manage in a better way:

Teacher A year 3: The girls are focused; they don't give up. They could solve some of the problems, they followed how they had lined up the cards. The boys, many of them just want to go on, something must happen.

Teacher B year 3: I think about the gender perspective, we have all seen the same, that the girls take over. They are the ones who have the energy to find the solutions. The boys think it's great fun to press the buttons and to invent, but the girls are more systematic in their ways of doing this. They quickly learned code, programming, and the other concepts.

Moreover, there is a view of students where "the silent ones who think get a place" /…/ "not those who scream, but those who sit in the back and work". This statement is expressed by one of the course leaders and reflects the norm, that is, the stereotype view of a programmer or computer scientist.

During teachers' presentations, a communicated strategy is about leaving the responsibility for student learning to the students themselves. This is expressed by teachers as "*It is possible to release control!*" and "*The students still learn!*":

Teacher, year 5: I cannot have the answers to everything, I can do it together with the students. I should not be stressed about that. My colleagues say: but how can you do this [programming]? I am with the kids, that's a lesson. If there are technical problems, in the worst case we must shut down, and then we can do it another day.

The programming lesson is summarised by one of the teachers in this way: "It was very successful without me being able to do anything!". The students are seen as skilled, and they manage without the teacher.

5 Discussion

In this study we have identified highly valued content but also norms, values and steering strategies (Öhman, 2007, 2010) related to programming that emerge as important in a professional development course. Below we discuss the results and relate them to both current research when the study was conducted, and studies carried out later. We also discuss the results in relation to technology education.

A norm that emerges in the professional development discourse is that the content itself, such as the decomposition of the problem, debugging, etc., was taken for granted, which was observed also by Öhman (2010). The specific content is similar to the content highlighted in previous studies (Jaipal-Jamani & Angeli, 2016; Sullivan & Heffernan, 2016; Yadav et al., 2014; Piadade et al., 2020; Rose et al., 2017). The consensus of the content that seems to be established in the observed context appears not to be questioned, a view yet also argued in later research by, for example, Kjällander et al. (2018). However, the artefacts in terms of robots and animations tend to take away the focus from the specific knowledge related to coding. The tools and artefacts seem to stay in focus instead of what students are supposed to learn about coding. This is also found in earlier research, for example, in Berglund and Lister (2010), but also discussed in later research related to technology education, for example, in Kjällander et al. (2018) and in Cederqvist (2021).

In the professional development course, there is an intention regarding learning progression to start with programming robots (blue bots and Lego robots) and then move on to coding with computer, using the Scratch program. But when using Scratch, the participants still focus on the artefact, for example, the cat on the display, rather than on the knowledge content related to coding. Developing more complex code knowledge, learning more qualified and efficient commands, etc., remains in the background.

Within the teaching practice of the participating teachers, a lack of clear progression in the teaching was mostly identified (also found in Sullivan & Heffernan, 2016). Moreover, problems with relating the use of programming to relevant and meaningful technology contexts were also identified (cf. Thuné & Eckerdahl, 2009). A thematic review conducted by the Swedish Schools Inspectorate in 2019, on how programming has been adapted to the two subjects of technology and mathematics, shows that the teaching has focused on the coding itself at the expense of the application of programming in the subjects' content, such as steering, controlling, regulating, and simulating processes. In this present study we find some examples of coding with an aim to steer the artefact, for example, the teacher in grade 7. However, the knowledge related to coding was not made explicit. Moreover, the teacher in grade 4 focuses on coding with the intention to relate to science but without a clear connection to relevant science content.

We also found that the ambitions concerning teaching programming with a clear learning progression and using relevant contexts were in conflict with the idea that computing must be fun (also found by Black et al., 2013). Moreover, as the content and the highlighted abilities in the course are presented in terms of thoroughness, patience and structure, programming is regarded by the participating teachers as suitable for the stereotypical image of a female student. At the same time, studies of students' attitudes show that male students have greater confidence in computer science and programming than female students, and that these gender differences increase when the students get older (Mason & Rich, 2020). The teachers highlighting of girls' competence in programming can be explained by the fact that most of the teachers were teaching rather young students (year 1–7) where female students have confidence in structured activities generally. This is interesting and worth highlighting in the context of programming teaching in terms of starting in early years and to keep in mind when the students get older.

Within the discourse, there is also an acceptance of the notion and a highly esteemed value, that the students can learn to program by themselves without any formal technology teaching by the teacher. In this way the teachers step away from the usual teacher role and its responsibility for student learning. The idea about stepping aside from the teachers' main task of promoting learning could be understood as a consequence of the discourse that has been studied and the values and the steering strategies identified in the training course (Öhman, 2010). Nouri et al. (2020) have also found such a result. Related to the view that students can learn by themselves, our study points out how collaboration among students is highly valued by the teachers (also found in Nouri et al., 2020). There are other results in technology education research that point out that technology teachers focus on fun and engaging student-led activities instead of the teaching goals (Jones et al., 2013).

In this study, we wanted to investigate what content and values concerning programming emerge as important in the discourse of a teaching practice of a professional development course and the participating technology teachers' teaching practices. The content seems to be obvious and fixed, which has been elicited through the course instructors' guidance strategies and appears important to the teachers. We have seen how the discourse of programming identified in the teacher training course, both dealing with content and values, is transferred by the participating teachers to their classroom practice mainly in technology teaching. Norms and values from the programming discourse within this professional development course become more important than the participating teachers' professional view on how classroom teaching should be conducted. We have seen how the participating teachers' actions are restricted in teaching situations as a consequence of how the course teachers govern the teachers' ways of acting and being (Öhman, 2010). The overall aim seems to be that the students should find programming fun and interesting. This intention has a powerful impact within the teaching practice.

References

Barker, L. J., & Garvin-Doxas, K. (2004). Making visible the behaviors that influence learning environment: A qualitative exploration of computer science classrooms. *Computer Science Education, 14*(2), 119–145.

Berglund, A., & Lister, R. (2010). Introductory programming and the didactic triangle. In *Proceedings of the 12th Australasian conference on computing education-volume 103* (pp. 35–44). Australian Computer Society, Inc.

Black, J., Brodie, J., Curzon, P., Myketiak, C., McOwan, P. W., & Meagher, L. R. (2013). Making computing interesting to school students: Teachers' perspectives. In *Proceedings of the 18th ACM conference on innovation and technology in computer science education* (pp. 255–260). ACM.

Börjesson, M., & Palmblad, E. (2007). *Diskursanalys i praktiken*. Liber.

Cederqvist, A.-M. (2021). *Seeing the parts, understanding the whole: A technology education perspective on teaching and learning in processes of analysing and designing programmed technological solutions* (Diss.). Gothenburg Studies in Educational Sciences.

Digitization Commission. (2014). *Summary of report on ICT in schools – report to the Swedish Government published in March 2014 (SOU 2014:13)*.

Fimyar, O. (1980). Using governmentality as a conceptual tool in education policy research. *Educate~ Special Issue, 2008*, 3–18. http://www.educatejournal.org/index.php/educate/article/view/143/157

Foucault, M. (1982). The subject and power. In D. Faubion (Ed.), *Essential works of Foucault 1954–1984. Volume 3, power* (pp. 326–348). Penguin Books.

Gee, J. P. (1990). *Social linguistics and literacies: Ideology in discourses. Critical perspectives on literacy and education*. Taylor & Francis.

Gee, J. P. (2014). *An introduction to discourse analysis: Theory and method*. Routledge.

Jaipal-Jamani, K., & Angeli, C. (2017). Effect of robotics on elementary preservice teachers' self-efficacy, science learning, and computational thinking. *Journal of Science Education and Technology, 26*(2), 175–192.

Jones, A., Buntting, C., & de Vries, M. (2013). The developing field of technology education: A review to look forward. *International Journal of Technology and Design Education, 23*(2), 191–212.

Mason, S. L., & Rich, P. J. (2020). Development and analysis of the elementary student coding attitudes survey. *Computers & Education, 153*. https://doi.org/10.1016/j.compedu.2020.103898

National Agency for Education. (2017). *Curriculum for the compulsory school, preschool class and the recreation centre*. Stockholm.

National Agency for Education. (2022). *Curriculum for the compulsory school, preschool class and the recreation centre*. Stockholm.

Nouri, J., Zhang, L., Mannila, L., & Norén, E. (2020). Development of computational thinking, digital competence and 21st century skills when learning programming in K-9. *Education Inquiry, 11*(1), 1–17.

Öhman, M. (2007). Styrning i "jumpasalen". In M. Börjesson & E. Palmblad (Eds.), *Diskursanalys i praktiken* (pp. 128–148). Liber.

Öhman, M. (2010). Analysing the direction of socialisation from a power perspective. *Sport, Education and Society, 15*, 393–409. https://doi.org/10.1080/13573322.2010.514735

Östman, L. (1995). *Socialisation och mening. No-utbildning som politiskt och miljömoraliskt problem [Socialization and meaning. Science education as a political and environmental ethical problem]*. Uppsala University, Uppsala Studies in Education 61.

Piedade, J., Dorotea, N., Pedro, A., & Matos, J. F. (2020). On teaching programming fundamentals and computational thinking with educational robotics: A didactic experience with pre-service teachers. *Education Sciences, 10*(9), 214. https://doi.org/10.3390/educsci10090214

Rose, S., Habgood, J., & Jay, T. (2017). An exploration of the role of visual programming tools in the development of young children's computational thinking. *Electronic Journal of E-Learning, 15*(4), 297–309.

Swedish Schools Inspectorate. (2019). *Digitala verktyg i undervisningen - Matematik och teknik årskurs 7–9.*

Sullivan, F. R., & Heffernan, J. (2016). Robotic construction kits as computational manipulatives for learning in the STEM disciplines. *Journal of Research on Technology in Education, 48*(2), 105–128.

Thuné, M., & Eckerdahl, A. (2009). Variation theory applied to students' conceptions of computer programming. *European Journal of Engineering Education, 34*(4), 339–347.

Yadav, A., Berges, M., Sands, P., & Good, J. (2016). Measuring computer science pedagogical content knowledge: An exploratory analysis of teaching vignettes to measure teacher knowledge. In *Proceedings of the 11th workshop in primary and secondary computing education* (pp. 92–95).

CHAPTER 10

Student Teachers' Experiences of Programmed Technological Artefacts

Range of Understanding and Ideas for Development

Anna Perez and Maria Svensson

Abstract

The starting point for developing teaching abilities is the experiences that you have of the object of learning. To prepare student teachers for teaching technology in a digital society, research related to their experiences of phenomena such as programmed technological artefacts is important. As future teachers, they must be able to prepare teaching about programmed technological artefacts as objects of learning. Therefore, more knowledge is needed about student teachers' understanding of them in everyday life.

To describe the different ways in which student teachers experience programmed technological artefacts, such as elevators, tumble dryers, traffic lights, and keyboards, we used a phenomenographic approach. Semi-structured interviews with eight student teachers in primary teacher education were conducted to find the variation in their experiences. This resulted in experiences described as: (1) the physical interface, (2) components as parts of a process, (3) connected, controlled, and regulated components, and (4) components as and in a system.

Aspects essential for understanding programmed technological artefacts are related to three dimensions; *computational thinking, systems thinking* and *the dual nature of technology*. For student teachers to understand these artefacts, aspects from all three dimensions need to be discerned. A problem is that student teachers in this study discerned only a few aspects of the phenomenon that they themselves will be teaching about.

Keywords

technology education – student teachers – programming – phenomenography – dual nature of technology – systems thinking – computational thinking

© ANNA PEREZ AND MARIA SVENSSON, 2024 | DOI:10.1163/9789004687912_010

1 Introduction

In recent years, teacher educators and researchers in technology education in Sweden have been challenged by the government's decision to integrate programming into the subject of technology in compulsory school. Including programming in technology courses for primary teacher education is important, but it also needs to be done with the support of educational research into programming as an aspect of technology education. In both, guiding student teachers as they develop their knowledge about teaching and learning, and as technology education researchers, it is important to build upon previous research about programming as an aspect of technology education. Unfortunately, there is still a lack of research in this area. To develop student teachers' understanding of programming as part of the subject of technology, more knowledge is needed about their own understanding of programming, as well as their understanding of teaching about programming.

In this chapter, empirical results are presented from a study in which eight student teachers in primary teacher education describe their ways of experiencing programmed technological artefacts after finishing a five-week technology course.

2 Background

2.1 *Artefacts in Today's Technological Society*

In recent decades, society and its technology have been exposed to major changes. Since ancient times, people have used tools to simplify their lives, starting with sticks and stones, but soon tools, known as artefacts, were being produced. An artefact is an object designed to have a specific function and a physical structure which interact and are interdependent. This duality has been written about by many researchers in the field of technology education (Dakers et al., 2019; de Vries, 2016; Kroes & Meijers, 2006; Säljö, 2013). In all times and cultures, humans have developed artefacts to meet the challenges and desires they face (Säljö, 2013). Säljö (2013) highlights the importance of artefacts for human learning and development and refers to the physical artefacts that are immersed in our everyday lives. These artefacts are understood as tools and instruments for measuring and weighing, as well as various forms of information technology. Over time, artefacts have become more and more advanced, with the aim of making life easier (Norman, 2014). During the era of our grandparents, there were big changes in their daily lives. Electricity was brought into the kitchen, the washing machine was installed in the basement,

Grandpa got a motor mower, and more. In just a few decades, the everyday artefacts we use have gone from being hand-powered to being mechanically powered by electricity or petrol. Furthermore, the functions of artefacts are today to a larger extent controlled by programmed units.

The evolution of everyday artefacts to simplify our lives has continued, and over the last decade these artefacts have been exposed to major changes. These technological artefacts come in a variety of forms, such as products, tools, production systems, vehicles, clothing, furniture, services, etc. (Norman, 2014). It is also more and more common for notifications to be sent between artefacts, so they can 'talk' to each other. Some examples are connected 'wearables', such as phones, watches, and jewellery, which are useful artefacts of today that are more digital than in the past, and used by almost everyone (Thierer & Castillo, 2015). According to Säljö (2013), the human intellect and its functions have been transferred to physical artefacts that people use to enhance their ability to deal with various problems. As the world around us is changing rapidly, and in order to have an influence and make wise choices in a democratic society, citizens today need to have knowledge and understanding of the technological artefacts that surround us (Hallström, 2009). Teacher education has a responsibility to train teachers to be able to teach their pupils to act as democratic citizens in this technology-intensive society, and to understand and use technological artefacts in their everyday lives.

2.2 *Digitalisation in Technology Education*

The increased digitalisation of society has also brought about changes in school policy documents in Sweden (Skolverket, 2017), as well as in other countries. In 2018, Sweden introduced programming into school curricula, not as a separate subject but as part of existing subjects, such as technology and mathematics. The aim was to develop pupils' digital competence, i.e., the competence necessary for being able to learn and work in the digital society (Ilomäki et al., 2016). But the description of programming in the curriculum was open to various interpretations, which has made teachers uncertain about their mission. Teachers often feel that they have little experience of how to organise teaching regarding programming and lack knowledge about what pupils are expected to learn, or how to address their difficulties (Cederqvist, 2020, 2021; Vinnervik, 2021; Webb et al., 2017). Furthermore, there is a shortage of pedagogical educational research about what teachers should address in relation to subject-specific content, such as programming in technology (Rolandsson, 2015).

Understanding the digital world is often referred to as digital literacy. However, digital literacy is a concept that is not well defined and is still evolving, with meanings that vary according to each user's perspective (Ilomäki et al.,

2016). When users come from different disciplines and backgrounds, a common framework can make it easier to address the same concept, and therefore it can be helpful to define the concepts within digital literacy, such as programming. Ilomäki et al. (2016) suggest a definition of digital literacy with four elements: The first is close to technology studies, consisting of the technical skills needed to use digital technologies. The second element is the ability to use and apply the digital technologies in a meaningful and appropriate way in work and everyday life. The third element is the ability to understand the ethical issues, limitations, and challenges that come with the use of digital technologies. The last element is based on the previous three in order to generate the motivation to participate in digital culture and encompasses social and cultural issues. In short, these elements are: applying digital technology as a tool, understanding computational thinking, and social and cultural issues. Programming is included in digital literacy within the elements of the use of digital technology and applying digital technology as a tool when programming. But computational thinking is also needed for understanding what can be controlled and regulated with programming, and how this is achieved in the world around us.

2.3 *Computational Thinking*

Computational thinking is a widely used concept, especially in computer science, and is often seen in connection with programming skills. Denning and Tedre (2019) claim that the concept of computational thinking consists of the skills and practices needed for designing computations for the purpose of making an artefact do a job for us. Computational thinking also includes the way in which we interpret the world around us as processes of information. Denning and Tedre (2019) describe computational thinking as having two faces. One looks backwards, the back-facing side, to explain what happens during the mechanics of the computer's work, how codes are expressed in a programming language, and how software can be assembled into systems. The other face looks forward, the forward-facing side, towards what needs to be designed, and incorporates a sensibility and understanding of the context in which the user of the software will be working. Both faces of computational thinking help us to understand the programmed technological solutions and artefacts that surround us, and how they are designed to fulfil a purpose and solve our problems (Denning & Tedre, 2019).

2.4 *Programmed Technological Artefacts*

Programmed technological artefacts are artefacts that are controlled and regulated by a code in a programming language. To control and regulate the artefact, signals from a sensor (input) are needed and sent to a control unit that processes the input, according to the code in the programming language, to control, for example, a motor as a result (output). Research shows that both pupils in primary and secondary school as well as student teachers in technology, have difficulty understanding how components function between input and output and the flow of information within the system; the components become 'black boxes' of the artefact (Compton & Compton, 2013; Hallström & Klasander, 2017; Koski & de Vries, 2013; Mioduser et al., 1996; Svensson, 2011). Understanding these artefacts requires much more than perceiving them as an 'entity as such'. Instead, they must be seen as the embodiment of design and purpose and located within the complex socio-cultural milieu of their inception, development, and use. Philosophical attempts to guide critical interpretations have been supported by focusing on the interrelated and dual (structural and functional) nature of technological artefacts (de Vries, 2005; Kroes & Meijers, 2006). Since the various technological solutions that we encounter in everyday life may be difficult to understand from a structural and functional perspective, one way to facilitate understanding of the structural and functional nature of a programmed technological artefact is to see it as a technological system (de Vries, 2005; Hallström, 2022). This, in turn, requires knowledge of systems thinking, and computational thinking.

2.5 *Systems Thinking*

Systems thinking, according to Senge (1990), is a conceptual framework of knowledge, principles, and tools that enable us to observe the 'whole', understand the interrelationships between system elements, and identify patterns of change. In relation to programmed technological artefacts, the use of systems thinking becomes important because it is not possible to think of these artefacts in terms of single objects, but rather of systems (Bertalanffy, 1968). These systems consist of several components that are connected to each other in such a way that the flow of information, energy, and/or matter is controlled, transformed, stored, or transported via different processes directed towards a desired output.

When addressing programmed technological artefacts, a code is used to control the components and the flow of information between them. This code can be seen as a nonphysical component, involving information, stored in a physical component, the control unit. The physical components of the programmed technological artefact, such as buttons, sensors, etc., are connected to the control unit and have the function of directing the flow of information in a way that is determined by what is written in the code. The components are structured so that it is possible for them to interact and move information between themselves to fulfil the function of the programmed technological artefact.

Applying a systems thinking perspective to programmed technological artefacts may facilitate an understanding of their dual nature. That is, to see the structural nature as a set of parts that interact, and the functional nature as an entity within which inputs are transformed into outputs with a function (de Vries, 2005). In a study by Slangen et al. (2011), systems thinking is suggested as an important key knowledge that will facilitate pupils' understanding of structural and functional relationships, in the context of both programming material and everyday life. Their investigation was directed towards robots, which are classic programmed technological artefacts. In Cederqvist's (2022) study of pupils' understanding of programmed technological solutions, she concludes that systems thinking may help pupils to experience the part–whole structure of programmed technological solutions in a powerful way, as well as helping them to compare and see differences and similarities between parts and parts, and between whole and whole, in relation to both everyday technological artefacts and pupils own programmed technological solutions.

There is currently a lack of corresponding studies on student teachers' understanding of programmed technological artefacts. However, it is possible to find studies in nearby areas, such as student teachers' understanding of technological systems (Hallström & Klasander, 2017). In their results, they conclude that student teachers mainly understand the visible parts of a system, such as buttons, power lines, and hydroelectric plants, and that they have difficulties understanding the invisible parts, such as the flow of information, matter, or energy, or control operations. They emphasise that students need to be trained in systems thinking, particularly regarding how components work and connect to each other and how the flows of information are included in technological systems. In relation to this, we sought to investigate student teachers' experiences of programmed technological artefacts after finishing their five-week technology courses during their primary teacher education.

3 Research Design

3.1 *Phenomenographic Approach*

The theoretical approach chosen for this study is phenomenography because the aim is to describe the different ways in which student teachers experience the phenomenon of programmed technological artefacts in everyday life. Phenomenography is based on the assumption that people experience phenomena in somewhat different ways depending on their previous experiences

(Marton & Booth, 1997). The ways in which the student teachers experience the phenomenon under study are categorised as a set of qualitatively different and logically related categories. These categories are not descriptive of how individuals perceive the phenomenon, but, as a set, they suggest potential ways in which individuals could perceive the phenomenon.

Learning in the phenomenographic tradition refers to coming to discern phenomena in new and more powerful ways (Marton & Booth, 1997). Here, this means perceiving objects as programmed technological artefacts in new ways, which are more powerful in terms of understanding and acting within the digitised technological world.

As future teachers, it is of great importance to understand the programmed technological artefacts that exist in our environment, especially because they are required to teach about these in the subject of technology. For us as teacher educators and researchers, the variation in student teachers' understanding indicates aspects that may be important for their learning about the phenomenon.

3.2 *Data Collection*

In this study, semi-structured interviews with student teachers were used to collect data. The questions used in the interview guide were tested in a pilot study with two informants. No major changes were made to the questions after the pilot study and therefore the two informants are included in the data. The students were informed about the purpose of the study in advance and the interviews were conducted by the first author, individually (in Swedish), and recorded (audio and video) using digital meeting software. Each interview lasted approximately 45 minutes. The data material is rich and includes both a conversation about programmed artefacts in our environment and student teachers' reflections about their future teaching. In this chapter, we focus on the part of the interview where their experience of programmed technological artefacts from everyday life is made visible.

In the interviews, eight (six plus two from the pilot study) prospective primary school teachers of technology for grades 4–6 participated as part of a sample of 30 students from two universities. This group of students was selected because knowledge about how they themselves understand content that they will learn to teach can provide an important starting point for the design of teacher education. At the time of the interviews, the participants had recently completed a pedagogical education course in technology that is a mandatory part of their teacher education program. All students on the two courses (at the two universities) were asked to participate, and eight volunteered to take part in the study. The aim of the course is to provide student teachers with an understanding of the nature of the subject of technology and to prepare them for teaching in accordance with the primary school curriculum documents. Even though the empirical data has been collected from only eight informants, the rich volume of data can be analysed in a qualitative way. It provides in-depth insights into how this sample of student teachers experience programmed technological artefacts. However, further variation can arise at the same time, as existing variation can be strengthened or weakened with additional informants.

An interview guide was developed and used to support the interviews. The theme for these interviews was students' experiences of programmed technological artefacts in everyday life and teaching about programming as part of the subject of technology. To start the conversation, four pictures (see Figures 10.1–10.4) were displayed. These represent everyday artefacts: traffic lights, a tumble dryer, an elevator, and a keyboard, which involve processes that are controlled using computer code. Using a prepared context with real objects, or pictures, in an interview situation facilitates understanding of the interview questions and communicating about abstract things (Collier-Reed et al., 2009; Svensson et al., 2012). The artefacts were chosen as being easily recognisable and objects that can be clearly linked to systems. The traffic lights and the elevator are parts of the transport system, the tumble dryer connects to the water system, and the keyboard is one part of a computer's systems and the global information system. All these are examples of how the artefacts are part of external systems. An external system is something that interacts at a level outside the artefact itself. But the chosen artefacts are also systems in themselves, with an internal system, containing components that are connected so that a flow of information is possible within the artefact. Researchers in the field of programming were consulted in the selection of the four artefacts. The keyboard may be the most difficult to link to a technological system because it is wireless and seen as a part of the computer, but even though they are not physically linked, the keyboard is also part of the computer's internal technological system.

In the development of the interview guide, a phenomenographic approach was employed. The interview guide thus consists mainly of open questions, with follow-up questions chosen on the basis of the informants' answers, with the intention of encouraging the informants to develop their thoughts about the phenomenon. Examples of questions asked in the interview are: *When do you encounter programming in your everyday life? Can you give some examples? Could any of these artefacts be linked to programming?*

The kind of follow-up questions used included: *What do you think? What do you think about it? Do you have any examples? Can you make a connection between the concepts?* The student teachers were asked to describe if or how the programmed technological artefacts are programmed and to compare

FIGURE 10.1 Traffic light

FIGURE 10.2 Tumble dryer

FIGURE 10.3 Elevator

FIGURE 10.4 Keyboard

them with each other, but then they were free to choose which artefact they wanted to talk more about.

3.3 *Analysis*

Experiencing the phenomenon of programmed technological artefacts is about distinguishing certain aspects of the phenomenon and variations in these aspects. Through the interviews, we, as researchers, gained access to student teachers' descriptions of programmed technological artefacts and, with the help of their statements, were able to identify different ways of experiencing the phenomenon. These different ways of experiencing together form a whole, which in phenomenography is called an outcome space.

The analysis in the present study started with transcribing the interviews verbatim and anonymising all the data. The transcripts were then analysed by repeated read-throughs, getting to know them as a collective. During these read-throughs, the sections in which awareness of the phenomenon appears were highlighted; these relevant parts were seen as different meaning units. In the next step, similarities and differences were searched for in these meaning units in order to identify meaningful variation. These tentative subsets were carefully described in terms of their differences, and subsequently in terms of their defining qualities. Descriptions of what constituted differences and similarities between the categories were then formulated. These tentative categories were continuously challenged by the researchers to ensure their uniqueness, and during this process the categorisation was tested on fellow researchers (Erlingsson & Brysiewicz, 2017; Ryan & Bernard, 2003). The first draft of categories was refined into a final form by systematic categorisation of the complete empirical material, creating an outcome space.

The categories address different aspects of the phenomenon and were hierarchically ordered in such a way that the lowest ordered category includes the simplest and smallest number of aspects in which the phenomenon can be experienced (Wood, 2000). The difference between the categories is specified by what needs to be discerned in order to move on to the next category, and this is a critical aspect (Pang & Ki, 2016). Different ways of experiencing phenomena can be described as more or less complex, and thus considered more or less powerful than others in relation to certain criteria. The most complex way of experiencing a phenomenon usually means that several critical aspects of that phenomenon are discerned at the same time. These critical aspects together build up dimensions of variation. By paying attention to and focusing on the dimensions of variation that exist within a phenomenon, necessary qualities in the learning conditions can be picked out (Marton & Booth, 1997; Svensson, 2011).

The aspects that vary within the categories are described as dimensions of variation. The whole of the analysis process was carried out in a dialogue between the two authors.

4 Results

The analysis resulted in a set of four categories, each qualitatively distinct from the others. A list of these four categories is presented below, followed by a description of each category with extracts from the interviews. Since no single extract fully captures one category, the extracts have been chosen to illuminate the main thrust of the categories, especially in terms of the meaning and structure of the understanding of programmed technological artefacts that the different categories signify.

- Category 1: the physical interface
- Category 2: components as parts of a process
- Category 3: connected, controlled, and regulated components
- Category 4: components *as* and *in* a system

4.1 *Category 1: The Physical Interface*

In this first category, the student teachers experience the artefacts as an object with which they interact to achieve a certain output. They predominantly have a user perspective on the artefacts and focus on the interaction between themselves and the user interface: buttons, dials, and representations that indicate how individuals should use the artefact. The structure of the artefacts is in the foreground. Thus, the students' awareness is directed towards two aspects: the output and the physical interface of the artefact. The output of the artefact is seen in this category as an activity; something happens, moves, is started, etc. In the following excerpts, we see examples of how student teachers understand the artefacts, starting with Alice.

Interviewer: But would any or all these artefacts [keyboard, tumble dryer, traffic lights, elevator] be linked to programming? And if so, how do you see the connection?

Alice: […] you kind of control and that by turning the dial on the tumble dryer. I can set different programs and it's programmed; I can push a button for the elevator. […] Yes, it's programmed to follow what the button says, if I want to go up or down. But I think the keyboard is [programmed] too. I press a key there [on the keyboard] and then it becomes a symbol in the computer. And then I think the traffic light is programmed, yes, to follow a certain pattern, to prevent traffic accidents.

There are examples here of inconsistency in the use of the terms program and programming, as we can see when Alice talks about the tumbler dryer. It is not clear if, when referring to the program, she understands it as a computer program (a program written in a programming language) or as the drying program in the tumble dryer manual.

Interviewer: The tumble dryer, I want to know a little more about it. What is the purpose of the programming in the tumble dryer?
Alice: That I should be able to turn a dial and set different programs and that it should follow the program I've turned the dial to. So, it can vary its function as I turn the dial. So, there are different programs, that's what I think.

In the next excerpt, Greta describes how the elevator is connected to programming through the buttons as physical interface, and the output as an activity.

Interviewer: Okay, yes, the elevator then, is it programmed?
Greta: I know something about how elevators work, you press a button, and it goes up or it goes down, but it's controlled by something up there.
Interviewer: Do you think it's programmed then?
Greta: It almost has to be.
Interviewer: Why then, or why not?
Greta: Because otherwise it wouldn't have gone up or down, but if I press the buttons, it's activated.

In this category, student teachers do not mention the processes between the components and the output, such as the control and regulation of components or the flow of information, energy, and matter in and between components. Their awareness is not directed towards the processes between the physical interface and the output; the process is a black box. Whether the students do this consciously or unconsciously is not entirely clear, even though they were asked during the semi-structured interviews to develop their thoughts further.

4.2 *Category 2: Components as Parts of a Process*

In this category, unlike the first, the student teachers describe components that lie behind the physical interface. Thus, we can see that component not visible to the user, such as sensors, are part of their awareness. The structure of the artefact is still in the foreground. We interpreted this as an awareness of the processes between input and output, looking behind the interface, opening that black box. They do not explicitly connect the processes to programming, but their understanding of the processes indicates more awareness of technological parts of the artefacts, such as components. In this category as well, the student teachers describe the technological artefacts from a user's perspective, where they are pressing buttons and so on, and the output is a desired activity.

However, there are indications of experiencing logical output connected to consecutive activities, as we can see in Clara's excerpt below, where she talks about the elevator.

Interviewer: And what about this elevator? How does it work in its programming?

Clara: Yes, I guess it also has buttons and they're set to operate, close the door, or open the door. As soon as you press it [the button], it's pre-set that something is going to happen, you go in and the door is going to close, so it notices that I'm standing there and then it will open again. So, then it has a sensor or something that says that now we're going to open the door again.

In the following excerpt, Elsa also indicates that she is aware of processes that lie behind the user interface.

Interviewer: And what about this elevator? How does it work in its programming?

Elsa: There's a lot more going on behind, than you're pressing the elevator button, like, you get in the elevator, press the floor you're going to and then it pulls you up and knows when to open the door.

When Daniella describes the keyboard, her description indicates knowledge about the user interface and the output but, as we can see in her description of the traffic lights, she is also aware of more aspects that are related to the processes between input and output.

Interviewer: What is the purpose of the programming in these artefacts?

Daniella: The keyboard is that you press keys, and you get that command on the computer. And traffic lights should change colour either after a time interval or [with] some kind of sensor. I've understood that you can have different sensors in the ground, but that it should mark when you have to stop, you're allowed to go and it's programmed with colours so that you can see it.

In this category, the student teachers open the black box by displaying an understanding of the presence of components, such as sensors, in the process between the interface and the output.

4.3 *Category 3: Connected, Controlled, and Regulated Components*

In the third category, the student teachers' awareness is directed towards not only the user interface and the output, but also the various components in the process that need to be connected, controlled, and regulated to achieve a desired output. We interpret this as the students describing a need for feedback loops in the programming and, in relation to the previous categories, the connection between the function and the structure of the artefact is more developed. There are functions in the process that need to be foreseen, which can be achieved by someone programming the artefact. This indicates that the user perspective is expanded with a development perspective, understanding the influence of the programmer on the process.

In the excerpt below Elsa talks about the elevator and the tumble dryer, she integrates the fact that components are connected and there is a flow of information between the components.

Elsa: Yes, the elevator needs to talk to its own system more and sort of fix the buttons, the same way as the tumble dryer needs to talk to its parts because it has a certain order it has to work in and detect when it has to do things.

In the following excerpt, Clara describes the importance of regulation in programming, which we interpret as an understanding of feedback loops.

Interviewer: How is it [the tumble dryer] connected to programming then?

Clara: Yes, but if it's pre-set, that it [the laundry] could be this heavy, we will drive [rotate] this long. Or I don't know if a tumble dryer has this, but the washing machine usually has [...] So if it's more than 5kg we should drive a little longer or if it's this hot, then we should do... It's like this. If a requirement is to be met or something, then the programming changes, something like that.

In the following excerpt, Ben explains that there are connections between components that have a control function and programming. This excerpt indicates a more developed understanding than the previous two, but still focuses on the control, connection, and regulation of components.

Ben: [...] but shall we take it picture by picture, shall we? Maybe start with the tumble dryer? So, everything that controls is really programmed [...] All hardware parts in the products must be programmed so that they behave in the right way. And that it ends the program and that it keeps the right temperature and all that.

> It's the programming in it, and it actually applies to all of them [the four artefacts in the photos]. But in different ways.

In this category, the black box is opened further by combining the components in a specific way and seeing the components inside the artefacts as part of an internal system. We interpret this as a way of understanding the back-facing side of programmed technological artefacts, how the mechanics of the computer work.

4.4 *Category 4: Components as and in a System*

In this fourth category, a more developed understanding of codes, languages, and combining components into systems is demonstrated. Here, the student teachers experience more details of the processes between user interface and output, such as how the sensors, as components, are used in the internal system to direct the flow of information, and the use of feedback loops to change the output of a programmed artefact. The black box is opened, and programming is understood as a process involving physical components, sensors etc., and a code, a nonphysical component, that are connected to achieve the transformation of information.

In this first excerpt, Ben describes some aspects connected to an understanding of the back-facing side, with the internal system of the programmed technological artefacts. When it comes to the traffic lights, there are also indications of an understanding of the forward-facing side, with the external systems included.

Interviewer: Please describe a bit more how you see differences between these artefacts [keyboard, tumble dryer, traffic lights, elevator].

Ben: [...] the tumble dryers, you press start and then it's set to yes, now I'm going to run this program for an hour or if the laundry gets done early. There's like an 'or' in it if you talk about programming languages. On the other hand, when it comes to the traffic lights, if there are no cars then nothing happens, but it must adapt all the time to the flow of cars. [...] You have an input to the tumble dryer that you, as an individual, provide and then it [the tumble dryer] behaves according to its programming. It takes an hour or whatever for the program to finish. While the input to the traffic lights comes in all the time and changes all the time and is then dependent on the cars.

We also interpret the above excerpt as showing that Ben has an awareness of the differences between a program in a programming language and a program as a manual for the tumbler dryer. Hence, he expresses a more developed understanding of programming in the process than in the earlier categories.

In the next excerpt, Ben describes the interplay between the artefact and the environment and external systems. Hence, we can see that he is aware of more aspects of the forward-facing side.

Ben: [...] a car comes and then five minutes later another car comes and then during rush hour when everyone is going home from work, there are a lot of cars. There will be so much variety with the cars, it's not the same as the programs in an elevator or a tumble dryer.

In this category, components are understood as both physical and non-physical, where the programming language is a non-physical component that has a similar impact as a physical component.

We also interpret the student teachers' understandings of the components in a programmed technological artefact as being connected and interacting as a system, within an internal system. At the same time, the programmed technological artefact, with all its components, is connected and interacts with external systems.

5 Summary of Results

Within the data we collected about student teachers' experiences of programmed technological artefacts, there is a prevalence of excerpts that are categorised as categories 1 and 2, showing a less developed understanding of programmed technological artefacts. The more developed understanding is presented in categories 3 and 4 and is found less frequently in the excerpts in the meaning unit. In conclusion, in the categories there is a variation in student teachers' experiences described as dimensions of variation (see Table 10.1). The three dimensions that varies are *computational thinking*, *systems thinking* and *the dual nature of technology*.

A critical difference between categories 1 and 2 is the identification of the process between input and output. Between categories 2 and 3, a critical difference is the understanding of components' functions in the process. Between categories 3 and 4, there are at least two critical aspects. One is the understanding of what influence the programming language has on the process. The other is understanding that the artefact itself is a system and at the same time is a sub-system in a system, the traffic light as a sub-system in a transportation system.

TABLE 10.1 Categories and their relation to the three dimensions of variation

Categories/ dimensions of variation	Computational thinking	Systems thinking	The dual nature of technology
Cat. 1: the physical interface		User interface and the output	Structure of the artefacts interface
Cat. 2: components as parts of a process	Mentioning single sensors -> indicates back-facing side	User interface, opening the black box – components and processes -> output	Structure including the artefacts visible (interface) and invisible components
Cat. 3: connected, controlled, and regulated components	Sensors and feedback loops as parts of the computer's mechanics - > back-facing side	Describing the connection, control, and regulation of components in the black box -> output	Structure and function as two parts of the artefacts
Cat. 4: components as and in a system	Computer mechanics and code -> back-facing side User context -> forward-facing side	Connecting the components and describing a flow of information or matter in internal and external systems	Structure and functions intertwined

Two things are noteworthy in this study. Firstly, the critical aspects identified are strongly linked to the knowledge areas of computational thinking, systems thinking, and the dual nature of technology. Secondly, the student teachers in this study, and probably many other student teachers in technology right now, discern only a few aspects of a phenomenon that they themselves will be teaching about.

The identified critical aspects are important for teacher educators to understand when enabling student teachers' learning about programming as part of the subject of technology. When discerning more aspects of a phenomenon, a more developed understanding of that phenomenon is possible (Marton & Booth, 1997). For student teachers, to be able to teach about a phenomenon, a more developed understanding of that phenomenon is necessary.

6 Discussion

In our study, the student teachers' experiences of programmed technological artefacts vary. Ranging from students who describe the artefacts as programmed, but do not really understand what this entails, to students who express an understanding of the various processes when programming controls functions of an artefact. This is in line with Hallström and Klasander's (2017) investigation of student teachers' understanding of technological systems, where the structure, such as buttons and other visible devices, are part of the students' awareness, while the more invisible devices, such as a circuit board where there is a transformation of information, are more difficult to understand. We interpret this as student teachers, who have finished their technology courses (5 weeks), discern only a limited number of aspects of programmed technological artefacts. This limited number of aspects indicates that their understanding of the phenomenon is not well developed and to a large extent is similar to compulsory-school pupils' understanding of programmed technological artefacts (Cederqvist, 2020). It is thus important to remember that the student teachers in this study were not taught in or about programming during their own compulsory schooling. Therefore, the level of their understanding might not be more developed than that of current pupils in compulsory school, who today are taught programming. An implication of this is that, during the student teacher's education, specifically during courses in technology, they have to develop their own understanding of programmed technological artefacts. At the same time, they have to develop their knowledge of teaching about programmed technological artefacts. Teacher educators' awareness about the need of addressing both content knowledge and teaching knowledge is crucial for developing student teachers' understanding of programmed technological artefacts. The problem is that to fulfil the policymakers' desire to add programming to already-established subject areas, such as technology, there is a need for increased research and pedagogical education knowledge. It may also be important to discuss how extensive, in terms of time and content, the technology course in primary teacher education should be in order to develop relevant knowledge (Vinnervik, 2021). Our suggestion for technology education is to emphasise the problems that can be solved by using programming as a tool in processes and not on understanding and writing codes in programming languages.

The result in this study contributes with knowledge about three dimensions of variation, computational thinking, systems thinking, and the dual nature of technology, needed for understanding programmed technological artefacts. What we have found is that for student teachers to understand artefacts that are programmed, aspects from all three dimensions need to be discerned. When someone discerns more aspects, it enables new and more nuanced ways of understanding a phenomenon (Marton & Booth, 1997). Therefore, it is important for student teachers, during their technology courses, to learn about all three dimensions and how aspects of these are connected and are understood as a whole. This would help them develop their understanding of programmed technological artefacts.

Another interesting result was how aspects of, and relations between, the three dimensions—computational thinking, systems thinking and the dual nature of technology—become visible to different extent in the hierarchically ordered categories (see Table 10.1). Starting with computational thinking, the back-facing side is a term that is used to describe the mechanics of computers' work, programming languages, and how they are understood as a system (Denning & Tedre, 2019). This is similar to what is explained in systems thinking as the internal system, which can be described as the flow of information, energy, and matter between components in a delimited system, such as an artefact (Svensson et al., 2017). On the other hand, the forward-facing side is described as an awareness of the context in which the programmed technological artefacts are going to be used. This is connected to systems thinking and the perspective of external systems, described as the interactions between the artefact and the surrounding environment. Concerning the dual nature of technology, the structural and functional perspectives are identified in the categories, together with the back-facing side and the internal system of a programmed technological artefact. This opens opportunities for learning about all three dimensions, computational thinking, systems thinking and the dual nature of technology when focusing on programmed technological artefacts as a learning object in the technology subject. However, to make learning in technology happen, we interpret the following critical aspects: the process between input and output, the function of components in the process, the function of the programming language in the process, and that the artefact itself is a system and at the same time part of larger systems. This needs to be discerned and varied, by primary teachers in technology education as well as by teacher educators in technology courses.

References

Bertalanffy, L. V. (1968). *General system theory: Foundations, development, applications.* G. Braziller.

Cederqvist, A.-M. (2020). Pupils' ways of understanding programmed technological solutions when analysing structure and function. *Education and Information Technologies, 25*(2), 1039–1065. https://doi.org/10.1007/s10639-019-10006-4

Cederqvist, A.-M. (2021). *Seeing the parts, understanding the whole-A technology education perspective on teaching and learning in processes of analysing and designing programmed technological solutions*. University of Gothenburg.

Cederqvist, A.-M. (2022). An exploratory study of technological knowledge when pupils are designing a programmed technological solution using BBC Micro:bit. *International Journal of Technology and Design Education, 32*, 355–381.

Collier-Reed, B. I., Ingerman, Å., & Berglund, A. (2009). Reflections on trustworthiness in phenomenographic research: Recognising purpose, context and change in the process of research. *Education as change, 13*(2), 339–355.

Compton, V. J., & Compton, A. D. (2013). Teaching the nature of technology: Determining and supporting student learning of the philosophy of technology. *International Journal of Technology and Design Education, 23*(2), 229–256.

Dakers, J. R., Hallström, J., & de Vries, M. J. (2019). *Reflections on technology for educational practitioners: Philosophers of technology inspiring technology education*. Brill.

de Vries, M. J. (2005). The nature of technological knowledge: Philosophical reflections and educational consequences. *International Journal of Technology and Design Education, 15*(2), 149–154.

de Vries, M. J. (2016). *Teaching about technology: An introduction to the philosophy of technology for non-philosophers*. Springer.

Denning, P. J., & Tedre, M. (2019). *Computational thinking*. MIT Press.

Erlingsson, C., & Brysiewicz, P. (2017). A hands-on guide to doing content analysis. *African Journal of Emergency Medicine, 7*(3), 93–99.

Hallström, J. (2009). Teknik som allmänbildning. Lärdomar av ingenjörers syn på teknisk kunskap under 1920-talet. In P. Gyberg & J. Hallström (Eds.), *Världens gång–teknikens utveckling. Om samspelet mellan teknik, människa och samhälle*. Studentlitteratur.

Hallström, J. (2022). The dual nature(s) of technology: Towards a unified conception of digital and analogue technology. *PATT, 39*, 20–27.

Hallström, J., & Klasander, C. (2017). Visible parts, invisible whole: Swedish technology student teachers' conceptions about technological systems. *International Journal of Technology and Design Education, 27*(3), 387–405.

Ilomäki, L., Paavola, S., Lakkala, M., & Kantosalo, A. (2016). Digital competence–an emergent boundary concept for policy and educational research. *Education and Information Technologies, 21*(3), 655–679.

Koski, M.-I., & de Vries, M. (2013). An exploratory study on how primary pupils approach systems. *International Journal of Technology and Design Education, 23*(4), 835–848.

Kroes, P., & Meijers, A. (2006). The dual nature of technical artefacts. *Studies in History and Philosophy of Science, 37*(1), 1–4.

Marton, F., & Booth, S. A. (1997). *Learning and awareness*. Routledge.

Mioduser, D., Venezky, R. L., & Gong, B. (1996). Students' perceptions and designs of simple control systems. *Computers in Human Behavior, 12*(3), 363–388.

Norman, D. (2014). *Things that make us smart: Defending human attributes in the age of the machine.* Diversion Books.

Pang, M. F., & Ki, W. W. (2016). Revisiting the idea of "critical aspects". *Scandinavian Journal of Educational Research, 60*(3), 323–336.

Rolandsson, L. (2015). *Programmed or Not: A study about programming teachers' beliefs and intentions in relation to curriculum.* KTH Royal Institute of Technology.

Ryan, G. W., & Bernard, H. R. (2003). Techniques to identify themes. *Field methods, 15*(1), 85–109.

Senge, P. M. (1990). *The art and practice of the learning organization* (Vol. 1). Doubleday.

Skolverket. (2017). *Läroplan för grundskolan, förskoleklassen och fritidshemmet 2011: reviderad 2017.* Skolverket.

Slangen, L., van Keulen, H., & Gravemeijer, K. (2011). What pupils can learn from working with robotic direct manipulation environments. *International Journal of Technology and Design Education, 21*(4), 449–469.

Svensson, M. (2011). *Att urskilja tekniska system: didaktiska dimensioner i grundskolan.* Linköping University Electronic Press.

Svensson, M., Zetterqvist, A., & Ingerman, Å. (2012). On young people's experience of systems in technology. *Design and Technology Education: An International Journal, 17*(1), 66–77.

Säljö, R. (2013). *Lärande i praktiken. Ett sociokulturellt perspektiv.* Prisma.

Thierer, A., & Castillo, A. (2015). Projecting the growth and economic impact of the internet of things. *George Mason University, Mercatus Center.* https://www.mercatus.org/system/files/IoT-EP-v3.pdf

Vinnervik, P. (2021). *När lärare formar ett nytt ämnesinnehåll: Intentioner, förutsättningar och utmaningar med att införa programmering i skolan* (Diss.). Umeå University.

Webb, M., Davis, N., Bell, T., Katz, Y. J., Reynolds, N., Chambers, D. P., & Sysło, M. M. (2017). Computer science in K-12 school curricula of the 21st century: Why, what and when? *Education and Information Technologies, 22*(2), 445–468.

Wood, K. (2000). The experience of learning to teach: Changing student teachers' ways of understanding teaching. *Journal of Curriculum Studies, 32*(1), 75–93.

CHAPTER 11

Swedish Technology Teachers' Understandings of Computer Programming as Modelling

Helen Brink

Abstract

Knowledge *about* computer programming and knowledge *in* computer programming are important features in technology education in Swedish compulsory schools, and therefore mentioned in the technology syllabus. Knowledge about computer programming is necessary to understand the contemporary world we live in, where much of the technology is connected and programmed. To develop knowledge in computer programming, pupils can be offered teaching which allows the pupils to program. In this chapter, results from a previous study (Brink et al., 2021) will be discussed, where interviews about models and modelling were conducted with lower secondary school technology teachers. The results reveal that teachers can understand computer programming as modelling in technology education and experience similarities between the two concepts, modelling and computer programming. Implications for teaching practice can be to teach more explicitly when modelling in different modes of representation and when different modes are preferable. This study contributes to development of the body of concepts relevant to technology education.

Keywords

technology education – teachers' conceptions – experience – modelling – computer programming

1 Introduction

In Swedish compulsory school, pupils are expected to develop digital competence, and in the technology subject pupils should, among other things, be given opportunities to work with product development, modelling, computer

© HELEN BRINK, 2024 | DOI:10.1163/9789004687912_011

programming, and digital tools (Skolverket, 2019). Teaching with models and modelling can be one area where digital competence is developed and digital tools are used. *Digital models* are mentioned in the Swedish technology subject syllabus, for year 7–9. However, digital models are not defined or explained in the syllabus or in the complementary material (Skolverket, 2017, 2019). Therefore, there is reason to believe that teaching with digital models can differ between schools and teachers and that the concept digital models can have different meanings for different teachers. A digital model can for instance be a model created with a digital design tool like Computer Aided Design (CAD). A recent study (Brink et al., 2021) shows that technology teachers can experience computer programming as modelling. It can be problematic if the two concepts computer programming and modelling are mixed or used synonymously. Treating the two concepts as synonyms can hinder comparisons between different technology educations (nationally and internationally) and make technology teachers' professional development more difficult, or the development of the technology subject more difficult. Failure to distinguish between the two can also make it impossible for the pupils to grasp the similarities and differences between models and computer programs.

Research in technology education for lower secondary schools is scarce and there are a great many areas where we know little about teachers' experiences of teaching the technology subject. However, Brink et al. (2021) conclude that teaching with digital models involves different contents and purposes and that pupils are given different prerequisites to learn. Examples of contents possible to learn are mathematical concepts, manufacturing methods, design processes, digital software handling, or general technology knowledge for everyday use. Modelling with digital tools such as CAD and computer programming are relatively new elements in technology education, and research shows that teachers need to develop their own knowledge about how to teach with new technologies and use them in a pedagogical manner (Koehler & Mishra, 2005; Mishra & Koehler, 2006). In another study from Sweden, teachers' pedagogical strategies for teaching computer programming were investigated, and one conclusion was that teachers need more content knowledge and pedagogical knowledge in the area (Bjursten et al., 2022). More knowledge about teaching computer programming and teaching modelling with digital tools can help teachers' professional development. The aim of this capter is to describe technology teachers' understanding of computer programming as modelling. The research question is:

What is the nature of technology teachers' understanding of computer programming as modelling?

2 Swedish Technology Education

In this part, technology education in Swedish lower secondary schools will be described. In Sweden, technology education should provide pupils with opportunities to develop technical knowledge and skills and knowledge *in* and *about* technology (Skolverket, 2022). That means that pupils should acquire knowledge to use and develop technology, acquire an understanding of different intentions with technology, and understand how technology affects humans, the environment, and society. Teaching *in* technology can involve designing, technical development, and technology-oriented methods (Svensson, 2011). Teaching *about* technology, on the other hand, can involve assessing and examining technology and working with moral and ethical issues in relation to humans, the environment, and society (Svensson, 2011). Svenningsson (2019) suggests that knowledge *in* and *about* technology is important for all pupils, not only for those pupils planning a future within the technology domain, since all pupils are part of a democratic society immersed in technology.

Further, technology education is naturally holistic (Gibson, 2008) and authentic education is important to achieve holistic learning (Gibson, 2009). Models and modelling are also important in technology education (Gilbert et al., 2000), and Hallström and Schönborn (2019) show that modelling can serve as a meaningful route to authentic education. Teaching about models and modelling can also help pupils transfer knowledge and skills between different contexts (Hallström & Schönborn, 2019). An ability to consider new insights in relation to different contexts can be useful to achieve holistic and authentic learning.

In addition, teachers' attitudes to education are also important, and it is known that a teacher's attitude influences the presentation of the subject content (Dakers, 2018). Thus, it is important to gain more knowledge about technology teachers' experiences of teaching different areas in the technology subject to support pedagogical choices and to develop a common body of concepts related to the subject. With a common body of concepts, collegial discussions can increase and the subject and profession can develop.

The core content for the technology subject in Sweden, defined in the syllabus Lgr22 (Skolverket, 2022), is divided into three parts; (1) *Technology, humans, society and the environment*, (2) *Technological solutions*, (3) *Working methods for developing technological solutions* and for year 7–9, the third part includes the following:

- Different phases of technical development: identifying needs, investigating, proposing solutions, designing, and testing. How different phases in the work process are interlinked.

- Pupils' own constructions in which they apply control and regulations, including with the aid of programming.
- How digital tools can provide support in technical development work, for example when producing drawings and simulations.
- Documentation in the form of manual and digital sketches and drawings with explanatory words and terms, symbols and measurements, as well as documentation with physical and digital models. Simple, written reports describing and summarizing work carried out on construction and technical development (Skolverket, 2022).

The study preceding the present one (Brink et al., 2021) was carried out under the former curriculum Lgr11 (Skolverket, 2019). However, the changes regarding the part *Working methods for developing technological solutions* are minor and the latter and former syllabuses are comparable in this matter. There is reason to believe that the results from the 2021 study by Brink et al. are continuously relevant under Lgr22. Teachers need to interpret the steering documents and transform the required core content into their own teaching. In this transformation, teachers have a great deal of autonomy with little guidance from the steering documents (Bjurulf, 2008). And more, the Swedish technology subject lacks a commonly agreed body and identity (Fahrman, 2021; Sjöberg, 2013) which can affect teachers' pedagogical choices and make the transformation of core contents more difficult.

3 Modelling in Technology Education

Models in technology education have been used for a long time for different purposes. They are for instance used as tools for pupils to examine, predict and optimize the artefacts or systems under development and for communicating and assessing the outcomes (de Vries, 2013). In technology education, models can also support pupils' product development through allowing manipulation and mental exploration. Nia and de Vries (2016) argue that models have a dual nature, with both intrinsic and intentional aspects. Intrinsic aspects can be physical aspects like the form or material of the model, but for models created with digital tools, physical aspects can be problematic. So, to avoid misunderstandings, Nia and de Vries (2016) suggest discussing intrinsic and intentional aspects to include models created with digital tools. Further, since models in technology education can be a tool for reflecting and examining ideas, the model is important for pupils to develop technological knowledge. One example of this potential for knowledge development is when a pupil is optimizing a computer program. Citrohn et al. (2023) found in their study that teachers' pedagogical choices in a teaching activity affect what possibilities pupils have to learn about models and modelling. With a more open instruction to the activity, more complex learning could occur and pupils put more effort in planning for modelling and for demonstrating for instance model functions. In the study, examples of models are pedometers and greenhouses and the pupils made computer programmed models of pedometers and greenhouses, with Micro:bit. The authors conclude that the models and the model functions help develop pupils' knowledge when testing different solutions to problems (Citrohn et al., 2023). Put in other words, the pupils in the study were reflecting ideas and examing ideas (Nia & de Vries, 2016).

3.1 *Models as Representations*

Models in technology settings are often described as a representation of something (Boulter & Buckley, 2000; Nia & de Vries, 2016), where idealizations (simplifications) and abstractions can be made depending on the purpose of the model (de Vries, 2018). Idealizations of the model are made to fit the given criteria and can be seen as a bridge between for instance collected data and mathematical representations like an equation or formula. One example is when a wobbly, graphical curve of measured values is smoothed to fit ideal settings. Idealization can be a prerequisite for mathematical calculations. Abstractions of a model are made when some aspects of reality, not relevant for the present situation, are left out (de Vries, 2018). Such aspects can be friction or air resistance, not included in the model of a free fall motion. Another example of abstraction is when a model of a car is made of simple material like paper, wood and cardboard, but remains a representation of a car. Idealizations and abstractions distinguish a model from the original. When something is represented with a model, inherent simplifications need attention in the teaching situation (Nia & de Vries, 2016), which requires teachers' awareness of these aspects.

When discussing models and describing models as representations, it is important to distinguish between internal and external representations (Gilbert, 2010). An internal representation is the idea or thought still not expressed and only accessible to the person herself. The idea is still in the person's mind. The external representation, on the other hand, is expressed and can therefore be shared and communicated with others. In school and technology education, pupils use both internal and external representations when finding technological solutions to a problem or developing an artefact.

In different contexts, different modes of external representation can be suitable, depending on what features are highlighted and what approximations are made in the model. The external representations can be described in terms of modes as concrete or material, verbal, visual, mathematical or symbolic, and as gestural representations (Gilbert et al, 2000; Gilbert, 2005). Concrete or material representations can be three-dimensional physical and material objects, like a car made of paper and cardboard. Verbal representations can be spoken or written texts where the product, technical solution, code or phenomenon is explained. Metaphors or analogies are common. Examples of visual representations are pictures, sketches, diagrams, animations or simulations. Mathematical or symbolic representations can for instance be formulas, graphs or chemical equations. Gestural representations are made with the body as a tool, for example when motions of the hands are used to explain a natural phenomenon or a technical solution.

However, models of something are not just static representations like diagrams, written texts or concrete models. They can also be representations of events, different situations or processes that need to be explained from a time perspective, that is, dynamic representations (Boulter & Buckley, 2000). Simulations or animations are examples thereof. Abstraction and idealization, as discussed earlier, can also be used when modelling objects and events (de Vries, 2013). In the process of modelling objects and events, the pupils can explore the intentions and functions and a simplified version of reality can be created (de Vries, 2013). Since a computational program can represent an event or process, it can be seen as a dynamic representation.

3.2 *Modelling for Learning Computational Thinking*

According to Elmer and Davies (2000), modelling activities in technology education are more than just activities to learn technological skills, they are also about learning analytical thinking. Other concepts that can be compared to analytical thinking are computational thinking and algorithmic thinking. There seems to be an agreement that computational thinking consists of capabilities like problem-solving, iterative thinking, algorithmic thinking, creativity, reasoning, decomposition and debugging (Sen et al,. 2021). Algorithmic thinking involves problem-solving through different steps in a particular order to attain the intended outcome. In both modelling and computer programming, problems are to be solved and different steps need to be taken into account. The modelling process can be described through different steps such as identifying a problem, undertaking research, developing solutions, producing solutions and evaluating solutions (Middleton, 2005). However, it is not always that all steps are used, or that the steps will follow in order. It is suggested that the iterative process of modelling should be described in terms of different aspects instead of described as a series of steps (Williams, 2000), in order to make the iterations more prominent. Citrohn et al. (2022) describe that teachers can use models in their teaching to connect to and teach about design processes. A conclusion can be made based upon this argumentation, that teaching with modelling can also include teaching about algorithmic thinking and computational thinking. Since there are similarities between the process of modelling och the process of computer programming, it is important to support teachers' pedagogical choices and help develop the body of concepts in this area of teaching.

3.3 *Modelling for Learning to Assess and Value*

Technology education in Sweden aims at developing students' technological awareness and understanding of the digitalized, programmed and connected world, so they can orient themselves and act in a technologically intensive and digitalized world where more and more information and communication are digital (Skolverket, 2022). Technology education should thus enable pupils to become responsible consumers and producers of technology. Teaching where pupils can value and assess models or computer programs can help them raise questions about the use of and intentions with technology (Ankiewicz, 2019). Models and computer programs need to be investigated from both functional and intentional perspectives and the pupils need to develop a critical mind for assessing models and computer programs (as representations of something real) to understand how the models and computer programs affect both them and society (cf. Elmer & Davies, 2000; Norström, 2013; Seery, 2017).

4 Computer Programming in Technology Education

When young children spontaneously create programming code, playing with robots in a school context, they create code in three different representations, visually through pictures, verbally through their speech, and symbolically through (drawn) arrows (Anderhag et al., 2021). As a result of having access to these alternatives, children choose a mode of representation that they themselves find suitable. Similarly, different models can be represented in different modes depending on the situation as discussed previously.

Computer programming is one way of demonstrating an event or process digitally. In technology education, objectives can be to examine natural and constructed events or processes through tools that are programmed (Cuperman & Verner, 2013). Computer programming can thereby be a tool for reflections and a way to stimulate pupils to investigate artifacts and understand operational principles. Cuperman and Verner (2013) give two examples thereof, computer programming of a robot representing a fly-trapping flower, and computer programming of a pace-maker for helping maintain the pace of a heart. In these examples, the authors highlight that digital manipulation is an integral part of the robots (the models of the flower and of the heart). The robots (physical models) are supposed to imitate the functions and behavior of the natural world. Digital manipulations like computer programming include development of thinking skills and can be seen as a tool for thinking (Elmer & Davies, 2000; Resnick et al., 1998). However, computer programming and modelling in robotics education have more outcomes; technological concepts are given a mental representation and the creation of robotic models gives different concepts a physical representation (Resnick et al., 1998).

As discussed earlier, both mental (internal) and expressed (external) representations are used in technology education. A tool that can be used to give pupils access to the transition from a mental model to an expressed model is to ask them in connection with an unplugged activity to describe different steps that a computer program is supposed to do with ordinary words and sentences, orally or in writing, before starting the computer programming activity (Bjursten et al., 2022; Carmichael, 2000). This enables discovery of the logical sequences within the program, that is, algorithmic thinking. The written descriptions of what a computer program is supposed to do can be understood as a visual and textual representation of the computer program. Further, schools have for many years now used simple programming tools with simple interfaces (like block programming) that enable pupils to try and learn about computer programming (Carmichael, 2000). Carmichael (2000) describes this as computer-based models.

5 Interviews with Technology Teachers, the Sample

During 2018 and 2019, 12 semi-structured interviews were conducted with in-service technology teachers, teaching in Swedish lower secondary schools grade 7–9 (pupils age 13–15), to gain more knowledge and describe the teachers' experiences of teaching digital models using digital tools (Brink et al., 2021). The interviewed teachers had been working as technology teachers between 1 and 19 years. Seven of them were female and five were male. The teachers also taught other subjects in addition to the technology subject, such as mathematics, science, crafts, music, or art. Some of the teachers had colleagues to collaborate with, others working in smaller schools did not.

The semi-structured interviews included questions that were open in order for the interviewees to answer in their own ways, as they themselves found necessary and relevant. However, an interview guide was used to help frame the phenomenon *teaching digital models* (see Brink et al., 2021). The interviews were then analyzed thematically (Braun & Clark, 2006) with four resulting themes. During the thematic analysis different aspects of the teaching emerged, like teaching contents and methods for all four themes. One of the themes describes that technology teachers are experiencing teaching digital models for *Enabling digital modelling* (Brink et al., 2021). In this theme the results show that teachers teach to give pupils opportunities to learn contents like the process of modelling and to handle computer software. The teaching activities include modelling in a digital design software (CAD) and computer programming. The next section will present results from the theme *Enabling digital modelling*, to show that teachers can understand computer programming as equivalent to modelling.

6 Computer Programming as Modelling, from a Teacher Perspective

In this section, the research question *What is the nature of technology teachers' understanding of computer programming as modelling?* will be answered. Previous results from Brink et al.'s (2021) study will be discussed with focus on teaching that *enabling digital modelling*, one of the resulting themes from that study. Excerpts from the interviews are presented to underpin conclusions drawn. The argumentation starts with describing vague concepts in technology education, continuing with linking problem solving to computational thinking. Thereafter a comparison between computer programming and different modes of representation is given. This section ends with a discussion on simulations of computer programs.

6.1 *Vague Concepts*

The interviewed teachers' views of models created with digital tools within technology education differ, and they talk for instance about movies, games and models created using computer aided design software. The teachers describe the concept of models as vague and that the steering documents offer little help in interpreting the concept. Other things are also experienced as vague in technology education according to the interviewed teachers. Teaching concerning computer programming is another example. Teacher 9 says: "they [the pupils] may not think about programming as building or creating." In this case, the teacher was talking about teaching technological development and about pupils creating and executing their own ideas. The teacher points out that teaching can be improved concerning instructions about what the pupils should do and learn in order to make the intentions behind computer programming more clear.

However, the interviewed teachers also talk about computer programming as models. Teacher 1: "Well, a robot you control with code, that can be a digital model." Teacher 1 described that the concept *digital model* is not very well defined and that each and every teacher needs to interpret different concepts mentioned in the syllabus. This statement is confirmed by other interviewed teachers in the study. Previous research also points out that teachers have great freedom when interpreting the steering documents (Bjurulf, 2008), wich can be related to the interviewed teachers' experiences of vague concepts.

6.2 *Problem-Solving and Computational Thinking*

Problem-solving is experienced as part of computer programming, according to the interviewed teachers. One excerpt from teacher 10 states that: "in school, much is about pupils trying, thinking outside of the box. It is very much about problem-solving. I think programming is just that." Problem-solving is also described as part of computer programming and algorithmic thinking by Sen et al. (2021) and as part of modelling (Citrohn et al., 2023; Middleton, 2005). Hence, problem-solving can link the two concepts together.

The modelling process is also described when the interviewed teachers talk about computer programming activities in technology education. Here, it is prominent that teachers experience computational thinking and algorithmic thinking as parts of the modelling process (Elmer & Davies, 2000; Sen et al., 2021). Teacher 2 says: "And I think that you should see it as a digital model, do a, like a sequence, to say that it is a digital model." Teacher 7 describes the modelling process and the efforts the pupils are making: "It is about patience and to be able to plan your work." In other words, when modelling pupils should structure their work sequentially and logically. Teacher 7 goes on to describe algorithmic thinking as a prerequisite for successful computer programming.

6.3 *Computer Programming and Modes of Representations*

In the interviews, different modes of representations of a computer program are given. According to Gilbert et al. (2000) and Gilbert (2005), modes of representations of a model can be concrete or material, verbal, visual, mathematical or symbolic, and gestural. By comparing these modes with the interviewed teachers' descriptions of computer programming, it can be argued that teachers can experience computer programming as modelling.

Pupils need to visualize and represent their computer programs, according to the interviewed teachers. This is explained by a teacher, describing a teaching activity where pupils are beginning to learn how to make computer programs. The teaching activity starts with pupils planning their programs. The pupils produce a description of the program and also a description of how the program should function. This teaching strategy is common as an unplugged activity among technology teachers for grade 4–6 in Sweden (Bjusten et al., 2022). Teacher 10 says on one occasion that: "It could be a model of a program." Each pupil's description of the computer program (the *model of a program*) is a verbal or textual representation of the computer program. This could exemplify pupils transferring their internal, mental models to external, expressed models (Carmichael, 2000). The value of making a computer program or model expressed, is for others to see the potential or weakness in the program so it can be improved (Carmichael, 2000).

In another interview, teacher 8 explains one computer programming activity as: "It is controlling a figure on the screen, a figure you created yourself." According to the model typology of Gilbert (2000) and descriptions by Boulter and Buckley (2000), the controlled figure on the screen can be seen as a dynamic and visual model. Teacher 8's explanation of controlling a figure shows that computer programming can be experienced as a dynamic model. Furthermore, teacher 7 confirms experiences of computer programming as modelling:

> Now, when we have the programmed games, you could say that you create a digital model of reality, so to speak. If you are a figure catching apples falling down from a tree, you create a picture of your idea. You are making a model, a digital image out of it. (Teacher 7)

The model of the figure catching apples is a visual representation of what is programmed.

6.4 *Simulations of Computer Programming*

Other teaching activities from the interviews are for instance programmed lawn-mowers or programmed traffic lights. These two teaching activities exemplify and visualize digital manipulations and simulations. Simulations carried out in computer programmes are interpreted as models by the interviewed teachers, here shown with an excerpt from teacher 11. "Well, the traffic lights, we are modelling them too, using Micro:bits." Micro:bit is a small computer, programmable from a web-page with block-programming or textual code. At the web-page, pupils can simulate their programs. Simulations of a computer program often have a time perspective, that is, a model of dynamic mode (Boulter & Buckley, 2000). Several teachers describe computer programming and modelling where simulations unite the two concepts, as shown in the next example from teacher 10.

Teacher 10 explained that the pupils should program and simulate lamps to be turned on and off, to steer the lamps with numbers of zero and one. The purpose of the activity was to increase understanding among the pupils about the need of turning things off when programming. Otherwise, the lamp would light up forever:

> It is about digital signals, you talk about them. Turning the signal on and off. Pupils are familiar with the fact that something happens when they turn things on, but they do not think about turning things off, for instance a lamp. (Teacher 10)

Teacher 10 talked about the lack in pupils' computer programs of mechanisms for turning things off. In ordinary life, pupils take this for granted. The model of the computer programmed lamp is a simple but important example of teaching where simulations of a computer program are needed.

All examples and excerpts above, taken from the interviews, show a broad range of understandings that teachers can have related to computer programming and models. To conclude, even if the interviewed teachers experience the concept digital models as vague, they can still experience computer programming as modelling. Problem-solving is represented in both computer programming and modelling activies. Computer programming can be presented with similar modes of representation as models, and simulations on a screen unite the two concepts of computer programming and modelling. However, experiences of computer programming also include working with physical material for concretizing the computer programs.

7 Physical Material for Concretizing Computer Programming

Computer programming in technology education can be one example of teaching in and about technology. In this teaching, teachers can use physical material to concretize computer programming, and pupils are thereby given opportunities to examine and assess the models they are programming. As discussed earlier in this chapter, examining and assessing models are important ways of understanding contemporary society (Seery, 2017). The models in this teaching can be both physical and digital, since the physical parts are being controlled by computer programs (see for instance Citrohn et al., 2023). This section will continue with examples from the interviews, illustrating how teachers describe their teaching in terms of concretizing computer programming.

Teacher 12 used LEGO-robotics in a teaching activity based on two cases. One case is when pupils are programming the robotics to act like a lawnmower. The other case is when pupils are programming the robotics not to fall off a table when moving around the surface of the table. The pupils were supposed to program sensors connected to the robotics, for detecting changes in the surroundings. "Yes, and there the pupils are programming a LEGO-car, or a LEGO-robotics. They should make it move in a certain way. And that is a kind of model of reality." The sensors can help pupils understand the programming activity and make it more concrete.

Computer programming of traffic lights seems to be a common activity, according to the interviewed teachers, for concretizing programming with physical material. Several of the interviewed teachers describe activities like this. Teacher 9 says: "So then we switch platforms, and mix the digital and the physical. [...] But there, pupils can simulate the code inside the software." Teacher 9 continues by describing that it is of great advantage if pupils work systematically and start describing what the computer program should do and in what order as an unplugged activity (Bjursten et al., 2022). When this preliminary step has been carried out, the pupils continue with programming in the software, simulate the program and finally, test the program with physical objects. Teacher 9 again: "It is another way of teaching the same thing, and maybe then, more pupils will learn." Teaching including different representations of the same thing can give pupils knowledge of when to use different modes, and when different representations of the model are preferable (Gilbert, 2000).

Several of the interviewed technology teachers use the subject mathematics to concretize computer programming. In one of the interviews, teacher 3 explained that: "[Pupils] should work with programming using Micro:bit. They are going to program them to measure temperature. They will measure both inside and outside temperature. After that, they will investigate the temperature in math and write and make diagrams." The diagrams in mathematics are created digitally, and compose a symbolic and visual representation of a mathematical model (Gilbert, 2000). The data from the programmed Micro:bit creates the mathematical model. However, even though the teacher emphasizes that the programming creates the data, the focus in this teaching is also to make pupils understand the use of programmed objects in everyday life. Above all, all these examples of concretized material used when computer programming show teaching both *in* and *about* technology. The examples also show teaching where technology teachers use computer programmes as models that represent some part of the programmed reality. In addition, teachers point out that physical material makes the teaching varied, and that pupils experience the teaching as more fun when for instance working with electrical components. According to the teachers, only computer programming on a screen is not sufficient for effective learning.

8 Concluding Discussion

In this chapter, Swedish technology teachers' understandings of computer programming as modelling have been discussed. I have also discussed that the concepts of computer programming and modelling can be conceived as closer to one another than often realized, even though not construed as entirely synonymous. Often during the interviews, the teachers mix the concepts and use them synonymously. Two explanations for this could be that the steering documents in Sweden do not give teachers much guidance and that the technology subject lacks a common and agreed body of core concepts. Other explanations can be that problem-solving is present in both computer programing and modelling, and that similar modes of representation can be applied in computer programs and models alike.

However, teaching *in* and *about* technology calls for different and multiple representations when modelling, with the aim of giving pupils opportunities to examine, predict, and optimize the outcomes (de Vries, 2013). The interviewed teachers in this study use visual and dynamical representations, mathematical models, or physical robotics when teaching computer programming. They are also using simulations to visualize computer programming. On some occasions, the teachers describe using textual or verbal representations when the pupils are working with planning their programs, that is, modelling. Teaching for planning computer programming is one way to teach algorithmic and computational thinking. Problem solving, as a part of computational thinking, is also mentioned when teachers describe their teaching of modelling and computer programming.

A common conception, expressed in the interviews, is that computer programming can be understood as modelling by the teachers and that they can mix the concepts. It can be problematic if these two concepts are mixed or used synonymously, for comparisons between different technology educations (nationally and internationally), for technology teachers' professional development, or for the development of the technology subject. It is suggested that it would be beneficial to use a more nuanced and precise concept apparatus in this area of the technology subject (Brink et al., 2021).

To conclude, one way of teaching computer programming and modelling in a more nuanced way can be to include different modes of representations explicitly in teaching, and to make the pupils aware of the differences between the modes, when and in what context different modes are preferable. A teacher can teach about something which can be represented through several models in terms of different available modes, and discuss pros and cons of the models together with the pupils. Finally, modelling that includes creating code, visually, verbally, symbolically, or using block-programming or textual code, can be described and taught as computational programming. These suggestions could create greater awareness among the pupils about models and the programmed and connected contemporary society we all are part of.

References

Anderhag, P., Björn, M., Fahrman, B., Lundholm-Bergström, A., Welland, M., & Wållberg, T. (2021) Kod som teknisk lösning: en studie om grundskoleelevers uppfattningar av ändamålsenlighet i deras spontana programspråk. *Nordina, 17*(1), 113–129. https://doi.org/10.5617/nordina.7020

Ankiewicz, P. J. (2019). Andrew Feenberg: Implications of critical theory for technology education. In J. R. Dakers, J. Hallström & M. J. de Vries (Eds.), *Reflections on technology for educational practitioners: Philosophers of technology inspiring technology education* (pp. 115–130). Brill publishers.

Bjursten, E.-L., Nilsson, T., & Gumaelius, L. (2022). Computer programming in primary schools: Swedish technology teachers' pedagogical strategies. *International Journal of Technology and Design Education.* https://doi.org/10.1007/s10798-022-09786-7

Bjurulf, V. (2008). *Teknikämnets gestaltningar: En studie av lärares arbete med skolämnet teknik* (Thesis). Karlstad University.

Boulter, C. J., & Buckley, B. C. (2000). Constructing a typology of models for science education. In J. K. Gilbert & C. J. Boulter (Eds.), *Developing models in science education* (pp. 41–57). Kluwer Academic.

Braun, V., & Clarke, V. (2006). Using thematic analysis in psychology. *Qualitative Research in Psychology, 3*(2), 77–101. https://doi.org/10.1191/1478088706qp063oa

Brink, H., Kilbrink, N., & Gericke, N. (2021). Teaching digital models: Secondary technology teachers' experiences. *International Journal of Technology and Design Education.* https://doi.org/10.1007/s10798-021-09659-5

Carmichael. (2000). Computers and the development of mental models. In J. K. Gilbert & C. J. Boulter (Eds.), *Developing models in science education* (pp. 177–190). Springer Netherlands.

Citrohn, B., Stolpe, K., & Svensson, M. (2023). The use of models and modelling in design projects in three different technology classrooms. *International Journal of Technology and Design Education, 33*, 63–90. https://doi.org/10.1007/s10798-022-09730-9

Citrohn, B., & Svensson, M. (2022). Technology teacher's perceptions of model functions in technology education. *International Journal of Technology and Design Education, 32*, 805–823. https://doi.org/10.1007/s10798-020-09632-8

Cuperman, D., & Verner, I. M. (2013). Learning through creating robotic models of biological systems. *International Journal of Technology and Design Education, 23*(4), 849–866. http://dx.doi.org/10.1007/s10798-013-9235-y

Dakers, J. R. (2018). Nomadology: A lens to explore the concept of technological literacy. In M. J. de Vries (Ed.), *Handbook of technology education* (pp. 17–32). Springer.

Elmer, E., & Davies, T. (2000). Modelling and creativity in design and technology education. In J. K. Gilbert & C. J. Boulter (Eds.), *Developing models in science education* (pp. 137–156). Springer Netherlands.

Fahrman, B. (2021). *To know a subject – Teachers' views about the subject of technology* [Licentiate thesis]. KTH Royal Institue of Technology.

Gibson, K. (2008). Technology and technological knowledge: A challenge for school curricula. *Teachers and Teaching: Theory and Practice, 14*(1), 3–15. https://doi.org/10.1080/13540600701837582

Gibson, K. (2009). Technology and design, at key stage 3, within the Northern Ireland curriculums: Teachers' perceptions. *International Journal of Technology and Design Education, 19*(1), 37–54. https://doi.org/10.1007/s10798-007-9039-z

Gilbert, J. K. (2005). Visualization: A metacognitive skill in science and science education. In J. K. Gilbert (Eds.), *Visualization in science education* (pp. 9–27). Springer.

Gilbert, J. K. (2010). The role of visual representations in the learning and teaching of science: An introduction. *AsiaPacific Forum on Science Learning & Teaching, 11*(1), 1–19.

Gilbert, J. K., Boulter, C. J., & Elmer, R. (2000). Positioning models in science education and in design and technology education. In J. K. Gilbert & C. J. Boulter (Eds.), *Developing models in science education* (pp. 3–17). Springer Netherlands.

Hagberg, J. E., & Hultén, M. (2005). *Skolans undervisning och elevers lärande i teknik: Svensk forskning i internationell kontext.* Vetenskapsrådets rapportserie. Vetenskapsrådet.

Hallström, J., & Schönborn, K. J. (2019). Models and modelling for authentic STEM education: Reinforcing the argument. *International Journal of STEM Education, 6*(1), 1–10. https://doi.org/10.1186/s40594-019-0178-z

Kilbrink, N. (2013). *Lära för framtiden: Transfer i teknisk yrkesutbildning* (Thesis). Karlstad University.

Kilbrink, N. (2018). Technical vocational education: From dualistic to pluralistic thinking. In M. J. de Vries (Ed.), *Handbook of technology education* (pp. 193–204). Springer.

Middleton, H. (2005). Creative thinking, values and design and technology education. *International Journal of Technology and Design Education, 15*(1), 61–70. https://doi.org/10.1007/s10798-004-6199-y

Nia, M. G., & de Vries, M. J. (2016). Models as artefacts of a dual nature: A philosophical contribution to teaching about models designed and used in engineering practice. *International Journal of Technology and Design Education, 27*(4), 627–653. https://doi.org/10.1007/s10798-016-9364-1

Norström, P. (2013). Engineers' non-scientific models in technology education. *International Journal of Technology and Design Education, 23*(2), 377–390. https://doi.org/10.1007/s10798-011-9184-2

Sen, C., Ay, Z. S., & Kiray, K. A. (2021). Computational thinking skills of gifted and talented students in integrated STEM activities based on the engineering design process: The case of robotics and 3D robot modeling. *Thinking Skills and Creativity*, 42. https://doi.org/10.1016/j.tsc.2021.100931

Seery, N. (2017). Modelling as a form of critique. In P. J. Williams & K. Stables (Eds.), *Critique in design and technology education* (pp. 255–273). Springer.

Sjöberg, S. (2013). Grundläggande tekniska förmågor. In J. Hallström & C. Klasander (Eds.), *Ginners teknikdidaktiska handbok. Några teser om teknik, skola och samhälle.* Linköping University.

Skolverket. (2017). *Kommentarmaterial till kursplanen i teknik: Reviderad 2017.* https://www.skolverket.se/publikationsserier/kommentarmaterial/2017/kommentarmaterial-till-kursplanen-i-teknik-reviderad-2017?id=3790

Skolverket. (2019). Läroplan för grundskolan, förskoleklassen och fritidshemmet 2011: Reviderad 2019 (6 uppl.). https://www.skolverket.se/getFile?file=4206

Skolverket. (2022). *Läroplan för grundskolan, förskoleklassen och fritidshemmet.* https://www.skolverket.se/undervisning/grundskolan/laroplan-och-kursplaner-for-grundskolan/laroplan-lgr22-for-grundskolan-samt-for-forskoleklassen-och-fritidshemmet?url=-996270488%2Fcompulsorycw%2Fjsp%2Fsubject.htm%3FsubjectCode%3DGRGRTEK01%26tos%3Dgr&sv.url=12.5dfee44715d35a5cdfa219f

Svenningsson, J. (2019). Carl Mitcham: Descriptions of technology. In J. R. Dakers, J. Hallström, & M. J. de Vries (Eds.), *Reflections on technology for educational practitioners: Philosophers of technology inspiring technology education.* (pp. 13–24). Brill Academic Publishers.

Svensson, M. (2011). *Att urskilja tekniska system: Didaktiska dimensioner i grundskolan* (Thesis). Linköping University.

de Vries, M. J. (2013, December 2–6). Modeling in technology and engineering education. In P. J. Williams & D. Gedera (Eds.), *PATT 27, technology education for the future: A play on sustainability, Christchurch, New Zealand.* University of Waikato.

de Vries, M. J. (2018). Philosophy of technology: Themes and topics. In M. J. de Vries (Eds.), *Handbook of technology education* (pp. 7–16). Springer.

Williams, P. J. (2000). Design: The only methodology of technology. *Journal of Technology Education, 11*(2), 48–60. https://doi.org/10.21061/jte.v11i2.a.4

CHAPTER 12

Teachers' Experience of Science Centres as a Resource for Programming Education

Maria Sparf

Abstract

Digital technologies have increasingly become part of society, and teachers and other pedagogues need to face digitisation in constructive ways. Among other things, this can include the use of different types of personnel and technical resources for teaching, both formal and informal. In municipalities where there is a Science Centre (SC), this is used as part of the education offered in compulsory school in Sweden. When computer programming was introduced into the curriculum for all grades, SCs already had experience in delivering programming education as part of their programming courses. This experience was used as a resource by teachers who sought support with teaching programming to their students. In this context, SCs can therefore be said to have played a pioneering role in the delivery of programming education in compulsory school. This chapter outlines how teachers who visited SCs with their students view the teaching of programming in compulsory school, why they chose to visit the SC with their class, and what they need in order to be able to teach programming themselves.

Keywords

programming – science centre – teaching – design for learning

1 Introduction

This study was carried out just before programming became a mandatory part of the syllabus in Swedish compulsory schools. Many teachers did not know where to start or at what level programming should be taught. Teachers sought support in different ways. Some turned to Science Centers (SCs) for support, which were relevant since most SCs offer courses in programming for compulsory school pupils (Svenska Science Centers, 2018). SCs had experience of teaching programming through leisure courses for children and teacher

© MARIA SPARF, 2024 | DOI:10.1163/9789004687912_012

training. In that sense, SCs can be said to have played a pioneering role, as they offered teaching in programming early on, before the compulsory school system had had time to adjust to the content. It is always important to investigate how teaching is conducted by those who adopt educational technologies at an early stage, as it presents opportunities to spread pioneers' knowledge further and pave the way for those who are a little more hesitant (Jahnke & Kumar, 2014). The teachers in this study took advantage of this by visiting SCs with their students. Using SCs as a resource for teaching programming might also increase teachers' competence in delivering technology education in compulsory school.

1.1 *Design for Learning*

In this study, Design for Learning (DFL) is used as a theoretical framework (Kress & Selander, 2012; Selander, 2017). Using DFL contributes to a systematic understanding of how contextual aspects such as physical surroundings and previous (in-service) education influence the ways in which teachers perceive programming education at SCs.

Within DFL there are three aspects of educational design: design *for* learning, design *in* learning and recognition of knowledge. In educational contexts, lesson planning and access to different kinds of resources constitute examples of design *for* learning. These in turn influence what opportunities individuals have to design *in* their own learning (Leijon & Lindstrand, 2012). That is, how individuals make use of the various learning resources in relation to their own interests and previous experience. Finally, recognition of knowledge refers to what counts as knowledge in a certain area (Selander & Kress, 2017). The recognition of knowledge can influence the design *for* learning, i.e., the planning of teaching. For example, gradings can govern what kind of knowledge that is recognised as signs of learning in a formal school context. At Swedish SCs, there is an intention to adhere to the current curriculum during school visits. However, there are no formal knowledge requirements in the form of grades that govern the teaching.

Selander (2017) highlights that a design-oriented perspective can contribute to discussions about how new meaningful contexts for teaching and learning can be created. The DFL framework is therefore applicable to use in this context; however, it is mainly the aspects design *for* learning and recognition of knowledge that are relevant in this study.

1.2 *Aim and Research Questions*

The aim of this chapter is to describe how teachers who visited SCs with their students view the teaching of programming in compulsory school and why they chose to visit SCs.

The research questions to be answered are:
- How do teachers perceive the teaching of programming at SC?
- What do teachers consider is required to teach programming themselves?
- How can SCs support the school teaching of programming, according to the teachers?

2 Education at SCs

A Science Centre is a complement to the school's theoretical teaching in science and technology. Most SCs offer a wide range of STEM educational activities (science, technology, engineering, and mathematics). The present study focuses specifically on programming that can be linked to certain areas of STEM, primarily technology but also mathematics.

2.1 *STEM Education at SCs*

The aim of an SC is, among other things, to stimulate curiosity and interest in STEM education. A visit at an SC should be both entertaining and educational at the same time. SCs fall somewhere in the middle between being a fun leisure activity and a communicator of knowledge in science and technology (Kubacki et al., 2012). In this way, SCs constitute a semi-formal learning environment (Sparf, 2020) that can be compared with, for example, museums (Sasson, 2014).

At most SCs, different activities are arranged for pre-booked groups such as planned lessons as part of the school day (Sasson, 2014; Swedish Science Centers, 2018). SCs offer lessons and activities, especially in subjects that are sometimes considered difficult to understand and therefore may need to be experienced in other ways than what is possible in a normal classroom. An example of this is the use of visualisation technology at some SCs, which can be a way to make *difficult-to-explain phenomena* in science and technology clearer and more understandable. The education at SCs can also apply to areas where special equipment is required that can be expensive to provide at individual schools. With this compensated focus, Swedish SCs have in recent years among other things had a mission on demonstrating technologies aimed at sustainable energy production.

The use of visits to SCs by schools as part of school education has been described by among others Shaby et al. (2017) and Krange et al. (2020). During lessons, the educators at SCs are given an active role in guiding the learning content. Krange et al. (2020) emphasise the importance of the pupils receiving support from their teacher or an SC educator in order to get the most out of the visit.

However, there may be a difference in perceptions about what the outcome of a visit to an SC should be. Insulander and Öhman (2022) describe that museum and SC educators often consider the visit as successful when the pupils feel comfortable and find personal relevance in the learning content. Meanwhile, teachers from schools evaluate the visit in relation to the curriculum and the practical schoolwork (Insulander & Öhman, 2022). This means that SC educators and teachers do not always have a consensus on what the outcome of the visit should be. This is in line with findings by Kisiel (2005) from a study investigating teachers' motives for taking their students on a school trip to an SC. One reason given was that the activities offered at the SC lined up with learning objectives stated in the curriculum. Other motives were to give the students new learning opportunities not available at school, and to provide a fun and engaging day out. Kisiel (2005) argues that it is important that both schoolteachers and SC educators are aware of these different expectations and plan the visit with these in mind.

Morentin and Guisola (2014) have also drawn attention to this difference between what the school expects and what SCs can offer. They show in their experimental study what can be done to reduce the gap that can arise between these two stakeholders. Firstly, teachers need to carry out more preparation for the visits together with their classes. Secondly, SCs need to adapt more to what the schools are asking for in the form of support for teaching. At the same time, according to Morentin and Guisola (2014), it is the teacher's task to ensure that what the students learn at the SC is adapted to the current curriculum.

2.2 *Programming at SCs*

There are some studies on programming teaching and learning at Swedish SCs. These studies highlight various aspects of using SCs for programming education and are summarised in this section.

Insulander and Öhman (2022) conducted a study on how educators at SCs and teachers in compulsory schools develop a collaboration on programming education. The SC educators aimed to offer activities to support pupils' interest in programming. The schoolteachers, on the other hand, wanted to build up their own competence and use the visit to assist them in creating a plan with progression for teaching programming. At the same time, the pupils would get an introduction to programming. Insulander and Öhman (2022), similar to Morentin and Guisola (2014), call for a dialogue between SC educators and schoolteachers so that the education offered at SCs is designed not only to arouse pupils' interest, but also for lasting learning. Insulander and Öhman (2022) argue that this would make informal learning easier to transform into formal knowledge.

Awareness of the different ways in which students can be engaged can be important both for teaching at SCs and for teaching at school. In a study analysing students' engagement during programming education, Sparf (2021) provides insights into how behavioural, emotional, and cognitive engagement can be identified and understood. Engagement can be considered multidimensional, where the different dimensions all influence each other. To have the opportunity to be engaged, it is important to understand the learning content. Visual programming, for example, can give students an "aha experience" when they are able to see the codes and how things are connected (Sparf, 2021). However, whether students perceive programming activities as emotionally engaging or not may depend on whether they find the tasks meaningful and cognitively stimulating.

There are also discussions about whether programming education at SCs is designed more for boys or girls (Ivarsson 2019; Silfver, 2019). There are examples where it may be perceived that some of the activities are directed more towards the interests of boys than girls which in such cases might lead to maintain stereotypical gender patterns. However, Ivarsson (2019) considers that although there are challenges with moving programming education from school to SCs, it is important not to get stuck in stereotypes about male and female behaviour. Instead, it is important to see what students learn from different parts of the programming exercises. In that way, the very fact that both boys and girls are programming can be an example to challenge gender stereotypes.

Sparf et al. (2022), investigated pupils' different learning approaches when programming at SCs. The study identified five different approaches that pupils used to solve various programming tasks. The qualitatively separate approaches that pupils used during the observed lessons were: mathematical, trial and error, step-by-step, routine, and aesthetic. Each of these approaches allows pupils to use and practice different abilities that are important for programming. These abilities can be compared to, but are not equal to, computational thinking (CT), which in turn is linked to competencies that are important for active participation in a digital society. At the same time, each approach can provide a basis for learning programming. The ideas and strategies that characterise CT correspond to a number of abilities. In this study, CT is seen as an overall competence used in programming. CT can be summarised as including different ways of thinking when solving problems, something that is important to be able to program (Cetin & Dubinsky, 2017; Durak & Saritepeci, 2018; Wing, 2006). The term CT is not used in the Swedish curriculum, either in technology or in mathematics (Skolverket, 2022). See also the introduction of this book for further interpretations of CT.

In summary, there is some existing research into the learning of programming at Swedish SCs. At the same time, there is a gap in the current knowledge—namely, how teachers take care of the experiences at SC to develop their own teaching.

3 Context of the Study and Research Methodology

This section describes the lessons that the teachers took part in with their students at SCs. It also describes how the study was carried out and the methodological and analytical choices that were made.

3.1 *The Lessons at SCs*

All the teachers interviewed had recently—in some cases the previous day—participated with their students in an approximately two-hour programming lesson at an SC (for more information about the lessons see Sparf, 2020). All of the activities were initiated and planned by SC educators. The lessons started with a brief background to programming where the educators together with the students discussed and identified connections to programming in everyday life. The educators then explained how the equipment worked and the functions of the various programming blocks. The majority of each lesson was used for the students' own work on various programming tasks.

The lessons were all adapted for novices, although the content differed slightly depending on the students' age. All the software used in the lessons was based on so-called visual block-based programming. The youngest students in Grade One used the Scratch JR programming language on iPads. The task was to create animations to illustrate short stories or episodes using programming. The students in Grade Four used Scratch on computers. Their task was to program a computer game. The task for the students in Grades Three and Six was to program Lego robots. For the students in Grade Six, the robot assignments were linked to the theme of sustainability.

In spring 2017, when this study was conducted, programming was not yet mandatory. However, at the same time, there was an expectation on teachers, perceived or real, to introduce programming to their pupils. The teachers had mixed feelings about the new subject content, as will be discussed in the next section.

3.2 *Interviews with Teachers*

The data collection was carried out in a context where teachers together with their students participated in programming lessons at SCs. A few weeks before the booked lessons, the teachers were contacted to confirm if they were willing to participate in the data collection together with their class. Written permission from legal guardians and students was collected to be able to observe the lessons and converse with the students. The teachers who accepted to participate were also informed about the nature of the project and that the data would be treated anonymously. This is in line with good research practice according to the Swedish Research Council (Vetenskapsrådet, 2017). All data are handled in line with the General Data Protection Regulation (GDPR) (European Data Protection Supervisor, n.d.).

In total, three SCs were visited on 10 separate occasions. The data collection was based on observations during programming lessons, conversations with students, and finally interviews with teachers who participated in the observed lessons (see Sparf, 2020). The interviews that form the basis of this chapter were conducted either directly after the lesson was completed or by telephone the following day, depending on the time available during the SC visit. The interviews lasted 15–20 minutes. Using interviews provided an opportunity to obtain explanations and clarifications from the teachers in relation to the events that occurred during the lessons. A flexible interview guide was used which included questions about i.e. teachers' experience of programming, opportunities for in-service education, and what the teachers knew about concepts such as CT. Five teachers working in Grades One to Six were interviewed in this way (see Table 12.1). The names used are pseudonyms.

3.3 *Data Analysis*

The interviews were transcribed and then analysed with the support of thematic analysis inspired by Braun and Clarke (2006).

TABLE 12.1 Teachers who visited SCs with their pupils

Teacher	Male/Female	Grade	Previous experience of participating in programming workshops for teachers
Lisbeth	Female	4	No
Ulla	Female	3	2–3 h
Ellen	Female	3	2–3 h
Ulf	Male	6	2–3 h
Viola	Female	1	No

The steps used in the analysis were:

1. Repeated reading to get acquainted with the material.
2. Initial coding of relevant sections that were consistent with the research interest in programming education.
3. Themes were formulated and codes were merged and sorted out.
4. In order to minimise overlap between the themes, they were matched with each other and analysed according to the research questions.
5. The themes, including sub-themes, were further analysed with the DFL theoretical framework in mind.
6. Excerpts from the dataset were selected to represent the results of the analysis. The excerpts should be seen as representative examples, not as quantitative representations.

The analysis can be said to be inductive as no predetermined themes guided the analysis. At the same time, there are deductive elements where there was an interest in finding out more about what the teachers knew about certain concepts such as CT, as well as the teachers' opinions and attitudes regarding SCs. The DFL framework contribute to the analysis by focusing on how meaningful contexts for teaching and learning can be created. It is mainly the aspects design *for* learning and recognition of knowledge that are relevant in the analysis.

4 Results

There may be a number of reasons why teachers use SCs to support their teaching of programming, and it is therefore interesting to hear teachers' own perspectives. Various explanations emerged in the interviews and are presented in relation to the lessons attended at the SCs.

4.1 *A Revised Curriculum*

All the teachers interviewed were aware that a revised curriculum, which included programming, was about to be implemented. At the time of the interviews, the information from the government was still very limited, which made it difficult for teachers to know in which subjects or to what extent they were expected to teach programming:

> We have heard that it is coming up and that it is a part of the curriculum, we have got that from our principal, but not much more. (Ellen)

Teacher Viola felt that there was pressure to implement the changes rapidly:

> This is what I think. It is a bit hard when you end up in this position as an educator that something (literally) falls from the sky and then you suddenly have to implement it in the teaching. You do not have the tools. You do not have the education. (Viola)

The upcoming curriculum change left Viola feeling stressed, and as she figuratively put it:

> We felt the knife at our throats, we just had to get started. (Viola)

One of the ways in which Viola started to teach programming was to take her students to the SC. There, she believed, there was an opportunity to give the students what she herself had not yet mastered.

At the time of data collection, the curriculum change had not been implemented, but all teachers were aware that it was underway, which led them to behave as if it was already a fact.

4.2 *Support for Teachers*

The five teachers in the study said they needed more training and support to teach programming themselves. One thing that teachers wanted to see being offered as part of in-service training was hands-on programming knowledge:

> I think that we need to get an education in the first place, we teachers, so that we feel safe. (Lisbeth)

Lisbeth suggested that teachers could gain this knowledge through workshops with, for example, staff from SCs. Viola, on the other hand, felt that more rigorous education was needed:

> We need to have education, absolutely. This is not something you can educate yourself on; I think you should be allowed to take proper university credits. [...] you need deeper knowledge. So I definitely think we should get to go to university... Mandatory! For everyone. (Viola)

In addition to in-service training, the teachers talked about how they needed time to discuss and organise the work at the school:

> But I think we need to flesh out these goals, what is it that we are going to teach. Well, like, what is it the kids need, [...]. I think that is really important. It is good to learn together with the children, but at the same time

> you must have a deeper knowledge to understand, what we want and how far we should reach. (Ulla)

Ulla felt it would be useful to sit down together with her colleagues to figure out exactly what it was they were going to teach—something she said had not yet been possible.

All teachers in the study expressed a desire for support with teaching programming, and felt they had not yet received such support, even though it would soon be compulsory for all students to receive instruction in the subject.

4.3 *The Visits to the SCs*

The teachers gave various reasons for why they chose to visit the SCs with their class. There are two things in particular that the teachers highlighted in this regard: their own lack of competence and a lack of equipment at the school.

Lisbeth saw programming as something the students should learn and wanted to give them the chance to try out educational tools that she herself was unable to provide:

> Yes, partly I don't have the knowledge of programming myself and partly we haven't had access to computers [...] so it has been both. (Lisbeth)

Since Lisbeth felt that she herself had limited opportunities to teach programming, she chose to bring her students to SC:

> to do things that the curriculum suggests but that we don't really have the opportunity to do during school time. Even if programming is a bit on the periphery for now, we believe it is still an important tool. (Lisbeth)

Ulf said that even if there were opportunities to work with programming at the school, it was something else entirely when the students got the chance to try more student-friendly equipment at the SC:

> There's a difference between doing programming online, there are lots of programming sites and training in programming or like in this case robots. It's not everyone who has access to do it, so it's great. (Ulf)

The teachers wanted to give their students the opportunity to try new things and looked for different occasions to do so. These statements show that when there was a lack of knowledge or adequate equipment, teachers searched for other options.

All five of the interviewed teachers were engaged during the lessons observed at the SCs, either in helping their students or in trying out the various programming tasks themselves:

> Yes. I also saw this a bit as a learning opportunity for me as well, because I feel like I'm quite, well a little new to this. And it's really exciting to see what you can do with Lego robots. I have not worked with robots before, so it was exciting. (Ulla)

In this context, the teachers said that they also saw the visit as a learning opportunity for themselves—something they also appreciated.

However, in some cases, teachers had more implicit pedagogical purposes in mind for visiting the SC. Ulla's class had won a drawing contest that sponsored a trip to SC. The initiative came from a local company and the price went to the class that could best illustrate what a robot can be used for. This meant that the trip to the SC was paid for by the local company:

> We usually go to the SC in Grade Three, but we might not have chosen this lesson with programming, it is not certain that we would have done that. (Ulla)

Ulla explained that her class often went to the SC on a school trip at the end of spring as a fun activity. Usually, to motivate the trip, they signed up for a lesson about the solar system or the star constellations, something that gives the visit an extra dimension in addition to the visit at the SC experiment hall. This time, as the visit was paid for—both travel and admission fees—they were not able to choose the activity themselves. However, Ulla also said she felt that programming was relevant for her pupils, as it was about to become mandatory.

Although the background and the funding for each visit differed somewhat, the teachers' general opinion was that the most important reason for the visit was in order to give their students opportunities that did not exist at school.

4.4 *Lesson Outcomes*

There were variations in terms of how much the teachers had prepared for the programming lessons at the SC. Some had participated in one or two workshops on programming and had let their students try out programming at school. There were also teachers in the group who had not made any preparations at all. This meant that the students came to the SCs with different degrees of preparation.

When the teachers were asked what they thought about the lesson, all responses were positive. Ellen's remarks about her class's visit to the SC focused less on the students' learning and more on the fact that both she and the students had enjoyed the visit:

> The kids appreciated it very much; they thought it was great fun. (Ellen)

All teachers felt that the students had kept up and coped with the tasks well, and that most of them had found it fun. The main thing that the teachers highlighted was that the students got to try out programming, without specifying what that meant. However, when asked directly about what they thought the students had learned from the programming lessons, the teachers highlighted collaboration:

> Yes, for one thing, I see the collaboration, i.e., student collaboration, it's good, and games, it appeals to children. (Lisbeth)

Ulf also emphasised collaboration:

> Yes, they are learning, they are learning collaboration, they are working in pairs on this. (Ulf)

In addition, he stressed that the students had to engage in logic and abstract thinking:

> they practice logical thinking, and it also means that they practice abstract thinking […], they have to figure out what the robot should do to reach the goal, there are actually a lot of different parts that are practiced. (Ulf)

However, when asked directly which school subject Ulf primarily connected to programming, he answered:

> I see mostly technology. It's more about technology than mathematics, I think. (Ulf)

It was thus difficult for the teachers to specify what the students had actually learned from the lessons. Collaboration and abstract thinking are important parts of programming, but other aspects of programming such as problem solving or creativity did not become visible in the teachers' comments. However, as there was an interest in knowing how the teachers related programming to CT, the teachers were asked one direct question about this:

> Yes, [CT] is something that permeates everything [...]. If you learn one thing, you can draw conclusions and then build on the knowledge, so it's part of everything. (Viola)

Only one of the teachers said she was familiar with the concept of CT. Her view was that it is included in everyday school activities.

Thus, despite having attended the SC with their students, the teachers had rather vague ideas about what they thought the students should be learning about programming. In their answers, they mainly described a fun day and that the students had collaborated well. The teachers who mentioned connections to school subjects highlighted technology and mathematics. Technology was made visible by students operating robots, and mathematics in the use of logical thinking to solve tasks. The term computational thinking (CT) was not familiar to more than one of the teachers, whose opinion was that it is included in almost all learning. This may be true, but as this teacher did not develop her description, the understanding of the term still seemed somewhat vague.

5 Conclusions

Learning can take place in different environments and contexts, in leisure time as well as in school. The lessons observed as part of this study are not delivered in school, although the students participated in the lessons as part of their school day. It can therefore be valuable to study how teachers who bring their students to SCs to try out programming reflects about the teaching activities. By analysing the lessons given at SCs with the support of DFL (Kress & Selander, 2017; Selander, 2017), it becomes visible how SCs' focus on interest and engagement permeates the teachers' perceptions of the lessons. What teachers recognise as knowledge within programming in particular is not as clear, for which there may be various explanations, as will be discussed later.

There may be various motives for schools using SCs in technology education. Teachers' documented lack of competence in programming (Skolinspektionen, 2019) may be one reason, something that is also confirmed by the teachers in this study. Although it is not possible to draw any general conclusions from interviews with only five teachers, it appears that the teachers in this study felt that they themselves had too little knowledge to deliver programming education. That is why they chose to take their pupils to the SCs. Some of the teachers also reported that they did not have enough digital equipment at the school. This is something that Samuelsson (2014), among others, has also written about and which could be somewhat compensated by the equipment available at the SC.

An important element of the overall learning situation, according to the DFL theoretical framework (Kress & Selander, 2017, Selander 2017), is what is recognised as knowledge. Sparf (2022) found that during programming lessons, students were challenged to use several different abilities, and that they solved programming assignments in different ways. As the schoolteachers in this study said that their knowledge of programming in general and CT in particular was limited, it may have been difficult for them to notice these abilities. The teachers had not yet formed their own interpretation of what was recognised as knowledge in the field. Rather, the teachers focused on how the students collaborated with each other and whether they found programming enjoyable.

It may also be important that both teachers and SC educators understand the different ways in which students' engagement can manifest (Sparf, 2021). Thus, it is not always the student who is the most emotionally engaged who comes up with the solution to, for example, a programming problem. In the teachers' statements, it becomes visible that the teachers did not pay attention to different types of engagement, equally: they say that the students had fun, but none of them describe how the students engaged cognitively while they solved their programming assignments.

It is seldom valuable to discuss how one way of learning is better than another (Kress, 2015). Nevertheless, it can be important to be aware that the intentions of the design for learning may differ between different types of educators (Insulander & Öhman, 2022). For example, the SC educators in this study had the primary aim of stimulating curiosity and interest in programming. However, several studies show that if students are to get as much learning as possible from a visit to an SC, both preparatory and follow-up work should be done together with their regular teachers at the school (e.g., Kisiel, 2005; Morentin & Guisola, 2014). In this case, this may be difficult, as the teachers explained that one of the main reasons for visiting the SC was that they did not themselves had the competence to teach the subject. There is therefore a risk that the lesson at the SC will be a one-off occasion that is not followed up. Morentin and Guisola (2014) argue that it is the teachers' task to ensure that the curriculum is followed, even when visiting the SC. In the present case, this becomes twice as difficult, in addition to teachers feeling that they lacked competence, the new curriculum was not yet in force and there were no formal guidelines to rely on. In this context, it is important not to blame the teachers. They all expressed a desire for more in-service training in programming. Rather, it can be seen as a positive bonus that the students got to try out programming for a few hours at the SC even before it became mandatory.

The findings indicate that the participating teachers had an interest in learning more about programming. Some of the teachers saw workshops at SC as a way to get in-service training, others wanted to take courses at the university. All teachers in the study hoped that in time, they would receive more education through their employers – and not least, time to plan for the teaching.

At the time of the study, it may be that goals such as arousing curiosity and interest in programming or getting students to collaborate during lessons at SCs were good enough. Learning can be followed up later on. It has now been a few years since programming was included as mandatory content in the technology (and mathematics) syllabus. As programming becomes something that more and more teachers are learning to master, there will also be opportunities to develop the collaboration between SCs and schools – so that, despite having somewhat different starting points, they are working towards the same goal of enabling more students to learn the basics of programming in a fun and engaging way.

References

Braun, V., & Clarke, V. (2006). Using thematic analysis in psychology. *Qualitative Research in Psychology, 3*(2), 77–101. http://doi.org/10.1191/1478088706qp063oa

Cetin, I., & Dubinsky, E. (2017). Reflective abstraction in computational thinking. *Journal of Mathematical Behavior, 47*, 70–80. http://doi.org/10.1016/j.jmathb.2017.06.004

Durak, H. Y., & Saritepeci, M. (2018). Analysis of the relation between computational thinking skills and various variables with the structural equation model. *Computers & Education, 116*, 191–202. https://doi.org/10.1016/j.compedu.2017.09.004

European Data Protection Supervisor. (n.d.). *Data protection.*

Insulander, E., & Öhman, L. (2022). The touring science centre- an example of collaboration between a museum and a school. In L. Björklund Boistrup & S. Selander (Eds.), *Designs for research, teaching and learning: A framework for future education* (pp. 123–136).

Ivarsson, J. (2019). Responsible intra-action in a science center. *Cultural Studies of Science Education, 14*, 169–176. https://doi.org/10.1007/s11422-018-9876-y

Jahnke, I., & Kumar, S. (2014). Digital didactical designs: Teachers' integration of ipads for learning-centred processes. *Journal of Digital Learning in Teacher Education, 30*(3), 81–88. http://doi.org/10.1080/21532974.2014.891876

Kisiel, J. (2005). Understanding elementary teacher motivations for science fieldtrips. *Science Education, 89*, 936–955. https://doi.org/10.1002/sce.20085

Krange, I., Silseth, K., & Pierroux, P. (2020). Peers, teachers and guides: A study of three conditions for scaffolding conceptual learning in science centers. *Cultural Studies of Science Education, 15*, 241–263. https://doi.org/10.1007/s11422-018-9905-x

Kress, G. (2015). Semiotic work: Applied Linguistics and a social semiotic account of Multimodality. *AILA Review, 28*. https://doi.org/10.1075/aila.28.03kre

Kress, G., & Selander, S. (2012). Multimodal design, learning and cultures of recognition. *Internet and Higher Education, 15*, 265–268. http://doi.org/10.1016/j.iheduc.2011.12.003

Kubacki, G., Elsholz, M., & Trefzger, T. (2012). Science center im Spannungsfeld zwischen Wissenvermittlung und Freizeitspaß. *Didaktik der Physik*. Früjahrslagung-Mainz.

Leijon, M., & Lindstrand, F. (2012). Socialsemiotik och design för lärande: två multimodala teorier om lärande, representation och teckenskapande. *Pedagogisk Forskning i Sverige, 17*(3–4).

Morentin, M., & Guisasola, J. (2015). The role of science museum fields trips in primary teacher preparation. *International Journal of Science and Mathematics Education, 13*, 965–990. https://doi.org/10.1007/s10763-014-9522-4

Samuelsson, U. (2014). *Digital (o)jämlikhet. IKT-användning i skolan och elevers tekniska kapital* (Diss.). Högskolan för lärarutbildning och kommunikation.

Sasson, I. (2014). The role of informal science centre in science education: Attitudes, skills and self-efficacy. *Journal of Technology and Science Education, 4*(3), 167–180. https://dx.doi.org/10.3926/jotse.123

Shaby, N., Ben-Zvi Assaraf, O., & Tal, T. (2017). The particular aspects of science museum exhibits that encourage students' engagement. *Journal of Science Education and Technology, 26*, 253–268. https://doi.org/10.1007/s10956-016-9676-7

Selander, S. (2017). *Didaktiken efter Vygotskij. Design för lärande.* Liber.

Selander, S., & Kress, G. (2017). *Design för lärande- ett multimodalt perspektiv.* Studentlitteratur.

Silfver, E. (2019). Gender performance in an out-of-school science context. *Cultural Studies of Science Education, 14*, 139–155. https://doi.org/10.1007/s11422-017-9851-z

Skolinspektionen. (2019). *Digitala verktyg i undervisningen – matematik och teknik i årskurs 7–9*. Granskningsrapport Dnr 400-2018-6938.

Skolverket. (2022). *Läroplan för grundskolan, förskoleklassen och fritidshemmet 2022.* https://www.skolverket.se/getFile?file=9718

Sparf, M. (2020). *Grundskoleelevers design i lärande: en studie om lärprocesser i programmering* (Licentiate thesis). Linköpings universitet.

Sparf, M. (2021). "I am magic!" Pupils' engagement when designing in learning programming. *Designs for Learning, 13*(1), 35–43. http://doi.org/10.16993/dfl.168

Sparf, M., Löfgren, H., & Kreitz-Sandberg, S. (2022). Design for learning programming: Approaches taken by novice learners. *NorDiNa, 18*(1), 6–22. https://doi.org/10.5617/nordina.8251

Svenska Science Centers. (2018). *Presentation av Science Centers 2018.*

Vetenskapsrådet. (2017). *God forskningssed* [*Elektronisk resurs*] (Reviderad utgåva). Vetenskapsrådet.

Wing, J. M. (2006). Computational thinking. *Communications of the ACM, 49*(3), 33–35. http://doi.org/10.1145/1118178.1118215

PART 3

Computational Thinking and Programming in Technology Teaching

CHAPTER 13

Introducing Programming in an Early Primary Technology Classroom

The Distinction between Human and Robot

Astrid Berg and Cecilia Axell

Abstract

Since 2018, programming is a content in the technology subject in Sweden. Thus, teachers must develop new subject-specific competence to be able to realize their teaching in and about programming. This is especially challenging for primary teachers since primary technology education is a young subject and lacks a common professional base of proven experience. Research focusing on the classroom practices that are now taking form, and which are based on teachers' use of tutorials provided from different resources, is scarce. Hence, our understanding of which programming-related knowledge is possible to develop through participation in these practices is very limited.

As a novice, understanding the meaning of programming assumes an understanding of what a computational device may—or may not— 'understand' in relation to a human. When it comes to introducing early primary pupils to the concept of programming, there are examples of tutorials describing activities that focus on this very issue. In the study reported in this chapter, we explore an activity during an introductory lesson in programming in an early primary classroom, where the teacher used such a tutorial aimed to prompt reflections about *the differences between a human and a robot.* The aim of the study was to explore what content is constituted and hence what knowledge pupils are enabled to develop during this introductory activity.

The results showed that the constituted content focused on a central difference between human and robot; humans, as opposed to robots, have own will and ability to think. However, the analysis also showed that the pupils had ideas beyond this rather narrow content, and that classroom conversations with the youngest pupils about *the differences between a human and a robot* are, in several ways, challenging to orchestrate.

Keywords

programming – technology education – primary education

© ASTRID BERG AND CECILIA AXELL, 2024 | DOI:10.1163/9789004687912_013

1 Introduction

Today, society is increasingly dependent on and shaped by computer technology. In response to this ongoing development, countries across the world have introduced computational thinking and, as a part of this, programming into their national curricula. Since 2018, programming is a content in the technology subject in Sweden, ranging from school year 1 (age 7) to 12 (age 19). However, although programming is often emphasised as new content in Swedish primary school education, it has existed, with different purposes and to varying degrees, since the latter part of the 1960s. An important difference from previous attempts to include this content in schooling is that programming instruction in Sweden today includes all students, from the lower ages in primary school to upper secondary school (Rolandsson, 2019; Swedish National Agency for Education, 2022).

As a result of the curriculum change, teachers must develop new subject-specific competence to be able to realize their teaching in and about programming (Sentence & Csizmadia, 2017). The task of designing teaching activities that enable early primary pupils to develop technology-specific knowledge in terms of programming is a specifically challenging enterprise. Not only is programming a new content area, but primary technology education is a young subject and lacks a common professional base of proven experience (cf. Stein et al., 2007). Accordingly, resources such as the internet and in-service training provide primary school teachers with tutorials on how to stage programming activities in the primary classroom. When it comes to introducing early primary pupils to the concept of programming, there are examples of activities focusing on the difference between humans and robots.

As a novice, understanding the meaning of *programming* assumes an understanding of what a computational device may—or may not—'understand' in relation to a human. This understanding may be formulated as being able to discern the differences between the artificial and the human mind. Spektor-Precel and Mioduser (2015a, 2015b) describe this as understanding "Theory of Mind" (ToM) and "Theory of Artificial Mind" (ToAM).

Prior research shows that children form several different conceptions about computers. The interdisciplinary literature review by Rücker and Pinkwart (2016) reveals that children with little experience of computers are unsure whether computers can think for themselves; e.g., whether they see them as intelligent entities, unlimited and potentially omniscient databases, or as programmable artefacts. There is also research indicating that experience of programmed technology, such as robots, may yield a more highly developed viewpoint that reflects a distinction between biological life and intelligent machines (Bernstein & Crowley, 2008). In other words, the conception of 'computers' covers a dimension ranging from something being understood in purely psychological to purely physical terms. In the perspective of teaching and learning programming, this conceptual pattern that students bring to the technology classroom may offer significant challenges for the teacher (Rücker & Pinkwart, 2016).

Research focusing on the classroom practices that are now taking form, and which are based on teachers' use of tutorials provided from different resources, is scarce (Stigberg & Stigberg, 2020). Hence, our understanding of which programming-related knowledge is possible to develop through participation in these practices is very limited. In the light of this, it appears to be urgent to explore early primary classroom practices (Kjällander et al., 2016).

In the study reported in this chapter, we were interested in exploring an activity during an introductory lesson in programming in an early primary classroom, where the teacher used a tutorial provided during in-service training. The tutorial aimed to prompt reflections about *the differences between a human and a robot*. The aim of the study was to explore what content is constituted and hence what knowledge pupils are enabled to develop during this introductory activity.

2 Theoretical Starting Points and Central Concepts

2.1 *Constituted Content and Learning—A Sociocultural Perspective*

Since the aim of this study is to explore what content is constituted in the early primary technology classroom, we choose a theoretical starting point in sociocultural knowledge theory. From such a perspective it is assumed that, by participating in a social practice, one will have the opportunity to appropriate the competencies, values, communicative patterns, and ways of solving problems, etc., that are characteristic of the studied practice. In other words, the content—what we traditionally refer to as 'knowledge'—is embedded in practice. The goal of instruction is therefore to help students to engage with the forms of thinking and doing that distinguish the practice to which they are being introduced; i.e., to engage in the activities, ways of talking, and using cultural tools that, for example, engineers or computer scientists do (Säljö, 2014). With such basic assumptions, one can, through analysis of pupils and the teacher's tool-mediated and goal-oriented actions distinguish the content of an ongoing classroom activity and say something about what is possible to learn through participation (Wertsch, 1998).

2.2 *Theory of Mind and Theory of Artificial Mind*

As a novice, understanding the meaning of programming assumes, to some degree, an understanding of what a computational device may, or may not, 'understand' in relation to a human. In other words, this understanding may be formulated as being able to discern the differences between the artificial and the human mind. Spektor-Precel and Mioduser (2015a, 2015b) describe this understanding as understanding Theory of Mind (ToM) and Theory of Artificial Mind (ToAM). Theory of mind, as a personal capability, refers to the ability to imagine and understand other people's thoughts; i.e., knowing that they know, feel, or believe things which are different from our own, and that other people's decisions are based on desires and beliefs which are different from our own, and dependent on environmental conditions (Premack & Woodruff, 1978; Spektor-Precel & Mioduser, 2015a, 2015b). Theory of Artificial Mind, on the other hand, means understanding that a robot has no will; its actions depend on direct operation or programming. Spektor-Precel & Mioduser (2015a) further describe ToAM as understanding that "robots behave according to environmental conditions and make decisions accordingly (rule understanding), and that robots behave according to either direct operation or human programming that are dependent on environmental conditions" (p. 338).

Pea (1986) outlines a critical aspect of what it means to say that a computer reads a program when emphasizing that "programming is a formal system that interprets each part of a program (instructions to it) in terms of rules that are mechanistic" and adds that "while people are intelligent interpreters of conversations, computer programming languages are not"'. Hence, Pea (1986) explains how this violates human conversation: a computer "cannot infer what a speaker means if she is not absolutely explicit, whereas a listener in a human–human conversation can query the speaker for clarification" (p. 26).

Anthropomorphism is an aspect that can be connected to ToM. Anthropomorphism refers to a "tendency to imbue the real or imagined behaviour of nonhuman agents with humanlike characteristics, motivations, intentions, or emotions" (Epley et al., 2007, p. 864). This is a common feature in popular culture, such as fictional literature and films. Animals, or other representatives from nature, are depicted as behaving and thinking as if they were human. Also, inanimate technological objects are also often portrayed as having a human mind (Axell, 2015; Axell et al., 2022).

3 Methodology

This study was conducted within a qualitative research tradition, and it analyses the constituted content in the classroom from a sociocultural perspective

(Säljö, 2014). Hence, the study has a descriptive character, including detailed descriptions of who (teacher and pupils) does what with which tools and why.

3.1 *Data Collection*

The study was carried out in year one in a primary school in a medium-sized Swedish city. Data were collected through observations of a teacher-orchestrated activity that initiated an introductory lesson in programming in a technology classroom in early primary school (pupils aged 7). We requested a number of teachers if they were interested in participating in a study which aimed to investigate how programming is introduced in year 1 in primary school. The participating teacher is the one who gave a positive response to our request. She is an experienced teacher and has worked as a primary teacher for more than 30 years.

Data collection was conducted in a class with approximately 20 pupils. Since the teacher split the class in two separate groups—Group 1 and 2—and only one group at a time stayed in the classroom, the activity was carried out two times in a row: first with the pupils in Group 1 and then with the pupils in Group 2. The activity was documented by the researchers using three video cameras and three separate microphones. During the activity, one camera was placed at the back of the classroom, and the other two cameras and microphones along the right and left sidewall, respectively, of the classroom. The researchers stood behind, or moved between, the three video cameras to follow the teachers' actions including teacher–pupil interactions.

Ethical considerations were followed according to the recommendations outlined by the Swedish Research Council (2017). The teacher gave written consent to participate in the study. Also, written information was sent to the pupils' legal guardians, including a description of the study, a letter of consent regarding video and audio recording of the pupil in class, and information saying that they had the opportunity to withdraw their consent at any time. The names of the pupils are pseudonyms for the purpose of anonymising all participants. It was clearly formulated that the participation was voluntary, and that the participants had the right to withdraw at any time without giving any reason. The pupils' consent was handled by their parents, and the consent was given in writing. A few pupils did not get their legal guardians' consent. These pupils participated in the activity, but their statements and actions during the activity were disregarded in the study. The data was stored and protected in accordance with the General Data Protection Regulation (GDPR).

3.2 *The Activity*

We, as researchers, interpret the learning objective of the activity as prompting the pupils to reflect upon the difference between a human and a robot. The activity was carried out according to the tutorial: the teacher staged a sequence of events well in advance of the start of the lesson by turning off the light and opening a window in the classroom. Concurrently, it was cold and dark in the classroom when it was time for the pupils to enter. The teacher began the lesson by stating it was "somewhat dark and cold in here". In response to this, one pupil lit the lamp, and another closed the window. Then followed a conversation between the teacher and the pupils focusing on the difference between the human mind and the robot mind, and on the meaning of the concept of programming.

3.3 *Analysis*

The analysis of the video and audio recordings proceeded through three phases. Phase 1: We segmented the classroom conversation into episodes—independent units consisting of pieces of dialogue that shared the same focus (Gee & Green, 1998). Boundaries were set by shifts in the dialogue and/or activities. Phase 2: We transcribed each of the episodes from the video and audio recordings, and these were then translated to English. The transcription work focused on what was communicated by whom, verbally. Phase 3: The analysis of the empirical material focused on the constituted content of the classroom. The content was interpreted by studying who did what, using which tools and communicative strategies, and for what purposes.

4 Results

The analysis of the activity shows that the teacher led the pupils towards the difference between human and robot. Consequently, a ToAM perspective became central to make the pupils able to discern the role of programming. Hence, the content constituted in the classroom can be described as different aspects of this overarching content or object of learning. Since the constituted content in Group 1 and 2 was similar, we choose to let the results section be framed by a chronological description of how the activity evolved in Group 1. However, we supplement with an episode from Group 2, which in its content differs slightly from the corresponding episode in Group 1. The section is divided into three parts, A–C, corresponding to four aspects of content that are constituted during the activity and which follow the chronology of the lesson: Part A, Aspect 1 and 2: Humans understand implicit messages and act on them, and robots do not act on implicit messages due to a lack of emotions and physical characteristics of a human body; Part B, Aspect 3: We must tell robots what to do; and Part C, Aspect 4: We do what we want, and think for ourselves.

4.1 *Part A, Aspect 1 and 2: Humans Understand Implicit Messages and Act on Them, and Robots Do Not Act on Implicit Messages Due to a Lack of Emotions and Physical Characteristics of a Human Body*

Before the children enter the classroom, the teacher has posted a picture of a robot on the smartboard (Figure 13.1).

The teacher has also turned off the ceiling light in advance and opened a window. After the pupils have entered the classroom and sit down on their chairs, the teacher says: "it's horribly dark in here /.../ it's cold in here too." A pupil pays attention to the fact that the window is open, and the teacher confirms this with an "oops". One of the pupils—Alex—closes the window, and another—Kim—turns on the light. The teacher then says: "So...now...something happened here. What happened?" One pupil answers that it became bright, and that Kim turned on the light, whereupon the teacher responds: "But I didn't say that Kim should turn on the light, did I? I just said it was dark." The pupils face this claim with silence, and the teacher repeats the question: "Why did Kim turn on the light then when I said it was dark?" One pupil replies: "so that we can see" but the teacher questions this: "although I didn't say turn on the light." Another pupil then replies "but Kim *understood* that you meant turn on the light." In other words, the pupil's reasoning expresses an understanding

FIGURE 13.1
The sketched robot that was displayed on the smartboard in the classroom
SOURCE: HTTPS://WWW.FREEIMAGES.COM/VECTOR/CARTOON-ROBOT-4848433

of Theory of Mind—people understand what someone is thinking, even though this someone does not explicitly express it.

However, the teacher is not content with this answer, and asks the class: "She understood, how come she could understand that?" Nobody answers and the teacher repeats the question. However, it appears that the pupils do not know how to relate to the question—they are all silent. Finally, someone answers: "because you said it was dark" and "if you turn on the light then it won't be as dark, then it will be brighter". Obviously, the pupils instead focus on the logic of the event—there was a problem and what was done solved the problem—rather than on the ability of humans to 'read between the lines'.

The teacher then turns to the smartboard and the picture of the robot and asks a new question: "If Kim had been a robot, would she have turned on the lamp then?", whereupon the following line shift takes place:

Harper: No
Teacher: Why not
Harper: Because…it has no emotions.
Teacher: No emotions. What do the rest of you think?
Logan: The robot may not reach up to the lamp button or it may not be able to tell the difference between dark or light.
Teacher: It may not see a difference. Ok.

With her question, the teacher here contrasts human with robot. In this way, the characteristics and actions of a robot, and thus the pupils' perceptions of these characteristics and actions, can emerge in the light of what the pupils have just experienced about humans (humans turn on the light when someone says it is dark). With his answer, Harper pinpoints the critical aspect emotions; a crucial difference between human and robot is that the robot lacks emotions (cf. Turkle, 2005). However, there is no further exploration of what the lack of emotion means more explicitly, how it may explain why the robot would not turn on the light. When the pupil talks about the robot lacking emotions, one possible interpretation is that it means that the robot lacks human senses—sight, hearing, touch etc.—and another that it lacks sensors. Other possible interpretations are that the robot cannot empathize with the experiences and feelings of others (of darkness) or 'read between the lines.' What complicates the interpretation is that the Swedish words for emotion and sense (*känsla* och *känna*) are similar. Unlike Harper, Logan's reasoning focuses on the physical limitations of the robot—it cannot reach the switch or *tell* the difference between light and dark.

In Group 2, where the teacher asked the same question ("if NN was a robot…"), focusing on whether NN as a robot would have closed the window, the pupils' reasoning is in line with Logan's in Group 1; the robot does not close the window due to 'bodily' restrictions:

Riley: It might be too small.
Teacher: Maybe it's too small. But what if it's as big a robot as I am.
Jaden: It can rise.
Teacher: What if it's like me [as big as me].
Madison: Maybe it isn't cold.
Avery: It has poor hearing.
Teacher: It may have poor hearing. Yes.
Michael: It may not have any ears.

Madison's suggestion, that the robot isn't cold, may be interpreted as an expression of the fact that it cannot *perceive* coldness. The suggestions from Avery and Michael are instead about the robot not being able to *hear* (the teacher saying that it is cold in the room). This reasoning may originate in the robot picture on the smartboard; it lacks 'ears' and, thus, can't hear. Although the pupils answer in the negative to the teacher's hypothetical question, their arguments stem from what appears to be a view of robots as both physical machine and thinking human. The reason it doesn't close the window is because it (as a machine) lacks certain human qualities (emotion, touch, sight, hearing, tall 'body'), rather than due to it not being able to understand what the teacher is thinking and, through its own will, act accordingly (reasoning according to TOM). In other words, arguments that seem to stem from an understanding of ToAM are not expressed in pupils' reasoning. On the other hand, the pupils' answers can be interpreted as a reflection of the contrast inherent in the question; that is, the pupils focus on what distinguishes the physical robot body from the human one and 'lose' the robot's mind and its limitations. The cartoon robot on the smartboard can further contribute to focusing on the physical comparison. Another aspect in this context is the hypothetical nature of the question. Spektor-Precel and Mioduser (2015) argue that a robot's hypothetical actions such as "Can the robot help find a dog?" are harder for children to embrace compared to a robot's concrete behaviour "Which balloons will the robot pull?" Considering this, the pupils' responses could be interpreted as turning to what they can embrace; that is, physical differences.

In Group 1, the teacher continues to challenge the children when she asks why Alex closed the window:

Teacher: It was cold in here, too. And then you say the window is open. It was Alex who closed it. Why did he do that? Why did you [Alex] close the window?

Alex: You said it was cold.

Teacher: I said it was cold, yes. Why is he closing the window when I think it's cold then? What are you [Robin] thinking?

Robin: Because you were cold.

Teacher: Because I was cold. That was very nice of you [Alex] then, I think. [...]

Arden: So that we don't turn into ice cubes.

Teacher: We don't want to turn into ice cubes, no. [...] then you [Alex] understood that we should close the window.

This time, too, the conversation ends up in a kind of dead end, where the pupils' reasoning is about the logic of events. Conceptions that somehow describe that Alex, as a human being, can understand what other humans (the teacher) think, are not expressed. In other words, the teacher seems to have difficulties finding an entrance to introduce reasoning about ToM. However, Robin and Arden implicitly describe that human decisions are based on perceptions and desires, which in turn depend on surrounding factors (cold, dark) (cf. Spektor-Precel & Mioduser, 2015a).

In summary, during this part of the activity, a content is constituted that is made up of the pupils' verbally described ideas about the human mind and the 'mind' of the robot; humans can understand what someone else means without the latter explicitly expressing it. The pupils also express that robots lack emotions as well as certain physical human attributes. How this may explain why a robot would not close a window in a cold room is, however, not evident in the pupils' statements.

4.2 *Part B, Aspect 3: We Must Tell Robots What to Do*

After the teacher and pupils have established that Alex closed the window, because otherwise they would "turn into ice cubes", she shifts the focus back to the robot. She contrasts the robot's actions with those of humans and asks: "How do robots act, then?" One pupil interprets the question literally, focusing on the movement of the robot and replies that: robots "kind of walk like this" and represents a robotic movement with their arms. She justifies this by saying that "if they walk [...] as people walk, then their arms can be broken off." The teacher then takes the robot's action of walking as a point of departure, and links it to the question of what might govern such an action:

Teacher: Do they walk because they don't want to get cold...and close the window, or do they walk and turn on the light because they think it's too dark?

Scattered "no" answers from the pupils.

Logan: Then we have to tell them to do it.

Teacher: Then we have to tell them. Yes, that's right. That's the difference with a robot, and we tell robots, maybe not with our voice, but it's when we decide how a robot should work. Or a machine.

When the teacher relates the robot's actions to human actions, several pupils deny that the robot would close the window. Whether this denial is about them understanding ToAM, or is in line with previous reasoning, grounded in the notion that robots cannot perceive cold, is admittedly hard to know. Logan's reasoning, on the other hand, shows that he has distinguished a critical aspect of ToAM—the robot does nothing by itself, the human controls the robot's actions. Possibly it becomes easier for the pupils, as Logan here, to discern ToM when the teacher describes a robot's decision-making as identical to that of humans and asks pupils to relate to it. In other words, the pupils may regard it as unreasonable that a robot—like a human—does not want to "get cold" and therefore closes the window, when such a scenario becomes explicitly presented to them.

In the episode above, the teacher confirms Logan's answer and describes the relationship between robot and human: "we [humans] tell robots [...] decide how a robot should work". The teacher then introduces the concept of programming: "what will be our focus now is robots, it's programming". She then relates the concepts of programming to what has just happened in the classroom, and contrasts the relationship between humans and programming with robots and programming:

Teacher: Kim went and turned on the light even though I hadn't programmed her, and you [Alex] went and closed the window even though we hadn't programmed you, because you're [both] humans. But with robots, you program it, so it knows what to do. When you say different things or press different buttons. And that's what we're going to practice a bit.

Here, the teacher describes a robot as something that humans must program "so it knows what to do". The very act of programming is described as verbally expressing something, or "pushing buttons".

In summary, the content that is constituted during this part of the activity focuses the role of programming in human–robot relations; robots, unlike humans, must be 'programmed', and the humans program the robot by "telling" it what do (as described by both pupils and teacher) or by "pressing buttons" (teacher).

4.3 *Part C, Aspect 4: We Think for Ourselves, and We Do What We Want and What We Know Will Be Good*

Also in Group 2, the teacher leaves the hypothetical question ("if I was a robot") and tries another question: "Why does a robot do different things [...] why is he doing what he's doing?" One pupil replies "because he is programmed to do so", and thus introduces the term 'programming' into the classroom. The teacher confirms this using the same metaphorical formulation as in Group 1; "it's someone who has told him what to do." Then she asks if a robot "knows he should close the window if it's cold?". Just like in Group 1, several "no" answers are heard, but also some "yes" answers. After a brief pause, as if considering the mixed answers from the pupils, the teacher clarifies "*if* you have not programmed that, when you hear that someone is cold, then you should close the window. You must have programmed [the robot]."

The teacher then changes perspective and asks "humans, does anyone have to program us?" whereupon many of the pupils answer "no". One pupil says that "we do what we want" and another "we think for ourselves." In other words, the pupils implicitly express an understanding of ToAM—*unlike robots*, we do what we want and think for ourselves. Hence, it seems that the teacher's way of asking the pupils to relate to a description of human action as identical to that of the robot, helps them to discern and express ToAM and the relationship between ToAM and ToM. The teacher affirms the pupils, adding "we do what we want or what we know will be good." Here, the teacher adds another aspect of the difference between the actions of humans and robots; human actions, as opposed to robot action, can have a 'good' purpose. Presumably, 'good' refers to what the children described: so "we wouldn't get cold" and "turn into ice cubes". However, this aspect is not further elaborated.

In addition to the aspect that humans program the robot, the question "do we need to program us humans" also leads aspects focusing that we humans (unlike the robot) "think for ourselves", "do what we want" and "what we know will be good" being expressed.

4.4 *Summary of Results*

As shown in Table 13.1, the overarching content focus during the lesson was the differences between a human and a robot. Additionally, the aspects of this content constituted during the activity mainly concerned ToM. Aspects that the pupils expressed, and that we interpret as concerning ToAM, is *humans must program robots*. Lastly, the aspect *robots do not act on implicit messages due to lack of emotions and physical characteristics of a human body* (*such as ears to hear with*), as expressed by the pupils, may be interpreted as an expression of attributing the robot a human mind. However, we also acknowledge the possibility of other interpretations.

TABLE 13.1 Summary of the content that took form in the classroom and potentially critical aspects

Aspects of the overarching content *the differences between a human and a robot* constituted in the classroom
Differences related to ToM
– Humans understand implicit messages and act on them to satisfy other humans' (in this case) implicit desires
– We think for ourselves, and we do what we want and what we know will be good
Differences related to ToAM
– We have to tell robots what to do
– Robots lack emotions
Differences related to physical appearance
– Robots lack physical characteristics of a human body, such as ears

5 Discussion

In the following section, we discuss the main findings of the study, which includes the content constituted during the programming activity and the challenge of elaborating on critical notions as they are expressed in the classroom, what role the teacher's questions play, the ways the pupils express their meaning, as well as the aspects of ToM and ToAM in the discussions during the activity. Identified potentials for development of the classroom discussion are also presented.

5.1 *The Content Constituted in the Classroom*

The analysis of the classroom conversation shows that the teacher has difficulty finding questions that open the door to reasoning about the differences between human and robot minds. Eventually, however, aspects of the content *the differences between a human and a robot* is constituted in the classroom. These aspects can be summarised as follows: *humans, unlike robots, do what they want to do and think for themselves; and the human must therefore program or 'tell' the robot what to do.*

During the classroom conversation, other aspects of ToM and ToAM are also mentioned, but the meaning of these is neither confirmed nor further explored; one pupil believes that robots lack emotions, and another that we (humans) can understand what others mean without them saying it ("we understood that you meant 'turn on the light'"). These aspects could have enriched the lesson, because they constitute central aspects of TOM and ToAM. For example, Turkle (2005) suggests that the line between the artefacts and humans is drawn based on emotion: computers can think, but only humans can feel. When the pupils were pressed on the question why Kim turned on the light, and one pupil answered "we understood…", he expressed the aspect of ToM that is about being able to get into the thoughts of others. The teacher also mentions at one point that humans act with the aim of doing good ("we do what we know will be good.") This too is a central aspect of ToM and ToAM. A robot does not act with someone's best interests in mind. Rather, and as described by Mioduser et al. (2009), a robot's purpose originates from its function; e.g., its purpose is to execute activities that derive from humans' problems or needs. However, this aspect was not developed by the teacher and the pupils, who, therefore, did not have a chance to reflect upon it.

To conclude, the pupils had the possibility to develop knowledge during the activity that is about central aspects of ToM and ToAM in terms of one's own will and ability to think. However, the classroom conversation shows that the pupils have ideas beyond this rather narrow content. The challenge for the teacher is therefore to identify, pick up, and elaborate on such critical notions (as well as on his/her own contributions) to make it possible for the pupils to discern what e.g. *a robot lacks feelings* or *humans do good* may mean. This assumes, however, that tutorials and in-service teacher training focuses aspects of ToM and ToAM that are potentially crucial to early primary pupils' learning, and how these may be mediated. The analysis in this study implies that the teacher was not provided with such support.

5.2 *The Importance of the Questions*

From the described classroom conversation, it appears that the nature of the teachers' questions seems to play a role in pupils' ability to express themselves in relation to ToM and ToAM. Regarding ToM and the question why NN opened the window, only one pupil refers to aspects of ToM—"we understood what you meant." On the other hand, since the reasoning behind decisions such as, in this case, closing the window, happens "by itself", it is reasonable that the pupils have difficulties putting it into words. When asked the more concrete question of whether we humans need to be programmed the pupils answered "no" in unison. The question also spontaneously opened for further descriptions of ToM—humans have a will of their own and think for themselves. In other words, this question seemed to be more productive.

As for questions aimed at opening up conversations about ToAM, these were also variously productive. The hypothetical "if I were a robot, I would open the window" prompted pupils to focus on differences in physical attributes between robot and human, rather than on ToAM. However, the concrete question of robots closing the window because they do not want to get cold seemed to be easier for the pupils to embrace. A possible reason may be that it becomes easier to distinguish a robot as a robot when you do not have to follow the detour of a human transformed into a robot. Even the question of whether we humans need to be programmed was productive in relation to ToAM, as pupils' responses not only described aspects of ToM, but also implicitly of ToAM—we can think for ourselves; that is, robots *cannot* think for themselves.

In summary, the results suggest that hypothetical questions as well as open-ended why-questions (why did NN close the window?) were more difficult for pupils to grasp than concrete questions. This is in line with previous research on children's and adults' perceptions of robots (Banks, 2020; Fussel et al., 2008; Mioduser, 2015a, 2015b; Thellman et al., 2022). In other words, the nature of the question appears to be an important aspect for teachers to be aware of. We also find it interesting that the teacher's concrete questions painted a picture of the human mind with aspects that belong to ToAM (do people need to be programmed), respectively, and the robot's mind with aspects that belong to ToM (do robots close windows because they do not want to get cold). With this kind of contrasting, it becomes possible to distinguish the human mind against the background of the robot's mind, and vice versa. In other words, beyond the concrete nature of the question, this contrasting approach could have contributed to the fact that the question seemed to help pupils discern and express thoughts about TOM and ToAM.

5.3 *Awareness of ToM, ToAM and the Use of Anthropomorphic Language*

The classroom conversation not only illustrates the teacher's difficulties in finding entrances to conversations about ToM and ToAM. It also shows that pupils' answers can be difficult to interpret. Examples of this are the pupils' arguments that a robot would not close the window when it is cold—"it has no ears"—and that pupil X closed the window—"so that we wouldn't turn into ice cubes". This could be interpreted as the pupils lacking understanding of ToAM and ToM, and that terms such as 'ears' and 'hearing' are expressions anthropomorphising the robot (Fussels et al., 2007). As we described above, however, the pupils' answers can be understood to be due to the nature of the question. Another contributing factor may be that the pupils lack linguistic resources. Levy and Mioduser (2008) emphasise that children's tendency to anthropomorphise robots can be understood as an expression of a lack of language. The authors describe children's use of anthropomorphistic language as "the children bring it closer to their own world, linking it to their understandings of themselves as intentional and emotional beings, creating personal meanings with respect to the robot's" (p. 351). Seen from this perspective, the pupils' way of focusing on things they have words for and their own experience of (ears, hearing) can give them the opportunity to formulate a motive for why the robot would not close the window (lacking ears and therefore hearing). Similarly, talk in terms of NN performing an action (closes the window) because it solves the problem that has arisen (risk of turning into ice cubes) is a familiar narrative from everyday life. Talking about it in more abstract terms linked to ToM can be more difficult for the pupils.

Although the teacher must be able to assess the pupils' pre-understanding and use it as a continuing starting point in the classroom conversation, it is difficult if what the pupils are able to express does not correspond to the meaning of the questions asked. Hence, we argue that an awareness of both the nature of the questions and the pupils' limited access to a 'language about mind' can therefore be considered important didactic knowledge. We suggest that one possible way for the teacher to help the pupils to develop an ability to talk about the different 'minds', may be to challenge their ideas and ask them to further develop their thoughts.

An additional aspect of the classroom conversation concerns how the teacher and pupils talk about the robot, which in this case is an imaginary figure projected on the white board (Figure 13.1). The robot is most often referred to as "it" in the classroom, but the term "he" also occurs. The teacher uses formulations such as "the robot knows that…" Furthermore, both pupils and teachers talk about programming in terms of "telling the robot", and the teacher describes programming as the human addressing the robot with a "you". Although this is close to humanizing robots (using words that normally apply to human activity) and their "mind" in speaking about them, especially when those pictured in the classroom have a humanlike appearance (see Thellman et al., 2022), it can contribute to blurring the image of the robot as a machine (cf. Levi & Midouser, 2008; Spektor-Precel & Mioduser, 2015b). In other words, it may be more difficult for pupils to be able to distinguish and understand ToAM and ToM, as two distinctly different minds, and thus to understand the prerequisites for programming.

Another consequence of using anthropomorphic language when talking about the robot is that the context of the lesson itself is at risk at being clouded. For the pupils to experience that they are in a technology classroom, and to have the opportunity to develop the ability to use the linguistic tools of the programming practice, it can be seen as important that the teacher introduces and consistently uses practice-specific terms. Zhang & Nouri (2019), for example, emphasise that the ability to use a computing vocabulary leads to students' internal communication of ideas being as "accurate and comprehensible as possible" (p. 17). A difficulty in this context, however, is that robots, as everyday objects, are becoming increasingly common, and that in our everyday interaction with the robot, a non-technological language is formed. Research shows that we interact with and talk about e.g. the robot lawnmower as if it has a mind, even though we do not believe so (Thellman et al., 2022). Another aspect that contributes to this is media and children's culture, in which the typical robot has a human-like body, is autonomous and possesses a human rather than artificial mind (Axell & Berg, 2023; Slangen et al., 2011). On the other hand, and due to its abstract nature, it can be overwhelmingly difficult to talk about the mind of the robot and the role of programming, especially with 7-year-olds, without using words that normally apply to the human mind and human activities. This means that as a teacher you need to navigate between the parallel everyday and technological terms and concepts, which of course is a big challenge. This should be particularly difficult for primary school teachers who have not themselves been trained within a programming practice and learned to use its specific linguistic tools in an authentic context. However, to consciously create space for meta-conversations with the pupils about what words are used in the classroom, and how they relate to technological ones, can help pupils orient themselves in the content that is constituted in the classroom (cf. Aikenhead, 1996).

6 Conclusions

To conclude, this study illustrates that classroom conversations with the youngest pupils about the learning objects ToAM and ToM are not straightforward but require consideration of the words. In particular, it demands knowledge of what kinds of questions are productive, an ability to pick up and create conversations about aspects of ToAM, to interpret pupils' answers and to be aware of the pros and cons of using anthropomorphic language. In other words, the question of how the content the difference between humans and machine can be treated when teaching very young pupils appears to be something that needs to be investigated more closely. The common choice to use the robot as a symbol of the computational machine when programming is introduced in early primary school is, in this context, worth reflection. What opportunities and limitations does the robot offer in relation to children's learning in and about programming, and how can teaching be designed to meet these in the best way? As artificial intelligence (AI) becomes a larger part of our everyday life, the question also arises as to whether the robot can continue to be used as a tool to mediate learning that is about being able to distinguish the human and the machine's mind as two distinctly different entities. This dilemma can be sensed when the teacher asks the pupils if a robot "knows he should close the window if it's cold?" Well, does it?

References

Aikenhead, G. S. (1996). Science education. Border crossing into the subculture of science. *Studies in Science Education, 27*, 1–52.

Axell, C. (2015). *Barnlitteraturens tekniklandskap: en didaktisk vandring från Nils Holgersson till Pettson och Findus* (Diss.). Department of Social and Welfare Studies (ISV), Linköping University. http://liu.diva-portal.org/smash/record.jsf?pid=diva2%3A784127&dswid=-7451

Axell, C., & Berg, A. (2023). "You give a little bit more love to animals than to robots" – Primary pupils' conceptions of 'programming'. *International Journal of Technology and Design Education*. https://doi.org/10.1007/s10798-023-09838-6

Axell, C., Berg, A., Hallström, J., Thellman, S., & Ziemke, T. (2022). Artificial intelligence in contemporary children's culture: A case study. In D. Gill, J. Tuff, T. Kennedy, S. Pendergast, & S. Jamil (Eds.), *PATT 39: PATT on the edge technology, innovation and education* (pp. 376–386). Memorial University of Newfoundland.

Banks, J. (2020). Theory of Mind in social robots: Replication of five established human tests. *International Journal of Social Robotics, 12*, 403–414. https://doi.org/10.1007/s12369-019-00588-x

Bernstein, D., & Crowley, K. (2008). Searching for signs of intelligent life: An investigation of young children's beliefs about robot intelligence. *The Journal of the Learning Sciences, 17*(2), 225–247. https://doi.org/10.1080/10508400801986116

Epley, N., Waytz, A., & Cacioppo, J. T. (2007). On seeing human: A three-factor theory of anthropomorphism. *Psychological Review, 114*(4), 864–886. https://doi.org/10.1037/0033-295X.114.4.864

Fussell, S. R., Kiesler, S., Setlock, L. D., & Yew, V. (2008). How people anthropomorphize robots. In *2008 3rd ACM/IEEE international conference on Human-Robot Interaction (HRI)* (pp. 145–152). IEEE.

Gee, J. P., & Green, J. L. (1998). Discourse analysis, learning, and social practice: A methodological study. *Review of Research in Education, 23*(1), 119–169. https://doi.org/10.3102/0091732X023001119

Kjällander, S., Åkerfeldt, A., & Petersen, P. (2016). *Översikt avseende forskning och erfarenheter kring programmering i förskola och grundskola.*

Levy, S. T., & Mioduser, D. (2008). Does it "want" or "was it programmed to..."? Kindergarten children's explanations of an autonomous robot's adaptive functioning. *International Journal of Technology and Design Education, 18*, 337–359. https://doi.org.10.1007/s10798-007-9032-6

Mioduser, D., Levy, S. T., & Talis, V. (2009). Episodes to scripts to rules: Concrete-abstractions in kindergarten children's explanations of a robot's behavior. *International Journal of Technology and Design Education, 19*, 15–36. https://doi.org/10.1007/s10798-007-9040-6

Pea, R. D. (1986). Language-independent conceptual "bugs" in novice programming. *Journal of Educational Computing Research, 2*(1), 25–36. https://doi.org/10.2190/689T-1R2A-X4W4-29J2

Premack, D., & Woodruff, G. (1978). Does the chimpanzee have a theory of mind? *Behavioral and Brain Sciences, 1*(4), 515–526. https://doi.org/10.1017/S0140525X00076512

Rolandsson, L. (2019). Programmering i skolan. *Vägval i skolans historia.* https://undervisningshistoria.se/programmering-i-skolan/

Rücker, M. T., & Pinkwart, N. (2016). Review and discussion of children's conceptions of computers. *Journal of Science Education and Technology, 25*(2), 274–283. https://doi.org/10.1007/s10956-015-9592-2

Sentance, S., & Csizmadia, A. (2017). Computing in the curriculum: Challenges and strategies from a teacher's perspective. *Education and Information Technologies, 22*(2), 469–495. https://doi.org/10.1007/s10639-016-9482-0

Slangen, L., van Keulen, H., & Gravemeijer, K. (2011). What pupils can learn from working with robotic direct manipulation environments. *International Journal of Technology and Design Education, 21*, 449–469. https://doi.org/10.1007/s10798-010-9130-8

Spektor-Precel, K., & Mioduser, D. (2015a). 5–7-year-old children's conceptions of behaving artifacts and the influence of constructing their behavior on the development of Theory of Mind (ToM) and Theory of Artificial Mind (ToAM). *Interdisciplinary Journal of e-Skills and Lifelong Learning, 11*, 329–345. https://doi.org/10.28945/2332

Spektor-Precel, K., & Mioduser, D. (2015b). The influence of constructing robot's behavior on the development of Theory of Mind (ToM) and Theory of Artificial Mind (ToAM) in young children. *Proceedings of IDC 2015: The 14th International Conference on Interaction Design and Children*, 311–314. https://doi.org/10.1145/2771839.2771904

Stein, S. J., Ginns, S. I., & McDonald, C. V. (2007). Teachers learning about technology and technology education: Insights from a professional development experience. *International Journal of Technology and Design Education, 17*, 179–195. https://doi.org/10.1007/s10798-006-0008-8

Stigberg, H., & Stigberg, S. (2020). Teaching programming and mathematics in practice: A case study from a Swedish primary school. *Policy Futures in Education, 18*(4), 483–496. https://doi.org/10.1177/1478210319894785

Swedish National Agency for Education. (2022). *Läroplanen för grundskolan, förskoleklassen och fritidshemmet. Lgr22.* [*Curriculum for the Compulsory school, preschool class and school-age educare. Lgr22*]. https://www.skolverket.se/getFile?file=9718

Swedish Research Council. (2017). *Good research practice.* Swedish Research Council.

Säljö, R. (2014). *Lärande i praktiken. Ett sociokulturellt perspektiv.* Prisma.

Thellman, S., de Graaf, M., & Ziemke, T. (2022). Mental state attribution to robots: A systematic review of conceptions, methods, and findings. *ACM Transactions on Human-Robot Interaction, 11*(4), 1–51. https://doi.org/10.1145/3526112

Turkle, S. (2005). *The second self: Computers and the human spirit.* MIT Press.

Wertsch, J. (1998). *Mind in action.* Oxford University Press.

Zhang, L., & Nouri, J. (2019). A systematic review of learning computational thinking through Scratch in K-9. *Computers & Education, 141.* https://doi.org/10.1016/j.compedu.2019.103607

CHAPTER 14

Students' Conceptions of Programmed Technological Solutions

A Basis for Organising Teaching

Anne-Marie Cederqvist

Abstract

Educational content with respect to programming and Programmed Technological Solutions (PTS) is commonly contextualised in practical activities such as analysing and designing PTS using instructional materials, e.g. BBC micro:bit. However, we still know little of how teaching regarding this content could be organised in order for students to progress in learning. The aim of this chapter is to analyse the synthesised results from two phenomenographic studies on ways students, aged 10–14, conceptualise PTS in order to shed light on PTS as a learning object, as well as what and how to teach about PTS in technology education.

The two included studies are conducted in a Swedish school context. The first study investigates the different ways students, aged 11–12, conceptualise PTS in the process of analysing structure and function in PTS. The second study investigates the different ways students, aged 10 and 14, conceptualise PTS in the process of designing and coding PTS. In the studies, the instructional material BBC micro:bit was used together with PTS from everyday life. The synthesised results have been further analysed from a learning progression perspective based on the phenomenographic view of learning. The results show variations in students' ways of conceptualising PTS. In keeping with the phenomenographic perspective, learning can be described as students initially seeing PTS as an undifferentiated whole, which they gradually learn to discern from its structural and functional parts. However, in the processes, there is a complexity of conceptualising PTS in relation to the changing contexts of everyday life and the instructional material BBC micro:bit. Based on the results, it is suggested that students should initially learn to analyse the structure and function of existing PTS in different contexts. This will provide them with the necessary conditions to progress their learning in activities where they design and code PTS.

© ANNE-MARIE CEDERQVIST, 2024 | DOI:10.1163/9789004687912_014

Keywords

programming – programmed technological solutions – teaching – primary students – secondary students – BBC micro:bit

1 Introduction

In Swedish compulsory school, educational content with respect to programming and Programmed Technological Solutions (PTS) has been introduced as part of technology education in order to develop students' understanding of how PTS work and are controlled by programming (see Swedish National Agency for Education, 2019/2022). The content is commonly contextualised in practical activities by letting students analyse and design PTS using different instructional materials, such as BBC micro:bit. However, many teachers feel uncertain of what students are expected to learn, and how to address students' difficulties (Vinnervik, 2022a). In addition, we still know little of how students conceptualise PTS in these kinds of activities and what their progression in learning might look like, which is important knowledge when making pedagogical decisions about what and how to teach. In order to shed light on these issues, we need to learn more about students' conceptions of PTS, since this is an important source for understanding PTS as a learning object from a technology education perspective. By identifying what becomes critical for students to discern of a learning object, i.e., what the critical aspects of a learning object are, it is also possible to clarify what to focus on in teaching in order to provide better learning opportunities. This chapter presents the synthesised results of two phenomenographic studies which investigate the ways students, aged 10–14, conceptualise PTS, both when analysing and designing them, to contribute knowledge that can support teachers when organising teaching on PTS and programming.

2 Background

2.1 *Programming in a Swedish Technology Education Context*

In 2018, programming was implemented in the Swedish curriculum for compulsory school. However, programming was not introduced as a separate subject, but as part of existing subjects such as mathematics, technology, and social science. This implies that programming as content in teaching and learning situations needs to be reflected upon and discussed in relation to the subject specific content and expected learning outcomes within the particular subjects. In technology education, this involves directing attention to digital technology in everyday life and developing students' ability to identify and analyse the structure and function of technological solutions that are controlled by programming (PTS), but also developing their ability to design new ones and control them with programming (Swedish National Agency for Education, 2019/2022). However, previous research shows that there are few details revealed in the curriculum about what and how to teach (Vinnervik, 2022a).

In technology education, programming and PTS is commonly contextualised by letting students take part in design activities where they use instructional materials such as BBC micro:bit or Arduino to create their own PTS, which they control with programming. As such, programming knowledge within technology education has mainly been shaped for students to gain practical experience on how basic instructions can be produced to control a technological solution (Vinnervik, 2022b). However, previous research (Barak & Zadok, 2009; Mioduser et al., 1996; Slangen et al., 2011) show that students have difficulties conceptualising how to produce instructions for controlling PTS. Barak and Zadok (2009) argue that this is based on students lacking the necessary technological knowledge involved for solving the task.

Therefore, in these types of design activities, it is important to highlight the technological knowledge involved, such as the structure and function of the intended technological solution, i.e., its dual nature, which is necessary to understand in order to be able to produce it (de Vries, 2005). de Vries describes how the function refers to human intentions and the need to change one state into another, and the structure is the appropriate physical structure of components for realising the function. Further, to understand the dual nature of technological solutions involves four types of technological knowledge: knowledge of the physical structure, such as properties of the solution; knowledge about the function, such as the functions to be fulfilled; knowledge about the relation between the structure and function, such as setting up an appropriate physical structure of components to realise the function; and knowledge of technological processes, as ways to turn structure into function (de Vries, 2006). Therefore, when teaching and learning technology in processes such as analysing and designing PTS, it is necessary to address knowledge related to understanding the dual nature of PTS.

2.2 *The Processes of Analysing and Designing PTS*

An important part of being able to understand the technological solutions that surround us in everyday life depends on being able to identify and analyse technological solutions. Thus, this has become an important activity in technology education, where students are presented with existing technological solutions, which they practice analysing from their structural and functional aspects. Understanding the dual nature of existing technological solutions also constitutes an important basis for being able to design new ones. That is, the processes of analysing and designing can be seen as going hand-in-hand with technology education. There are several factors that need to be analysed in relation to the real-world conditions and taken into consideration during the design process concerning structural and functional aspects, and as de Vries (2005) puts it, success in the design process depends on the extent to which the designer is able to make a fit between these factors. Slangen et al. (2011) previously described how students have difficulties in converting real-world conditions into a technological solution in terms of a code, hence, these difficulties need to be addressed in teaching.

A way to facilitate understanding of structural and functional aspects, both when analysing and designing, is to apply the systems thinking perspective to a technological solution (Cederqvist, 2021; Slangen et al., 2011). That is, seeing the structural aspects in terms of a set of parts that interact, and the functional aspects in terms of an entity where inputs are transformed to outputs (de Vries, 2005). Moreover, systems thinking might also explicate the design process and the movement back and forth between functional and structural aspects (de Vries, 2019). However, to be able to organise teaching that helps students understand these processes and the structural and functional aspects they involve, we need to learn more about the ways students experience the processes and what becomes critical for them to discern.

In technology education, it is common to use instructional materials, representations, tools and activities to visualise and facilitate understanding of phenomena. For example PTS is commonly contextualised in practical activities using instructional materials such as BBC micro:bit. However, instructional materials, representations, tools, and activities bring contextual factors to the teaching situation, which indicates a certain way to experience a phenomenon (Marton et al., 2004). This implies that the context of the situation in terms of, for example, instructional materials or representations, provides certain access to important aspects of the phenomenon that learners are expected to understand. Thus, the teacher needs to be aware of what the important aspects are, as well as the extent to which the context in the teaching situation can bring the important aspects of the phenomenon to the fore. This knowledge is paramount for allowing teachers to make pedagogical decisions on what to address and how to organise teaching and learning situations. Hence, by identifying students' ways of conceptualising a phenomenon and what becomes critical for students to discern in the contexts used in teaching situations, it is also possible to clarify what to be focused on in order to provide better learning opportunities.

3 Aim and Research Questions

The aim of this chapter is to analyse the synthesised results from two phenomenographic studies on ways students, aged 10–14, conceptualise programmed technological solutions (PTS) in order to shed light on PTS as a learning object, as well as what and how to teach about PTS in technology education. The study is guided by the research questions:

- In what ways do students conceptualise PTS in the processes of analysing and designing PTS?
- What aspects are critical to discern in order to conceptualise PTS in a powerful way?
- In what way could teaching on PTS be organised in order for students to learn about PTS?

4 The Phenomenographic Perspective on Learning

In phenomenography, learning implies developing an ability to experience a phenomenon in a new way, different from the way it has been experienced previously. Experiencing in this sense implies being able to see important aspects of the phenomenon. These important aspects are referred to as the critical aspects that are necessary to discern in order to understand the phenomenon in a more powerful way (Marton & Booth, 1997). In phenomenography, attention is therefore drawn to the pedagogical value of identifying how learners experience a phenomenon, as well as identifying what learners need to experience in terms of critical aspects, in order to understand a phenomenon in a more powerful way (Åkerlind, 2015). The main research interest in phenomenographic studies is in the variety of possible ways of experiencing a phenomenon in a group of learners. By analysing learners' qualitatively different ways of experiencing the phenomenon, a set of hierarchical descriptive categories can be created; these are referred to as the outcome space. The categories describe more or less complex ways of experiencing the phenomenon, which are characterised in terms of how many aspects of the phenomenon a learner is capable of discerning and having in focal awareness simultaneously (Marton & Booth, 1997). Further, by investigating the critical differences between categories, the critical aspects of the phenomenon can be identified.

The results from phenomenographic studies in terms of critical aspects can be used in teaching to bring forward what to focus on and what to put into the background. The aim is to organise teaching situations that help learners to see critical aspects of a phenomenon (Åkerlind, 2015), and help them to recognise the meaning of the critical aspects in relation to each other, i.e., to understand a learning object from its part-whole relationship (Ingerman et al., 2009). Therefore, in this study, the aim is to bring together knowledge from two phenomenographic studies on students' ways of conceptualising PTS and what aspects are critical to discern in the processes of analysing and designing PTS. This will suggest what to bring forward and focus on in teaching, as well as a way of organising teaching that can help students to progress their learning about PTS.

5 Method

When conducting a synthesis of primary qualitative research studies, the content in the existing studies relevant to particular research questions are comparatively analysed and synthesised (Eisenhart, 2017). This is also what synthesising in this context means: to combine two or more parts to a new whole, and new ideas which go beyond what the initial parts provide (Kakos & Fritzsche, 2017). The primary studies in this chapter were selected based on their presumed contribution to addressing the research questions. The studies are *Pupils' ways of understanding programmed technological solutions when analysing structure and function* (Cederqvist, 2020) and *An exploratory study of technological knowledge when pupils are designing a programmed technological solution using BBC micro:bit* (Cederqvist, 2022).

An interpretative synthesis was conducted, based on finding new ways to interpret the results in the primary studies in an inductive way, to develop understanding of PTS as a learning object as well as to gain an understanding of what, and how, to teach about PTS in technology education. The empirical data and results from the two phenomenographic studies were first comparatively analysed and synthesised on what aspects are critical for students to discern in order to conceptualise PTS in a powerful way. This first part of the analysis resulted in a set of critical aspects that was further analysed from a learning progression perspective, in keeping with the phenomenographic perspective on learning. That implies finding ways of expanding students' ability to discern critical aspects as parts of PTS, the relationships between the parts, and how they relate to the processes of analysing and designing PTS.

There are methodological challenges to consider concerning synthesis of primary qualitative research studies. In the single studies, there is information which is important to provide in order to understand the studies and to interpret their results. However, when re-using existing data, it is not possible to provide all the information that was provided in the primary studies, with the consequence that there may be a lack of contextual knowledge for understanding the results with respect to context (Kakos & Fritzsche, 2017). Therefore, in the following section the contexts of the primary studies will be described and a summary of the results from each of the studies will be provided.

6 The Point of Departure—the Two Phenomenographic Studies

This study takes its point of departure in two phenomenographic studies, which are conducted in a Swedish school context. The studies present students' conceptions of PTS in two different situations. In the first study, the interview study, the students are analysing existing PTS (Cederqvist, 2020). In the second study, the design study, the students are designing and coding a PTS (Cederqvist, 2022). The instructional material BBC micro:bit was used in both studies to represent PTS (see Figures 14.1 and 14.2).

The BBC micro:bit material consists of a programmable microcontroller that includes a LED display, programmable buttons, light and temperature sensors, a compass, an accelerometer, and a Bluetooth for connection to other micro:bits.[1] The microcontroller can also be connected to external components such as a speaker, a motor or a sensor by using crocodile clips. The

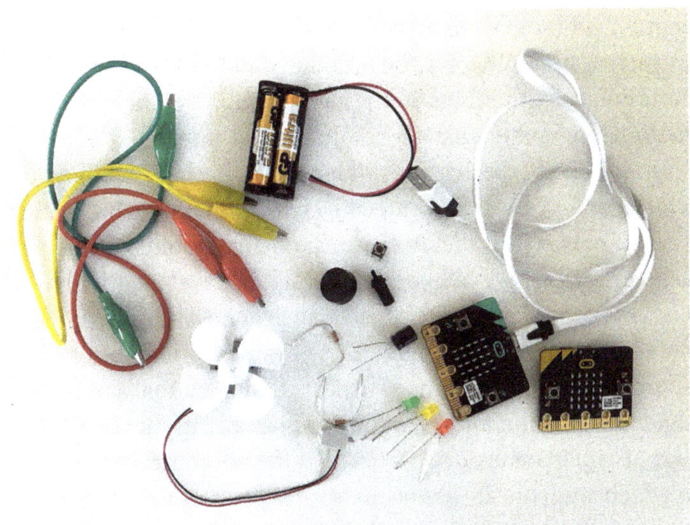

FIGURE 14.1 The BBC micro:bit microcontroller and external components

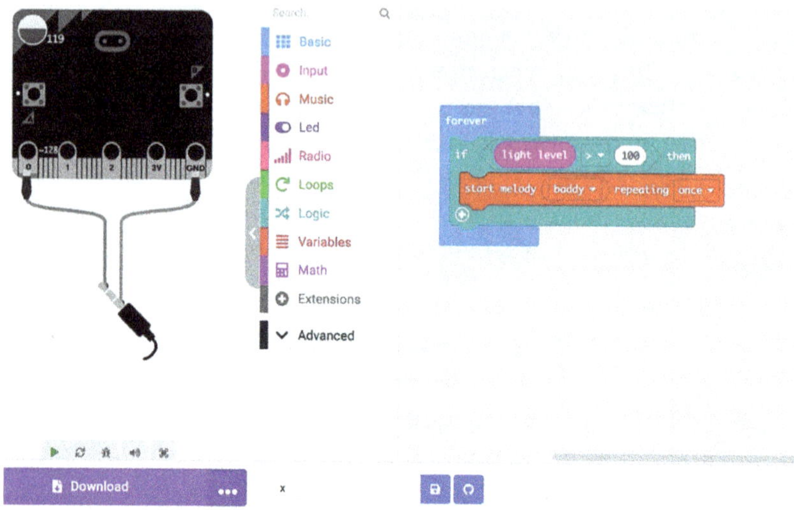

FIGURE 14.2 The block-based editor where blocks are snapped together to build a code

microcontroller can be programmed in either a text-based or a block-based editor. In the included studies, the block-based Microsoft MakeCode micro:bit editor was used (Figure 14.2). In the editor, students built their code by snapping blocks together.

6.1 *The Interview Study*

The interview study qualitatively investigates the different ways students conceptualise PTS when analysing structure and function. In semi-structured interviews students, aged 11–12, were asked to analyse structure and function of presented PTS such as BBC micro:bit constructions, and PTS in everyday life such as a TV remote control, a thermometer, a car key or a PTS of their own suggestion. The phenomenographic analysis resulted in an integrated outcome space (Figure 14.3), which shows that the students approach PTS differently, depending on which context PTS is presented in, i.e., the BBC micro:bit context or the context of PTS in everyday life. In the BBC micro:bit context, students initially approach the PTS as a whole device, which is developed based on their ability to discern increasingly more aspects, such as how the components work, or the flow of information. In the context of everyday life objects, students initially approach the PTS in terms of aspects that are visual and prominent, such as single components, and what the PTS does, i.e., its function. Thus, the different ways of conceptualising PTS in the different contexts are based on the extent to which students are able to discern aspects of PTS. In the outcome space (Figure 14.3), we can see that in Category 4 all aspects of PTS are discerned in each context, which indicates a more powerful understanding. At

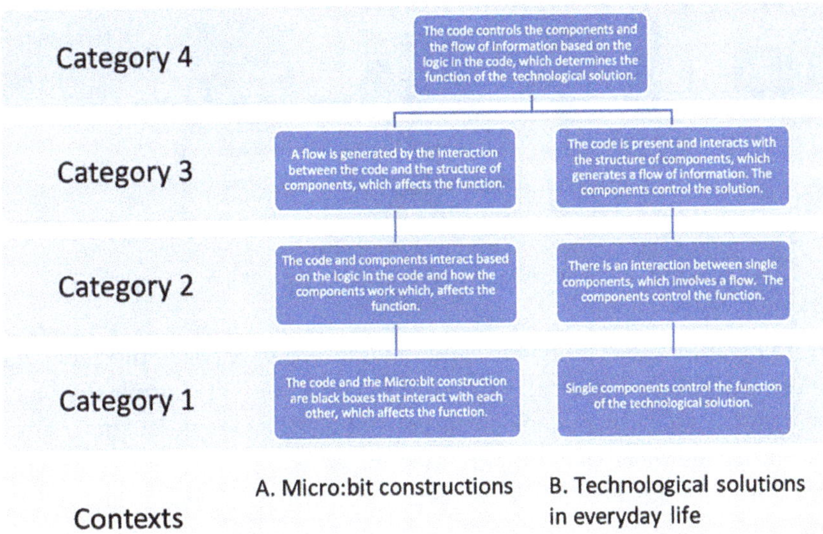

FIGURE 14.3 Pupils approach PTS differently, depending on which context PTS is presented in (from Cederqvist, 2020, figure 4, CC BY 4.0)

this stage, students are able to transcend contextual details and move towards a general understanding of the structure and function of PTS.

The result in the interview study can be summarised as follows. Independently of context, the aspects of PTS that are critical for students to discern in order to conceptualise PTS in a powerful way are:

- How individual components work and what their function is in PTS
- How the components are organised
- The logic in the code, and how the code controls the components
- The interaction between the code and the components that generates a flow of information that determines the function of PTS

6.2 *The Design Study*

The design study qualitatively investigates the different ways students conceptualise PTS in the process of designing and coding PTS. Data were gathered from students' sketches, video recordings and interviews, collected while students, aged 10 and 14, were working in pairs to design a burglar alarm using the BBC micro:bit. The phenomenographic analysis shows that the students experience two intertwined phenomena during the process that need to be understood in relation to each other: the dual nature of PTS and the BBC micro:bit material that represent aspects of PTS (Figure 14.4). The dual nature involves the logic in the code in terms of a control function, how the components work,

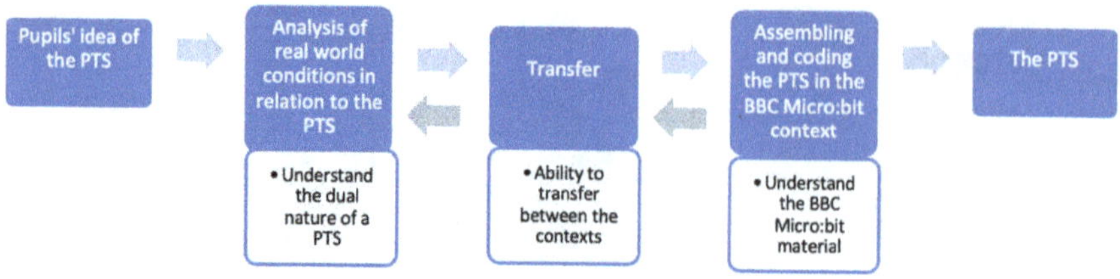

FIGURE 14.4 The students experience two intertwined phenomena during the process of designing a PTS (from Cederqvist, 2022, figure 7, CC BY 4.0)

how to organise the components, and the interaction between the code and the components that generates a flow of information that controls the function. The BBC micro:bit material involves what the blocks represent in terms of real-world conditions, what they represent in terms of programming concepts, the shape of the blocks, how the blocks are organised in the editor, the need for a control function in the code, and how to combine blocks into a code.

Initially, students come up with an idea for a PTS based on the real-world context in which it would be used. This implies they need to analyse the structure and function of PTS as experienced in the real-world context. The results from the design study show that students conceptualise the structure and function of the PTS in different ways, based on what critical aspects they are able to discern. In the "black box" approach, the students discern what components to use, the interaction between the components, and a code, which will affect the function. However, these parts are treated as black boxes, which is probably the result of a vague understanding of how the components work, how to organise them, and what the code represents. However, some students are able to unpack the black boxes. They are able to see what the code represents and how the components can be organised. Further there are students who also are able to describe the purpose of using specific components, and how these can be organised, as well as how these interact based on a code to generate a flow of information that controls the function in the PTS. Thus, in the first stage of the process, the students are challenged to discern the parts regarding the structure and function of the PTS.

In the next step, students need to transfer the structure and function into the BBC micro:bit context. The results from the design study show that the way students conceptualise the BBC micro:bit material affects their ability to transfer the structure and function of the PTS. Students conceptualise the BBC micro:bit material in different ways, based on what critical aspects they are able to discern. In the least powerful way, the novice user approach, students are not able to discern what the blocks represent in relation to structure and function of PTS as experienced in the real-world context. Further, they are not able to interpret the shape of the blocks and how to use this information.

These students work randomly in a trial-and-error way that results in producing non-functional codes. However, some students are able to discern what the blocks represent as real-world conditions and are able to interpret the shape of the blocks, which they can then use to produce a code. They also see the need for a control function. Further, there are students who discern what the blocks represents as programming concepts. These students are also able to navigate in the editor confidently to find the blocks, and are able to combine the blocks to control the function of the PTS. Accordingly, the BBC micro:bit material also presents students with a challenge to discern the parts with respect to how they represent the structure and function of PTS.

The results from the design study can be summarised as follows. In order to achieve the task, students need enough cohesion and detail with respect to the structure and function of the PTS in order to have anything substantial to implement in the BBC micro:bit context. This implies that students need to discern critical aspects of both the structure and function of PTS and the BBC micro:bit material, and recognise their meaning in relation to each other. The aspects that were found critical for students to discern concerning the structure and function of PTS are:

- The logic in the code in terms of a control function
- How the components work
- How components can be organised
- The interaction between the code and the components based on the control function and how this generates a flow of information that controls the function of the PTS.

The aspects found that are critical to discern concerning the BBC micro:bit material are:

- What the blocks represent in terms of real-world conditions and in terms of programming concepts
- The shape of the blocks
- How the blocks are organised in the editor
- The need for a control function in the code
- How the code can be combined in terms of blocks to control the function of the PTS.

7 Findings and Discussion

In this section, the analysed and synthesised results from the two phenomenographic studies, the interview study and the design study, are presented and discussed in relation to the research questions that guide this study.

7.1 *In What Ways Do Students Conceptualise* PTS *in the Processes of Analysing and Designing* PTS*?*

The results show that students' ways of conceptualising PTS in the processes of analysing and designing is affected by contextual factors. This implies that the contexts in which the students experience PTS, i.e., the micro:bit context and the everyday context, direct attention to certain aspects of PTS that appear as more or less important for the students. Further, the students conceptualise PTS from the way PTS appears as part-whole structures, which are initially conceptualised as an undifferentiated whole, which students gradually try to differentiate from its parts. Thus, students conceptualise PTS from discerned parts of its context. If the students are not able to differentiate the part-whole structure, and are not able to see the important aspects of PTS, it becomes difficult to conceptualise PTS in the processes. In line with what has previously been suggested by Marton et al. (2004) the findings of this study indicate that the context in which PTS is presented brings certain aspects to the fore, which indicates a certain way to experience PTS.

The difficulty for students seems to be the changing contexts. In the everyday context, PTS do not appear in the same way as they appear in the BBC micro:bit context. The results show that within each of the contexts, students are able to conceptualise PTS in a more powerful way the more they are able to see important aspects of PTS and the relation between these. Furthermore, the results indicate that the more powerful the conceptualisation in each context, the easier it is to transcend the contextual factors. This implies that the students need to learn to see the part-whole structure of PTS in different contexts in order to conceptualise PTS in a powerful way in any context. This also includes the ability to see the differences and similarities between the different contexts.

Accordingly, students need a strategy when approaching PTS in different contexts, which involves the ability to discern the structural and functional parts of PTS. The synthesised results indicate that systems thinking may be that strategy. In order to make sense of structural and functional parts of PTS, students seem to intuitively approach PTS using systems thinking. Results from both studies show that the more students are able to understand the structural parts of PTS and how they interact, and the relation between the structural parts and the functional aspects, the stronger their ability to conceptualise PTS. Systems thinking may also help students to compare the structural and functional parts of PTS and the relationships between the BBC micro:bit context and the everyday context. That is, systems thinking will help them to discern differences and similarities between different parts and different wholes. Seeing PTS in different but connected contexts, using the systems thinking perspective, may help students to develop a general understanding of PTS.

Similar approaches have previously been described by Cederqvist (2021), de Vries (2005/2019), and Slangen et al. (2011), who suggests systems thinking as a way to facilitate understanding of structural and functional aspects of technological solutions, both in analysing and designing.

7.2 *What Aspects Are Critical to Discern in Order to Conceptualise PTS in a Powerful Way?*

The results from the two phenomenographic studies, the interview study and the design study, show what aspects are critical for students to discern in order to conceptualise PTS in a powerful way. In Table 14.1, the critical aspects are summarised in relation to the two studies.

The analysed and synthesised results from the two studies reveal that there are certain aspects that are necessary to discern in relation to both being able to analyse and to design PTS. These aspects are related to the structure and function of PTS, and to the instructional material used for producing the PTS, which represents structural and functional parts of PTS. Similar findings have

TABLE 14.1 Aspects that are critical to discern in order to conceptualise PTS in a powerful way

The interview study	The design study	
PTS in BBC micro:bit context/ everyday context	PTS in everyday context	PTS in BBC micro:bit context
The logic in the code and how it controls the components	The logic in the code—a control function	What the blocks represent as real-world conditions and as programming concepts
How the components work and their function in the PTS	How components work	The shape of the blocks
The organisation of components	The organisation of components	The organisation in the editor
The interaction between the code and the components that generates a flow of information that determines the function of the PTS	The interaction between the code and the components that generates a flow of information that controls the function	The need for a control function in the code and how this can be combined in terms of blocks to control the PTS

SOURCE: DEVELOPED FROM CEDERQVIST, 2020, 2022

been presented by Cederqvist (2021) in relation to what key elements of technological knowledge is at play in the processes of analysing and designing PTS. In sum, the critical aspects that are necessary for students to discern in relation to being able to analyse and design PTS are:

The structure and function of PTS
- The logic in the code in terms of a control function
- How components work and what their functions are in the PTS
- How components can be organised to fulfil the function
- The interaction between the code and the components that generates the flow of information that controls the function

The instructional material (BBC micro:bit)
- How the structural and functional nature of PTS is represented by blocks as real-world conditions
- Programming concepts in order to control the flow of information in PTS
- How programming concepts are represented by blocks
- The shape of the blocks and how to use this information
- How the editor is organised
- How to combine blocks into a control function

The identified critical aspects can inform teachers about what to take as their point of departure and direct attention towards in the processes of analysing and designing PTS when teaching technology.

7.3 *In What Way Could Teaching on PTS Be Organised in Order for Students to Learn about PTS in a Powerful Way?*

In keeping with the phenomenographic perspective, learning about PTS can be described as students initially seeing PTS as an undifferentiated whole, which they gradually discern from its structural and functional parts, such as its components and how these work and are structured, as well as the logic in the code that controls the function in the PTS. In the processes, there is a complexity of conceptualising PTS in relation to the changing contexts of everyday life and the instructional material BBC micro:bit. This indicates the importance of bringing forward the ways PTS appear in these different contexts. Further, the findings indicate that before students take part in design and coding activities, technological knowledge in relation to structural and functional aspects of PTS should be addressed, which is in line with findings from previous studies (see Barak & Zadok, 2009; Cederqvist, 2021; Mioduser et al., 1996; Slangen et al., 2011). This involves knowledge of components and ways to control PTS in terms of a logical code, and how this could be represented by blocks, both as expressed in everyday language and as programming concepts. To this end, students also need to understand programming concepts and how these can be used to control the flow of information in the PTS. Thus, teaching should take departure in developing students' understanding of the structure and function of PTS. This might involve activities such as identifying and analysing the structure and function of existing PTS, both everyday objects and ready-made micro:bit constructions. Further, students need knowledge of how to interpret and control the flow of information in PTS, which implies that they need to understand programming concepts. This can be highlighted by the teacher both in theoretical and practical activities, which can be analogue or digital. Analogue activities, such as controlling a friend to move from one point to another using programming concepts, is a fun and fruitful way of learning programming concepts such as loops and conditionals. Also, instructional online materials, such as code.org, are useful when introducing programming concepts. However, it is also important to bring forward the programming concepts in relation to the situations they might be used in, i.e., the designing and coding of PTS, and to clarify the function of the programming concepts in these situations. Therefore, it can be fruitful for teachers to show examples of how to produce code in relation to different micro:bit constructions, in order to visualise the connection between the structural and functional parts of PTS and the blocks as expressed in everyday language and as programming concepts. In this way, students' difficulties in conceptualising how to produce instructions for controlling PTS, as previously described by Barak and Zadok (2009), Mioduser et al. (1996) and Slangen et al. (2011), might be overcome.

During the process of designing and coding, students can make use of and strengthen their understanding of the programming concepts they have learnt and how to design a PTS based on their knowledge of components and how these might contribute to functional PTS by interacting with a functional code. During the process of designing, students could also "learn by doing", e.g., how components might be organised, how to use the shape of the blocks, how the editor is structured, and how to combine blocks to a control function. These parts can, of course, be highlighted before taking part in the design activity but might be easier to conceptualise through hands-on activities.

In line with what Barak and Zadok (2009) argue concerning students lacking necessary technological knowledge when designing PTS, it can be summarised that there are certain aspects that are necessary to direct attention to before students take part in a design activity to produce and code a PTS. Further, there are aspects that may be learnt during the design activity. Below is a suggestion on how to organise teaching and what to focus on.

What to learn before the design activity:
- The logic in a code in terms of a control function
- Basic programming concepts, e.g., conditionals and loops
- How components work and what their functions are in the PTS, e.g., how does a light sensor work
- The relation between structure and function in PTS, i.e., the goal is to make components interact based on how they work and the logic in the code in order to generate a flow of information that controls the function of the PTS
- How the structural and functional nature of PTS is represented by blocks as real-world conditions
- How programming concepts are represented by blocks

What to learn during the design activity
- How components can be organised, i.e., how to assemble components to a functional PTS
- The shape of the blocks in the micro:bit editor and how to use this information
- How the micro:bit editor is organised
- How to combine blocks into a control function
- How the structural and functional nature of PTS is represented by blocks as real-world conditions
- How programming concepts are represented by blocks

7.4 *Final Remarks*

We have to bear in mind that teaching activities involving programming have to be aligned with what is stipulated in the curriculum (Vinnervik, 2022a). That is, in the syllabus for technology, to develop students' ability to identify and analyse the structure and function of existing PTS, and to develop their ability to design new ones and control them with programming (Swedish National Agency for Education, 2019/2022). This implies that teaching should be organised in a way that provides opportunities for students to discern the critical aspects in order to conceptualise PTS in the processes of analysing and designing PTS. Slangen et al. (2011) have previously highlighted students' difficulties in converting real-world conditions into a technological solution in terms of a code, and that this needs to be addressed in teaching. The findings of the current study indicate that these difficulties might be overcome. It is suggested that students should initially be taught to analyse the structure and function of existing PTS in different contexts such as everyday objects as well as ready-made micro:bit constructions. This will allow students to see the differences and similarities between the different contexts and help them develop a general understanding of PTS. Students should also learn basic programming concepts and how they could be used to achieve function in PTS. This will provide students with the necessary conditions to use programming as a tool in the design process to achieve their intended ideas of PTS.

In the process of designing PTS with the BBC micro:bit, the students initially need to analyse structure and function of their intended PTS. This implies that they need pre-knowledge that involves conceptualising the structure and function. Therefore, teaching should focus on this before involving students in the design activity. The more the students understand of the structural and functional parts of a PTS, the better opportunities they have to assemble and code the PTS in the BBC micro:bit context. The more of the micro:bit material they conceptualise, the better prerequisites they have to produce the PTS. In teaching, this implies presenting the design process as way of travelling between the real-world context and the BBC micro:bit context where the students are trying to make a fit between identified structural and functional aspects of PTS with aspects of the micro:bit material. As de Vries (2005) suggests, success in the design process depends on the extent to which the designer is able to make a fit between involved factors. If students have previously developed an ability to discern aspects of PTS in relation to the different contexts, they have greater opportunities to achieve the PTS, but also greater opportunities to conceptualise how PTS work and are controlled by programming, which can be applied in any context.

Note

1 https://www.bbc.co.uk/programmes/articles/4hVG2Br1W1LKCmw8nSm9WnQ/the-bbc-micro-bit

References

Åkerlind, G. (2015). From phenomenography to variation theory: A review of the development of the variation theory of learning and implications for pedagogical design in higher education. *HERDSA Review of Higher Education*, 2, 5–26. https://www.herdsa.org.au/herdsa-review-higher-education-vol-2/5-26

Barak, M., & Zadok, Y. (2009). Robotics projects and learning concepts in science, technology and problem solving. *International Journal of Technology and Design Education, 19*(3), 289–307. https://doi.org/10.1007/s10798-007-9043-3

Cederqvist, A.-M. (2020). Pupils' ways of understanding programmed technological solutions when analysing structure and function. *Education and Information Technologies, 25*, 1039–1065. https://doi.org/10.1007/s10639-019-10006-4

Cederqvist, A.-M. (2021). *Seeing the parts, understanding the whole – A technology education perspective on teaching and learning in processes of analysing and designing programmed technological solutions*. Acta Universitatis Gothoburgensis. http://hdl.handle.net/2077/68522

Cederqvist, A.-M. (2022). An exploratory study of technological knowledge when pupils are designing a programmed technological solution using BBC Micro:bit. *International Journal of Technology and Design Education, 32*, 355–381. https://doi.org/10.1007/s10798-020-09618-6

Curriculum for the compulsory school, preschool class and school-age educare: Revised 2019. (2019). Swedish National Agency for Education.

Curriculum for the compulsory school, preschool class and school-age educare: Revised 2022. (2022). Swedish National Agency for Education.

Eisenhart, M. (2017). A matter of scale: Multi-scale ethnographic research on education in the United States. *Ethnography and Education, 12*(2), 134–147. https://doi.org/10.1080/17457823.2016.1257947

Ingerman, Å., Linder, C., & Marshall, D. (2009). The learners' experience of variation: Following students' threads of learning physics in computer simulation sessions. *Instructional Science, 37*(3), 273–292. http://doi.org/10.1007/s11251-007-9044-3

Kakos, M., & Fritzsche, B. (2017). Meta-ethnography E&E. *Ethnography and Education, 12*(2), 129–133. https://doi.org/10.1080/17457823.2017.1278578

Marton, F., & Booth, S. (1997). *Learning and awareness*. Erlbaum.

Marton, F., & Tsui, A. (2004). *Classroom discourse and the space of learning*. Lawrence Erlbaum.

Mioduser, D., Venezky, R. L., & Gong, B. (1996). Students' perceptions and designs of simple control systems. *Computers in Human Behavior, 12*(3), 363–388. https://doi.org/10.1016/0747-5632(96)00014-3

Slangen, L., Keulen, H., & Gravemeijer, K. (2011). What pupils can learn from working with robotic direct manipulation environments. *International Journal of Technology and Design Education, 21*(4), 449–469. https://doi.org/10.1007/s10798-010-9130-8

Vinnervik, P. (2022a). Implementing programming in school mathematics and technology: Teachers' intrinsic and extrinsic challenges. *International Journal of Technology and Design Education, 32*, 213–242. https://doi.org/10.1007/s10798-020-09602-0

Vinnervik, P. (2022b). Programming in school technology education: The shaping of a new subject content. *International Journal of Technology and Design Education*. https://doi.org/10.1007/s10798-022-09773-y

de Vries, M. J. (2005). *Teaching about technology: An introduction to the philosophy of technology for non-philosophers*. Springer.

de Vries, M. J. (2006). Technological knowledge and artifacts: An analytical view. In J. R. Dakers (Ed.), *Defining technological literacy*. Palgrave Macmillan.

de Vries, M. J. (2019). Peter Kroes and Antonie Meijers: The dual nature of artefacts. In J. R. Dakers, J. Hallström, & M. J. de Vries (Eds.), *Reflections on technology for educational practitioners: Philosophers of technology inspiring technology education*. Brill Academic Publishers.

CHAPTER 15

Metaphors and Gestures in Programming Education

Andreas Larsson

Abstract

Programming is an inherently abstract activity that requires the use of metaphorical expressions. In this text, I intend to investigate the role of metaphor in programming education by attending to such expressions on a discursive and on a cognitive level. Research on programming metaphors and education shows that teachers often use metaphors that can be related to the programmer's intentions and to the function of a programming concept. Based on the classroom examples provided in the chapter, however, I argue that it is not always self-evident whether a teacher uses a metaphor in one way or the other. In contrast, the chapter indicates that the meaning of a concept changes in relation to the situation and that much of the teachers' metaphorical language is in fact based on programming conventions rather than on embodied experiences.

Keywords

metaphorical expression – programming metaphors – classroom observations

1 Introduction

At its core, computer programming is a very abstract activity. Consequently, speaking about, for example, programming and computing processes has always required the use of figurative language. Figurative language is sometimes referred to as the expression we use when we do not mean what we literally are saying (Gibbs & Colston, 2012), and we use figurative expressions all the time. Consider, for example, common expressions such as "the computer just died on me", or "just click here to kill the process". Most of us know that a computer is an inanimate artefact and is therefore unable to die. Nevertheless, these expressions make it reasonably clear to us that something is wrong

© ANDREAS LARSSON, 2024 | DOI:10.1163/9789004687912_015

with the machine or that we intend to stop the computer from performing an action.

Accompanying such expressions are often image-like gestures. Imagine, for example, someone speaking about a program that crashes while performing an explosion like gesture with both hands or try to envision a teacher who is explaining the inner workings of a database while "manipulating" and "moving" an imaginary object with one or two hands. Gestures such as these may probably go unnoticed to both speakers and listeners, still studies show that they generally exist (e.g., Larsson & Stolpe, 2022; Manches et al., 2020; Solomon et al., 2020).

1.1 *Metaphors and Programming*

Computer science can, in contrast to, for example, the natural sciences, be considered the inventor of its own object of study and also the formulator of its own laws. Also, computer science programming is centred around implementing a programmer's intentions rather than understanding how electrons function inside the computer (Colburn & Shute, 2007, 2010). Thus, programmers, educational facilitators, and copywriters have a constant need to 'invent' or add to a type of language that can provide an instant and initial understanding of what a computer is able to do, how we can accomplish this, and what tools to use (Colburn & Shute, 2008, 2017; Nofre et al., 2014). For example, according to Merriam Webster's dictionary, the word "computer" originates from the transitive verb "[to] compute", a word that was used as early as 1579. Thus, as the word was used around 400 years before the modern computer came into use, it is reasonable to suggest that the term 'computer', at least in the modern sense, is a metaphor that is based on connections between human abilities and the workings of the computer —and there are many more of these metaphors. A computer 'remembers' things, it is 'smart', and it is 'efficient problem solver'. Moreover, the computer 'listens' to what we say, it can 'fall asleep' and it can provide us with 'answers' to specific queries. I would go so far as to claim that many of these metaphors are essential for understanding programming on a general level.

1.2 *The Aim of the Chapter*

The aim of this chapter is to further explore the role of figurative language in programming and programming education. First, I intend to provide a scientific base illustrating how metaphorical expressions come to life. After this, I will provide concrete examples on how metaphors may appear in classroom situations and what we can learn from them.

2 Metaphor in Education

So far, I have referred to 'metaphor' as a rather generic linguistic phenomenon where one concept is referred to in terms of another. However, metaphor can also be understood as a cognitive phenomenon that structures the way we make sense of the world (e.g., Fauconnier & Turner, 1998; Gibbs, 2006; Lakoff & Johnson, 1999). Thus, metaphor analysis can be used to explore both *how* we speak around the computer and *why* we speak in the way we do (Larsson, 2023). To put it simply – the main rationale for studying programming education from this perspective is that since we know that people often use metaphors related to human action when, for example, speaking about a specific programming concept, said human action is probably important for understanding programming concepts (Larsson, 2022; Larsson & Stolpe, 2022; Manches et al., 2020; Solomon et al., 2020).

This observation aligns with research in other STEM-related areas (Amin et al., 2015; Larsson et al., 2019; Niebert & Gropengießer, 2014) and cognitive linguistics (Gibbs, 2006; Grady, 1997) suggesting knowledge about abstract phenomena is a result of physical interaction with the environment that surrounds us (see e.g., Lakoff & Johnson, 1999).

2.1 *Knowing How vs. Understanding Why*

There are two main ways to understand the role of metaphors in programming education:

1. Programming metaphors on a *discursive level*; i.e., the metaphorical expressions and gestures that we use when we engage in a programming-related event.
2. Programming metaphors on a *cognitive level;* i.e., the connections between a programming concept and the real world experiences by which it is made sense of.

To illustrate the two ideas, let us revisit one of the metaphorical expressions from the introductory section, "just click to kill the process". On a discursive level, the expression displays connections between the verb "kill" and stopping a computing process. This suggests that the metaphorical expression is grounded in the idea that a functioning machine is a living organism (e.g., a human being). Within cognitive linguistics, such a connection between concepts—in this case between life and function—is referred to as the *metaphorical meaning* evoked by the expression and is notated as a conceptual metaphor; in other words, A FUNCTIONING COMPUTER IS LIFE[1] or A MALFUNCTIONING COMPUTER IS DEATH. From this, it is possible to assert that the metaphorical expression demonstrates that we *know* that the computer 'will stop working' when we click the button, based on the concept that DEATH IS A MALFUNCTIONING COMPUTER.

Theorists Lakoff and Johnson (1980, 1999) and Grady (1997) claim that metaphorical expression such as the above would arise from recurring physical interactions between the environment and any abstract phenomenon. However, since this implies understanding that the metaphorical meaning of "just click to kill the process" would be underpinned by recurring experiences of death, there must be another way to determine why we use the expression the way we do.

Johansson Falck and Okonski (2022) argue that one way to approach this conundrum is by attending to how we use instances of real-world physical experiences in language in relation to abstract embodied concepts. For example, the reference to "death" ("a permanent cessation of all vital functions") implies that the concept 'death' is understood by means of experiencing "loss of all vital functions", i.e., in relation to the conceptual metaphor A MALFUNCTIONING COMPUTER IS (PERMANENTLY) WITHOUT VITAL SIGNS. In this way, the idea of analysing metaphorical expressions on a discursive and a cognitive level can allow us to assert what we know about programming and how we understand programming the way we do.

2.2 *Metaphor in Education*

Research about the role of metaphor in education have increased over the last decade. Generally, this type of research is focused on exploring on one 'level' at a time. Often, the research exploring metaphor on a discursive level concern, for example, metaphors used by teachers when speaking about programming, students' use of metaphors when engaging in programming activities, metaphors in programming textbooks, and experts' use of metaphors. Meanwhile, researchers interested in the cognitive aspects of programming focus their work on exploring the experiential grounding of different metaphors, teachers' embodiment, and enactment of metaphors. Following is a review of some of the relevant literature available today.

2.3 *Conceptual Metaphors on a Discursive Level*

In 1983, Howard Peele advocates the use of metaphors to help students develop general computational literacy. In the publication *Computer Metaphors: Approaches to Computer Literacy for Educators*, he argues that metaphors can be especially fruitful when it comes to learning a subject as complex and ever-evolving as programming. Also, he states, metaphors are extra helpful in establishing a starting point for new learners. Similarly, Colburn and

Shute (2008) argue that metaphoric language is central for understanding the abstract aspects of programming metaphors related to human actions (e.g., 'jump', or 'catch'), and can become a kind of 'shortcut' when it comes to learning programming. There is, however, a risk that anthropomorphic metaphors may result in misconceptions among students (Forišek & Steinová, 2012; Kohn, 2017; Qian & Lehman, 2017), and even though such metaphors may seem simple, research shows that teachers find them complicated to use in context (Munasinghe et al., 2021).

Another common computer metaphor is the tool metaphor (COMPUTER IS A TOOL OR PROGRAMMING IS A TOOL). There are many examples in advertising, where computers have been marketed as problem-solvers, word processors, or, for that matter, as a computer (Larsson, 2023). Larsson and Stolpe (2023) have identified that the tool metaphor can be related to many aspects of programming. For example, programming may solve an advanced mathematical problem, a programming language can have different tools built into it, or a specific sub-routine in a program can be used as a tool within the program. Consequently, the tool metaphor is highly dependent on the context in which it used.

Research also shows that teachers use different kinds of analogies when teaching (Larsson & Stolpe, 2023; Sanford et al., 2014). For example, several of the teachers compare programming concepts to ideas from Harry Potter or provide students connections between physical objects (e.g., a storage room) and the function of a computer. There are also a few studies addressing teachers' use of programming metaphors in the classroom. For example, Larsson and Stolpe (2022) conclude that teachers often use gestures related to physical objects and motion when lecturing about programming concepts. Also, Manches et al. (2020) identify these, and several other conceptual metaphors when investigating expert programmers' explanations of programming metaphors.

2.4 *Conceptual Metaphors on a Cognitive Level*

Studying metaphors on a cognitive level means exploring the connections between the source and the target domain of a metaphor; i.e., the meaning of a metaphor. Larsson (2022) studied textbook examples from four different sources, concluding that different textbooks conceptualise programming in different ways by emphasising different source domains. This implies that, depending on the textbook of choice, students may conceptualise programming and their roles as programmers differently. In another study, Larsson and Stolpe (2022) identified several events where teachers referred to data in relation to a physical object in both words and as re-enactments in gestures. From this, they conclude that DATA IS A PHYSICAL OBJECT should be seen as a core metaphor in programming. In turn, this implies that metaphors related to motion (e.g., "receiving data from a database", or to "place data in an array"), and physical space ought to be central in programming. Furthermore, the findings indicate that metaphors related to processing, moulding, or constructing larger objects would be plausible to find in relation to programming. Findings from Manches et al. (2020) and Solomon et al. (2020) show similar results, at least when it comes to expert programmers and programming teachers; three examples that indicate that teachers, regardless of their origin (Sweden, England, and the United States) display a similar conceptualisation of programming.

Research also shows that teachers gesture in different directions in relation to concepts. Larsson et al. (2021) noticed in a small case study that when their informants were speaking about the processes or functions of a program, the majority of gestures were performed in the sagittal direction (backwards and forwards). However, when referring to code or functions, their hands moved vertically. It was only when discussing the order of data in, for example, an array, that they gestured horizontally. Similar phenomena have been identified in learners' gestures, with the addition that native English speakers tend to gesture vertically from left to right, while native Arabic speakers tend to gesture vertically from right to left (Almjally et al., 2023).

Another aspect of teachers' metaphor use concerns the differentiation between homonyms (words that are syntactically identical but have different meanings) (see e.g., Kendon, 1988) in programming. One such example—"to drill"—has been identified by Larsson et al. (2022). In the study, a teacher's use of the concept has been analysed with respect to both speech and gesture. The results of the study indicate that teacher use the words as conventionalised programming terms when referring to the computer screen, and as a metaphor when speaking about the function of the code. Consequently, it is reasonable to suggest that the computer itself affects the way that the teacher reasons about programming and plays a vital role of the conceptualisation of code. Similar results—that teachers' use of metaphors is related to what may appear on the screen—has been identified by Stolpe and Larsson (in press) who have investigated the metaphoric nature of a teacher's language while engaging in a programming task.

2.5 *Metaphor in Action*

Based on the research above, one can be misled to believe that programming concepts are distinguishable from other instances of language use, however, that is not the case. Take for example the expression "I just *received* an e-mail *from* a student". The use of the word "receive" suggests that the expression is partly based around the idea of a user (me) coming into possession of an object (the e-mail) (AN E-MAIL IS A PHYSICAL OBJECT) and partly on the idea motion has taken place (sending an email is motion from location a to location B) (from the student's computer to my computer). The same logic would apply to the expression "I just *received* data *from* a database", where DATA IS A PHYSICAL OBJECT and DATA COMMUNICATION IS MOTION. Data can also be '*structured*' (ORGANISATION IS PHYSICAL STRUCTURES) into larger packages, printed on a screen (EVENTS ARE HUMAN ACTION), or stored in memory (COMPUTER AS HUMAN).

The above example implies the following: first, that two or more metaphors can be blended together to form a large 'network' of metaphors that can be employed when speaking about complex programming issues. Second, that programming is partly about function (e.g., DATA IS A PHYSICAL OBJECT or AN EMAIL IS A PHYSICAL OBJECT) and partly about intention (SENDING AN EMAIL IS MOTION FROM LOCATION A TO LOCATION B, DATA COMMUNICATION IS MOTION FROM LOCATION A TO LOCATION B) (Colburn & Shute, 2008; Larsson, 2023). In that sense, programming education differs from, for example, the natural sciences.

3 Examples from Practice

To illustrate the role of metaphor for programming education, I will provide examples where teachers' use of metaphors is in focus. The examples are based on videoed classroom observations of Swedish upper secondary programming education (see Larsson & Stolpe, 2022, 2023; Larsson et al., 2022 for details) and presents a variety of how teachers express themselves in the classroom. The metaphors presented below have been identified using three different metaphor identification procedures:

1. The Metaphor Identification Procedure (MIP) (Pragglejaz Group, 2007)
2. The Metaphor Identification Guidelines for Gesture (MIG-G) (Cienki, 2016)
3. The Procedure for Identifying Metaphorical Scenes (PIMS) (Johansson Falck & Okonski, 2022)

The three procedures follow similar protocols, and even though the procedures are based on different theoretical assumptions, they can be used complementarily to study the role of metaphor in programming education (Larsson, 2023). First, the analyst must establish an understanding of the context in which a concept/expression is used. Second, the analyst needs to determine how the concept is used in the expression and relate that to the basic/literal meaning of the concept/expression. Based on this information, the analyst will then be able to determine whether a concept/expression is metaphorical and, hopefully, also why a metaphor is used in said context.

The examples below have been chosen to represent a variety of teachers' programming metaphors. In addition, the examples demonstrate occasions when the same concept means different things. Finally, the section brings forward a discussion of the role gestures play in programming education.

3.1 *Metaphors Related to Motion*

3.1.1 Jump

The first classroom example concerns motion. During this event, a teacher is speaking about using a specific command to restart the execution of a program. When explaining the function of the command, the teacher says "one. next will get the program to move [corresponds to 'jump'] on" and makes a 'jumping' forward gesture with his left hand. Since 'jump' is a concept that is used to describe "a sudden upwards motion caused by muscular action of feet and legs", it is reasonable to suggest that the word "jump" is used metaphorically for an event that is caused by the command one.next. Judging by the gesture, the teacher envisions the jumping-process to take place in front of him and proceed away from him (Larsson & Stolpe, 2022). In that sense, the code is understood in terms of a 'force' that causes motion (Larsson, 2023), and it is the programmer who controls this force.

Another example of the same concept is when a teacher says, "you have been jumping between code lines, right?" while lowering his pen from a chest-height position (holding the pen in the right hand) down to his hip, while keeping his left hand in front of his chest as a point of reference. The difference between the two examples is located in the actor in the expressions. In the former, it was the command one.next that performs the motion, and as a program is an abstract 'force' the concept is used metaphorically. In the latter example, it is the students who are responsible for moving between the code lines. Thus, the expression is understood directly and is not a metaphor. In addition, the gesture in the latter example does not indicate that 'jump' is a programming process, but rather a change of location. Consequently, even though both teachers use the same expression in similar contexts, they still have different meanings.

3.1.2 Loop

Another common programming concept related to motion is 'loop'. In this example, a teacher is using 'loop' in two ways: (1) "place a loop" (i.e., a noun), and (2) "loop through" (i.e., a verb). Similar to results identified by Manches et al. (2020), the teacher uses circular gestures when speaking about the concept; inwards when speaking about 'loop' as an object and outwards when speaking about loop as a process (Larsson & Stolpe, 2022). Judging from the similarities between the circular gestures and the concepts, it is reasonable to suggest that the gestures are used as metaphors for 'loop' as a noun and 'loop' as a verb. This implies that the directions of the teacher's gestures are in some way related to the meaning of the programming concept 'loop' and that the loop-concept is related to motion. Attending to the verbal expressions, however, reveals that only the second loop-concept can be conceptualised in terms of motion. Even though a loop, when conceptualised as a physical object, is admittedly movable, it will have attained a firm position once the execution of the program has begun. Nevertheless, within the body of the loop, there is room for motion.

3.2 *Metaphors Related to Tool Use*

3.2.1 Making the Data Fit

There are of course other examples where the same concept is used in similar contexts but with different meanings. For example, one teacher spoke about making the data 'fit' the database. In order to do so, the students had to learn how to 'format' the data in the right way. Attending to the basic meaning of the concepts; "[to] put into a suitable state: made ready" and "to prepare (something, such as a computer disk) for storing data in a particular format"—suggest that the concept is based around two metaphors: (1) DATA IS A PHYSICAL OBJECT and (2) FORMATING IS ORGANISING PHYSICAL OBJECTS.

In another example—this time based on an interview with an upper secondary programming teacher (see Larsson & Stolpe, 2023)—programming and programming education are referred to in terms of "building toolbox". Similar to the above example, this example is based around two concepts: (1) 'building' meaning "to form by ordering and uniting materials by gradual means into a composite whole" and (2) 'toolbox' meaning "a chest for handheld devices that aids in accomplishing a task." This implies that programming is an activity that involves ordering objects into structures and that developing programmers need to accumulate a number of devices to accomplish this. However, the expression does not reveal what tools are considered essential for programming.

3.3 *Metaphors Related to Structures*

As described above, metaphors generally tend to associate and form larger structures on which we can map new metaphors. Structures like these can, at times, form analogy-like scenes on which a teacher may base a lecture.

Take, for example, an event where the teacher introduces a new concept—'function'—in terms of a kitchen scenario. The students are asked by the teacher to envision a scene in which their parents have asked them to clear the table, do the dishes, and take out the trash. At this stage of the event, the teacher has activated a FUNCTION IS CLEANING A KITCHEN metaphor where CODELINES ARE PARENTS' INSTRUCTIONS. During the event, the teacher acts out the different instructions given by the parents. Much like a theatrical performer, he uses his stage in front of the whiteboard as a physical source domain for the kitchen metaphor. Just before this event, the teacher had spoken about a function as a list of chores that one needs to do when cleaning the room (an activation of a FUNCTION IS CLEANING THE ROOM metaphor as well as CODELINES ARE A LIST OF CHORES metaphors), while later speaking about different functions in a program in terms of a house. From this, it is possible to infer that the teacher understands programming based on the following metaphors: (1) A PROGRAM IS A HOUSE, (2) A FUNCTION IS A SUBSECTION OF A HOUSE, and (3) A FUNCTION IS A LIST OF INSTRUCTIONS.

As seen in the example, all the metaphors used by the teacher seem to be held together by a general house metaphor. It follows that the teacher most probably understands 'the programmer' in terms of a builder (A PROGRAMMER IS A BUILDING) and that a skilled programmer needs to know how to, for example, use tools to build physical structures (PROGRAMMING IS BUILDING STRUCTURES), or manipulate objects (PROGRAMMING IS MANIPULATING). This shows how a well-designed metaphor can work as an anchor point for a whole metaphorical scene.

3.4 *Ontological Metaphors*

Lakoff and Johnson (1980) argue that there exists a set of metaphors that structures abstract phenomena in relation to embodied experiences. For example, we may employ experiences of handling a container—i.e., a physical object with an inside and outside, enclosed by a surface—to understand what it means to save 'file inside a folder' or to 'place a value in a variable' (Stolpe & Larsson, in press). Often, such structures are referred to as image-schemas (see Lakoff & Johnson, 1999). As seen in the above examples, programming is often communicated in terms of objects (a container-schema) and motion (source-path-goal–schema). Stolpe and Larsson (in press) argue that a metaphoric expression that is based on a container-schema needs to provide the listener with an element of inside, outside, and surface to be understood metaphorically. In the same manner, a metaphor that is based on a source-path-goal-schema requires an element that describes a starting point, an endpoint, and a path along which the object moves. Hence, an image-schema may become an analytical framework for determining to what extent a programming metaphor can be considered and what elements of a metaphor are 'taken for granted' by the teacher.

3.4.1 Conventionalised Language

Consider, once again, the teacher's expression "one.next will get the program to move [corresponds to "jump"] on" above. As seen in the analysis in Section 3.1.1., the verbal expression, in conjunction with the teacher's gestures, reflects a source-path-goal–schema (the program moves from one position to the next) and a description of the qualitative aspects of the motion ('jump'). Thus, the expression provides the listener with all the elements required to acquire a metaphorical understanding of the concept (and more). For the same reason, it is reasonable to argue that the concept is metaphorically understood by the speaker.

In contrast, the metaphorical expression presented in Section 3.1.2—"[to] place a loop"—lacks references to the starting point and the end point of the motion implied by the concept. Consequently, the expression is only partly understood as a metaphor. Instead, the expression is based on conventionalised language—i.e., metaphorical expressions that are so commonly used in a programming context that they are understood in direct relation to a specific activity.

Stolpe and Larsson (in press) have investigated several metaphorical expressions used by a teacher engaging in a programming activity. Their findings indicate that the teacher's metaphorical expressions often lack connections to real-world experiences and should therefore be considered conventionalised language. Hence, there is a risk that the 'everyday' language used by teachers may be misinterpreted by students.

3.5 *The Role of Gestures in Programming*

Gestures are usually defined as spontaneous movements of the hands and upper body and generally studied in relation to corresponding verbal expressions. Gestures have several functions in human communication (e.g., referring to an object, tracing a line or emphasising specific words). Moreover, gestures can be used as metaphors for abstract concepts. Kendon (1988) argues that gestures can provide verbal metaphors with a basic ontology, much like (verbal) ontological metaphors.

Research has shown that gesture is a big part of the way we understand and learn abstract concepts (e.g., Alibali & Nathan, 2012; Cienki & Müller, 2008; Dreyfus et al., 2015; Manches et al., 2020; Müller & Tag, 2010; Solomon et al., 2020). Larsson et al. (2022) have—in addition to the examples presented in this chapter—identified examples where a programming teacher uses gestures; e.g., shaping his hand as if he was holding an object or extending his hand to grasp 'data' from a 'database'. This indicates that teachers' gestures play a role when communicating the ontology of programming. Larsson et al. (2022) also concludes that gestures can function as anchor points for an overarching conceptual frame.

4 Conclusion

In this chapter I have explored the role of metaphors in programming education. The short review of current research in the area indicates that programming and metaphor are closely related in two ways: first, programming metaphors are necessary to use when speaking about the intentions behind using a specific function. Here, concepts such as 'jump', 'loop', and 'build' can provide a basic understanding of the intentions behind a concept (Colburn & Shute, 2008), and form a base from which one can illustrate one's programming knowledge. Nevertheless, the examples in the chapter have shown that the meaning of the concepts may differ depending on the situation. Hence, there is reason to believe that these types of metaphors need to be accompanied by elements that are otherwise lacking. Secondly, metaphors are needed to understand and communicate the function of programming concepts. Often such metaphors are understood in relation to real-world experiences of handling, manipulating, and structuring physical objects.

The research and classroom examples presented in this chapter indicate that programming metaphors are generally based around two or more 'metaphorical components'. I regard these components as conceptual elements that together form a metaphorical expression (see Johansson Falck & Okonski, 2022). Consequently, teachers' use of metaphors is important in the sense that the metaphors will affect the way that students think about programming and how they think while programming.

Note

1 All instances of conceptual metaphors will be notated using capital letters.

References

Alibali, M. W., & Nathan, M. J. (2012). Embodiment in mathematics teaching and learning: Evidence from learners' and teachers' gestures. *Journal of the Learning Sciences*, *21*(2), 247–286. https://doi.org/10.1080/10508406.2011.611446

Almjally, A., Howland, K., Good, J., & du Boulay, B. (2023). Investigating primary school children's embodied expression of programming concepts. *International Journal of Child-Computer Interaction*, 100574. https://doi.org/10.1016/j.ijcci.2023.100574

Amin, T. G., Jeppsson, F., & Haglund, J. (2015). Conceptual metaphor and embodied cognition in science learning: Introduction to special issue. *International Journal of Science Education*, *37*(5–6), 745–758. https://doi.org/10.1080/09500693.2015.1025245

Cienki, A. (2016). Analysing metaphor in gesture. In E. Semino & Z. Demjén (Eds.), *The Routledge handbook of metaphor and language* (pp. 131–152). Taylor & Francis.

Cienki, A., & Müller, C. (2008). Metaphor, gesture, and thought. In R. W. Gibbs (Ed.), *The Cambridge handbook of metaphor and thought* (pp. 483–501). Cambridge University Press.

Colburn, T. R., & Shute, G. (2007). Abstraction in computer science. *Minds and Machines*, *17*(2), 169–184. https://doi.org/10.1007/s11023-007-9061-7

Colburn, T. R., & Shute, G. M. (2008). Metaphor in computer science. *Journal of Applied Logic*, *6*(4), 526–533. https://doi.org/10.1016/j.jal.2008.09.005

Colburn, T. R., & Shute, G. (2010). Abstraction, law, and freedom in computer science. *Metaphilosophy*, *41*(3), 345–364. https://doi.org/10.1111/j.1467-9973.2010.01631.x

Colburn, T. R., & Shute, G. (2017). Type and metaphor for computer programmers. *Techné: Research in Philosophy and Technology*, *21*(1), 71–105. https://doi.org/10.5840/techne20174662

Dreyfus, B. W., Gupta, A., & Redish, E. F. (2015). Applying conceptual blending to model coordinated use of multiple ontological metaphors. *International Journal of Science Education*, *37*(5–6), 812–838. https://doi.org/10.1080/09500693.2015.1025306

Fauconnier, G., & Turner, M. (1998). Conceptual integration networks. *Cognitive Science*, *22*(2), 133–187.

Forišek, M., & Steinová, M. (2012). Metaphors and analogies for teaching algorithms. In *Proceedings of the 43rd ACM technical symposium on Computer Science Education* (pp. 15–20). https://doi.org/10.1145/2157136.2157147

Gibbs, R. W. (2006). *Embodiment and cognitive science*. Cambridge University Press.

Gibbs, R. W., & Colston, H. L. (2012). *Interpreting figurative meaning*. Cambridge University Press.

Grady, J. (1997). *Foundations of meaning: Primary metaphors and primary scenes* (Doctoral thesis). University of California.

Johansson Falck, M., & Okonski, L. (2022). Procedure for Identifying Metaphorical Scenes (PIMS): A cognitive linguistics approach to bridge theory and practice. *Cognitive Semantics*, *8*(2), 294–322. https://doi.org/10.1163/23526416-bja10031

Kendon, A. (1988, December). *How gestures can become like words*. This paper is a revision of a paper presented to the American Anthropological Association, Chicago.

Kohn, T. (2017). Variable evaluation: An exploration of novice programmers' understanding and common misconceptions. *Proceedings of the 2017 ACM SIGCSE Technical Symposium on Computer Science Education*.

Lakoff, G., & Johnson, M. (1980). *Metaphors we live by*. University of Chicago Press.

Lakoff, G., & Johnson, M. (1999). *Philosophy in the flesh: The embodied mind and its challenge to Western thought.* Basic Books.

Larsson, A. (2022). Reading the code between the lines—exploring the structure of metaphors in educational programming resources. *Nordic Studies in Science Education, 18*(3), 305–322. https://doi.org/10.5617/nordina.8774

Larsson, A. (2023). *Metaphor in mind—programming teachers' knowledge and beliefs in action* (Doctoral thesis). Linköping University. https://doi.org/10.3384/9789107523811

Larsson, A., Stafstedt, M., & Schönborn, K. J. (2019). Heat angels and paper cups: Pupils' use of metaphoric relations when engaging thermal cameras to investigate heat. In E. McLoughlin, O. E. Finlayson, S. Erduran, & P. E. Childs (Eds.), *Bridging research and practice in science education* (pp. 73–89). Springer.

Larsson, A., & Stolpe, K. (2022). Hands on programming: Teachers' use of metaphors in Gesture and speech make abstract concepts tangible. *International Journal of Technology and Design Education*. https://doi.org/10.1007/s10798-022-09755-0

Larsson, A., & Stolpe, K. (2023). Exploring the metaphoric nature of programming teachers' reflections on action—a case study with teaching in mind. *International Journal of Technology and Design Education*. https://doi.org/10.1007/s10798-023-09826-w

Larsson, A., Stolpe, K., & Falck, M. J. (2022). Analysing the elements of a scene—an integrative approach to metaphor identification in a naturalistic setting. *Cognitive Semiotics, 15*(2), 223–248. https://doi.org/doi:10.1515/cogsem-2022-2014

Larsson, A., Stolpe, K., & Johansson Falck, M. (2021). *A teacher's hands on programming: How orientations of gestures provide concrete dimensions to abstract thoughts.* 14th Conference of the European Science Education Research Association (ESERA 2021), Braga, Portugal.

Manches, A., McKenna, P. E., Rajendran, G., & Robertson, J. (2020). Identifying embodied metaphors for computing education. *Computers in Human Behavior, 105*, 105859. https://doi.org/10.1016/j.chb.2018.12.037

Munasinghe, B., Bell, T., & Robins, A. (2021). Teachers' understanding of technical terms in a computational thinking curriculum. In *Proceedings of the 23rd Australasian Computing Education Conference* (pp. 106–114). https://doi.org/10.1145/3441636.3442311

Müller, C., & Tag, S. (2010). The dynamics of metaphor. Foregrounding and activation of metaphoricity in conversational interaction. *Cognitive Semiotics, 10*(6), 85–120. https://doi.org/10.1515/cogsem.2010.6.spring2010.85

Niebert, K., & Gropengießer, H. (2014). Understanding the greenhouse effect by embodiment—analysing and using students' and scientists' conceptual resources. *International Journal of Science Education, 36*(2), 277–303. https://doi.org/10.1080/09500693.2013.763298

Nofre, D., Priestley, M., & Alberts, G. (2014). When technology became language: The origins of the linguistic conception of computer programming, 1950–1960. *Technology and Culture, 55*(1), 40–75. http://www.jstor.org/stable/24468397

Pragglejaz Group. (2007). MIP: A method for identifying metaphorically used words in discourse. *Metaphor and Symbol, 22*(1), 1–39. https://doi.org/10.1080/10926480709336752

Qian, Y., & Lehman, J. (2017). Students' misconceptions and other difficulties in introductory programming: A literature review. *ACM Transactions on Computing Education* (*TOCE*), *18*(1), 1–24. https://doi.org/10.1145/3077618

Sanford, J. P., Tietz, A., Farooq, S., Guyer, S., & Shapiro, R. B. (2014). Metaphors we teach by. In *Proceedings of the 45th ACM technical symposium on Computer science education* (pp. 585–590). https://doi.org/10.1145/2538862.2538945

Solomon, A., Bae, M., DiSalvo, B., & Guzdial, M. (2020). Embodied representations in computing education: How gesture, embodied language, and tool use support teaching recursion. In M. Gresalfi & I. S. Horn (Eds.), *The interdisciplinarity of the learning sciences, 14th International Conference of the Learning Sciences* (*ICLS*) *2020, Volume 4* (pp. 2133–2140). International Society of the Learning Sciences.

Stolpe, K., & Larsson, A. (in press). "Vi går in och tittar i mappen": En lärares strukturering av rumslighet i programmeringsundervisning. *Nordic Studies in Science Education.*

CHAPTER 16

Product or Process Criteria?

What Teachers Value When Assessing Programming

Lars Björklund and Charlotta Nordlöf

Abstract

Programming is increasingly being taught in schools in Sweden and around the world. Programming has been taught and assessed for a long time both in upper secondary schools and in universities, but in the lower grades, programming as part of technology education is still rather new. Therefore, there is a need for assessment criteria for programming in compulsory school. The aim of this chapter is to examine the criteria that individual teachers use when they grade and assess students in tasks related to programming. This study uses Q methodology. The informants, 28 programming teachers, were asked to sort and rank criteria regarding the assessment of programming according to their own subjective beliefs. The results were analysed using factor analysis to find groups of similar informants. The results reveal two teacher groups, where one puts greater emphasis on process criteria while the other primarily emphasises product criteria. Moreover, both groups place high value on students' ability to explain the program, which can be interpreted as a combination of product and process where the student uses the knowledge they have acquired and turns it into their own knowledge. At the end of the chapter, formative assessment is discussed, based on the results relating to process criteria, and how teachers can use these criteria to help students progress.

Keywords

assessment – technology education – programming – product – process

1 Introduction

Programming has been taught in elementary schools in Sweden since 2018, explicitly in mathematics and technology (see Chapter 1). In technology, core content related to programming is taught from grades 1–3 up to lower secondary school. There are several different ways in which programming as a subject content in

© LARS BJÖRKLUND AND CHARLOTTA NORDLÖF, 2024 | DOI:10.1163/9789004687912_016 technology can be taught and learned: from a perspective of technological solutions to understand how objects in everyday life are controlled with the help of programming; as working methods in technology, to enable students to create their own solutions to technical problems; and in terms of the relationship between technology, humans, society and environment and the consequences and risks involved (Skolverket, 2022). Thus, knowledge related to programming needs to be assessed in compulsory school. In upper secondary school and in universities, programming has been taught and assessed for a long time, but in the lower grades, programming-related content in technology education might cause some confusion for teachers, as in the following fictitious scenario:

> Jill has taught technology and mathematics in lower secondary school for the past 15 years. She enjoys being a teacher and is interested in her subjects. Overall, she has high self-efficacy as a teacher and is interested in developing her teaching practice. In her teacher training, programming was not included in either technology or mathematics. Two years ago, she participated in a course designed for teachers in lower secondary school on the basics of programming. Nonetheless, she is quite anxious about the parts of the curriculum related to programming.
>
> She is about to plan a lesson about traffic lights and how they work. What troubles her most is assessment. She is eager to use formative assessment and give her students feedback to help them to progress in technology. The knowledge requirements tell her to assess students' reasoning, their explanations of how technical solutions work, and how students carry out work involving technology and design. She realises that she needs to know more about how other teachers assess assignments related to programming in technology.

Like Jill, we wanted to know more about how teachers assess students' knowledge related to programming. Since programming-related content is still rather new to the technology syllabus in compulsory school, we investigated how teachers in different parts of the education system think about the assessment of programming, with the aim to use that knowledge to discuss and suggest assessment strategies for teachers in compulsory school, primarily teachers in grades 6–9.

2 Background

Given that programming is a relatively new subject content in elementary school, there is little existing research on assessment at that level. However, there is some research on assessment at higher levels of education (Daly & Waldron, 2004; Harland et al., 2013). Internationally, programming is often related to computational thinking. Computational thinking (Wing, 2006) can be described as "a problem-solving methodology that can be automated and transferred and applied across subjects" (Barr & Stephenson, 2011). There are various projects that aim to assess computational thinking skills. Bebras (Dagienė & Stupurienė, 2016) is an international initiative that started in Lithuania, with the aim to promote informatics (Computer Science, or Computing) and computational thinking in schools. The Bebras challenge is an international contest performed in schools where students complete different tasks on computers. Logical thinking, pattern recognition and simple programming are tested. Teachers sometimes use Bebras in their assessment of students' computational thinking. It was also used in a research study evaluating a Code Club, which found that learning at the Club did not have any impact on pupils' computational thinking over and above changes that would have occurred anyway (Straw et al., 2017). Initiatives like Bebras focus on computational thinking, but they may also be used as a teaching resource in technology education. Dr. Scratch is a web-based assessment tool that uses the concept of computational thinking and automatically grades scratch programs according to a special template and grading scale (Moreno-León et al., 2016).

In higher education, the emphasis is on the skill of writing a program using code. Simon et al. (2012) analysed the content of introductory programming exams and found that a large percentage of the marks (81%) were awarded for the product and the student's coding skills, and much less for other skills such as design, debugging, testing, or explaining and documenting.

It is not enough to know the syntax or semantics of a coding language to be able to create a program; one needs to be able to put the pieces together to create a program. In this sense, it is a kind of problem solving (Soloway, 1986): "Experts have built up large libraries of stereotypical solutions to problems as well as strategies for coordinating and composing them" (Soloway, 1986, p. 850). To be able to give formative support, teachers should also be able to assess the learning process in the classroom, which includes students tinkering, creating, debugging, persevering, and collaborating. In a previous study (Björklund & Nordlöf, 2021), we investigated the criteria teachers use when grading and assessing students in tasks related to programming. Six Swedish teachers were interviewed using Comparative Judgement and Repertory Grid techniques. Through this process, several criteria for assessment were identified. These criteria were divided into two groups, regarding the finished product (the program) and the work process, respectively.

Criteria related to the product (the program):
- The program is complete.
- The function of the program, it fulfils its purpose.
- The student utilises the various functions of the program language.
- The student can explain the code used in the program.
- The coding is advanced, on a high technical level.
- Readability is high, easy to follow and understand the program.
- Good coding.
- The program is accurate and not sloppy.
- The documentation is good.

Criteria related to the process and how the student works:
- Endurance, the student struggles and does not give up.
- The student pushes the work forward and moves on.
- The student identifies the problem and makes a plan to find an effective way of solving it.
- The student has a trial-and-error approach, tinkering.
- The student is creative.
- The student can learn from other people's code.
- Good team worker.
- The student shows enthusiasm, motivation.

In the present study, we wanted to know how a greater number of teachers valued and ranked the criteria identified in the first study.

3 Method

In this study, we use a method almost forgotten in education research, namely Q methodology (Stephenson, 1953). In this method, informants sort and rank criteria or propositions according to their own subjective beliefs. The results are analysed statistically and, using factor analysis, groups of different types of beliefs may be identified. Q methodology has been used mostly in psychology and sociology, but also to analyse teachers' and students' beliefs about things like physics education, simulations, and teacher training (Jurczyk & Ramlo, 2004). The method can be used with a small number of informants and still give useful results. In the current study, a list of altogether 19 different criteria was developed. It was based on the output of the previous study (Björklund & Nordlöf, 2021), and further developed and expanded by the research team. The criteria were used in a web-based Q methodology sort.

Twenty-eight programming teachers were invited to participate. The participating teachers included both men and women representing schools from different parts of the school system: middle school, lower secondary, upper secondary and university. Some of them had experience of working as programmers, others were ordinary teachers. Some were qualified technology teachers. The participants' age varied between 30 and 77 years, and their experience as programming teachers varied from one to more than 40 years.

The participants were asked to sort the 19 criteria according to how important they deemed them as assessors. They were also asked to motivate some of their choices. Background data on these groups of teachers were analysed to see whether age, education, expertise, subject, or other factors could explain the results.

It is worth noting that valuing of the statements is relative. If a statement is placed on the low end of the scale, this does not mean that the teacher finds the statement unimportant, but that it is less important in relation to the other statements.

4 Analysis and Results

The traditional way of analysing a Q methodology survey is not to look at separate elements (criteria) and how they are valued, but to look for groups of informants sorting the elements in a similar way. Using this method, the relatively small number of participants is more than enough to get useful results. Thus, when we performed a proper Q methodology analysis, i.e., factorisation, two qualitatively different profiles of assessors were discovered. In other words, among the participants there are two groups of teachers (with 16 and 8 members respectively). Within the two groups, the teachers have not sorted the statements identically, but in a similar way.

The two groups are illustrated in Figures 16.1 and 16.2.

4.1 *Comparing Teachers in Groups 1 and 2*

When comparing Group 1 and Group 2 (Figures 16.1 and 16.2, respectively), we can see there are several statements that both groups have valued in a similar way (see Table 16.1).

First, we will look at which criteria both groups have given high value, i.e., rated as important, when grading and assessing students in programming. The statement *Student can explain the code,* meaning that the student can describe and explain the program, is valued highly by both groups. Both groups gave this statement the highest ranking (3). The statement *Readability of the code*

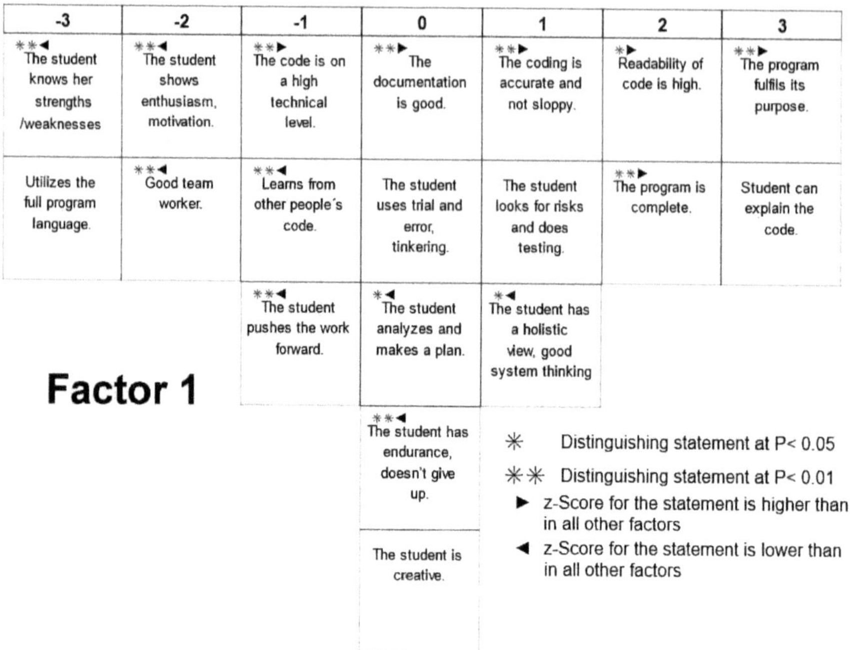

FIGURE 16.1 Illustration of Group 1 and how the teachers sorted the statements

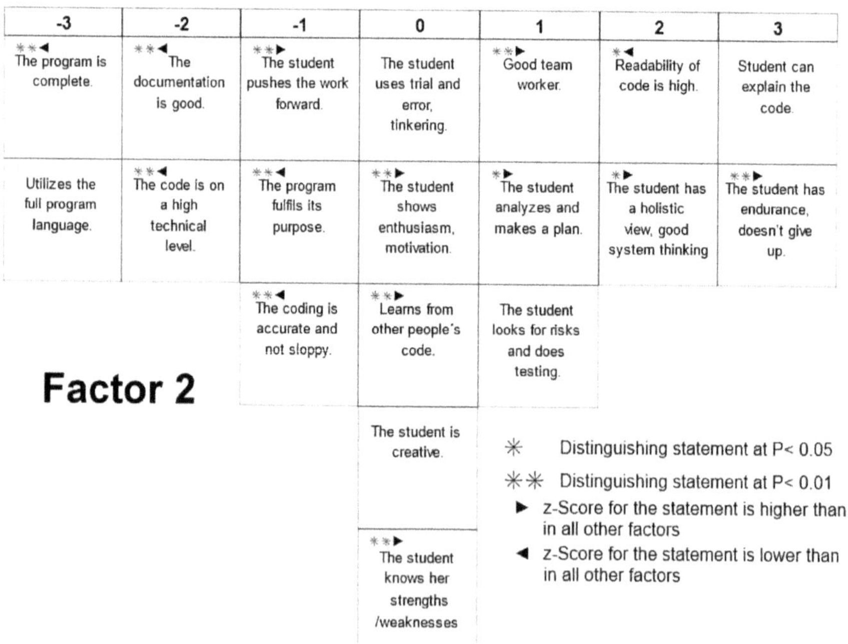

FIGURE 16.2 Illustration of Group 2 and how the teachers sorted the statements

TABLE 16.1 Statements valued similarly by both groups

Statements given high value by both groups	Statements given low value by both groups
Student can explain the code	(The student) utilises the full program language
Readability of the code is high	

is high is also ranked highly by both groups (2). This means that overall, the participating teachers believe that readability is important when assessing students, and for students to be able to explain the code and the program is ranked highest among the teachers in the study. This result is also confirmed by the mean value of the rankings by all the participants.

Further on, there are also interesting findings among the statements that are valued as not so important. There is one statement that both groups valued as the least significant statement in assessment, and that is *The student utilises the full program language.* Both groups valued this statement as least important (−3). This result is also confirmed by the mean value of the rankings by all the participants.

As seen in Figures 16.1 and 16.2, there are also several statements that both groups put in the middle of the scale, in a neutral position. In this chapter, we focus on the statements placed at the positive or negative end of the scale, i.e., those that teachers most clearly took a stand on. In all, most of the teachers find it important for the students to explain the code and for the code to be readable. Further, to utilise the full program language is not found to be important. Next, we will take a closer look at the differences between the two teacher groups.

4.1.1 What Is Significant for Teachers in Each Group?

There are two statements that are rated as important when grading students in programming in Group 1 (see Figure 16.1). *The program fulfils its purpose,* meaning that the program works and solves the problem, is ranked highest (3). The second most important criterion (2) is *The program is complete.* Both statements are focused on the program itself and that the programming is correct. In addition, Group 1 places three statements at the bottom of the scale: *The student knows her strengths and weaknesses* (−3), and further *Good team worker* (−2), and *The student shows enthusiasm, motivation* (−2). These statements all concern the student's way of working and their attitude.

Teachers in Group 2 meanwhile valued *The student has endurance, does not give up* as most important (3) (see Figure 16.2). The second most important statement in this group is *The student has a holistic view, good system thinking* (2). At the other end of the scale, we find *The program is complete* (−3) as least important. Second least important (−2) are *The documentation is good* and *The code is on a high technical level.* Thus, the teachers in Group 2 give high value to statements regarding the student's way of working and their attitude, and low value to statements concerning the program itself and that the programming is correct.

4.1.2 Product and Process—What Differentiates and Unites the Teachers

The criteria featured in this study can be divided into two categories, or themes, based on *what* is being assessed. The first is *Product criteria,* including statements regarding the outcome and the result; in other words, the program itself. The second is *Process criteria,* including statements regarding students' progression and their process of working and learning. Table 16.2 lists the criteria based on these themes.

Additionally, two criteria stand apart. Firstly, the criterion *The student can explain the code used in the program* is interpreted as including both product and process. Secondly, the criterion *The student has a holistic view, good system thinking* we consider as related to overall assessment and therefore is not included in Table 16.2.

In Figures 16.3 and 16.4, we have included the themes by adding colours to the diagrams. This illustrates the differences between the two groups, and further, what they have in common. As the figures show, Group 1 values product criteria (green) as most important, while Group 2 values the process criteria (yellow) as most important. Both groups value the ability to take a holistic approach (white) highly.

This indicates that the teachers in the study exemplify two different perspectives on assessment in programming. Group 1 mainly values product criteria, while Group 2 mainly values process criteria. The two groups do have some aspects in common, e.g., they both value a holistic approach to programming. Both groups also value the criterion *Readability of the code is high* as important, and *Utilises the full program language* as less important. As a result, Group 1 is named *Product assessors* and Group 2 is named *Process assessors.*

In the Q-sort questionnaire, the participants were asked to sort the statements according to importance. In addition, they were given the opportunity to motivate and comment on their choices via optional open questions. The responses help to explain the themes and give further insights.

TABLE 16.2 The statements divided into product and process criteria

Product criteria	Process criteria
The program is complete.	Endurance, the student struggles and does not give up.
The function of the program, it fulfils its purpose.	The student pushes the work forward and moves on.
The student utilizes the various functions of the program language.	The student identifies the problem and makes a plan to find an effective way of solving it.
The coding is advanced, on a high technical level.	The student has a trial-and-error approach, tinkering.
Readability is high, easy to follow and understand the program.	The student is creative.
The program is accurate and not sloppy.	The student can learn from other people's code.
The documentation is good.	Good team worker.
The student can explain the code used in the program.	The student shows enthusiasm, motivation.
	The student knows her strengths/ weaknesses.
	The student looks for risks and does testing.
	The student can explain the code used in the program.

Some of the teachers who valued product criteria highly gave comments which further explain their views. One teacher said that "If the program does not solve the problem, it is difficult to make a fair assessment also of other aspects (of programming)". This example illustrates how some teachers see the outcome, the program, as the main goal of the learning process. Meanwhile, the statements from the teachers who valued process criteria highly reveal a different position regarding the main purpose of programming and learning. For example, one teacher said that "Errors occur when writing code; different types of errors are part of learning". Another example of a process-oriented view is "Programming is problem solving, which means that active work where the student uses trial and error and refines their approach is an important aspect". One teacher refers to the statement *The student analyses and makes a*

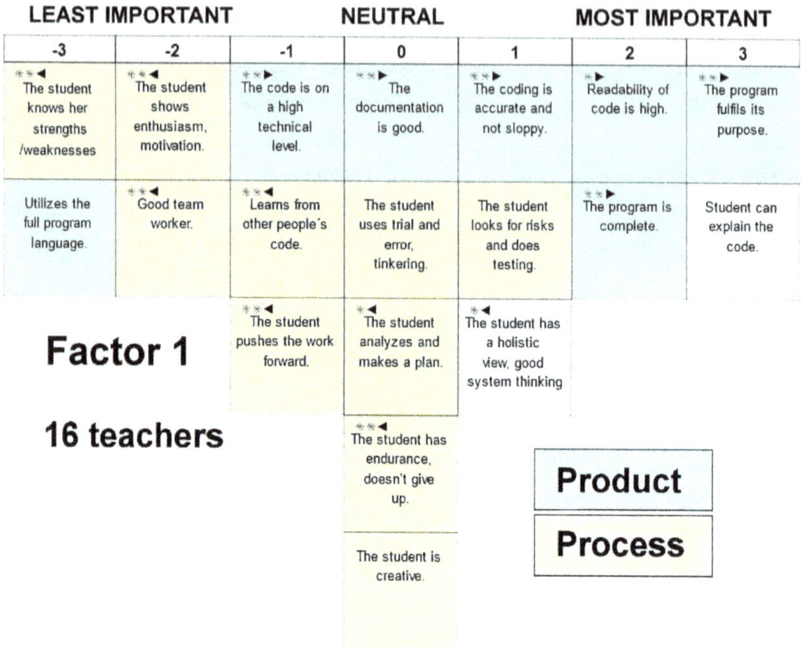

FIGURE 16.3 Illustration of Group 1 (*Product assessors*) and how the teachers in this group sorted the product and process statements

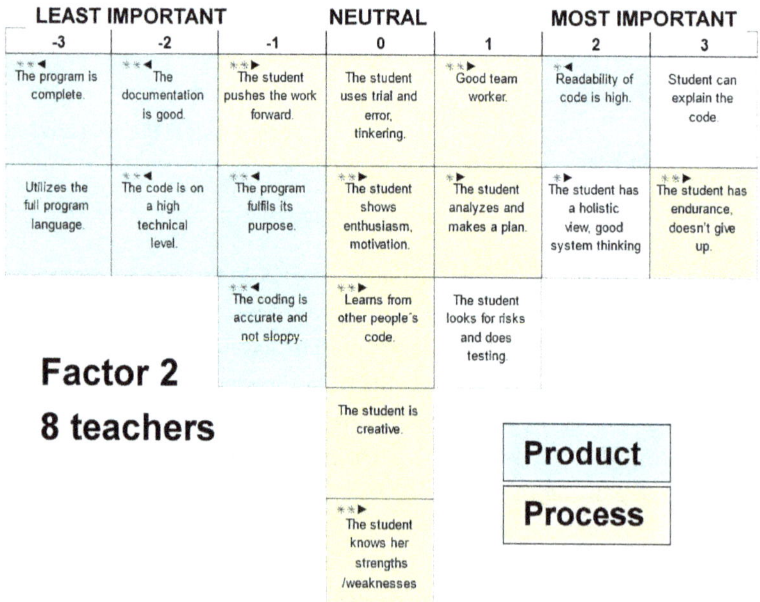

FIGURE 16.4 Illustration of Group 2 (*Process assessors*) and how the teachers in this group sorted the product and process statements

plan to be as "the foundation of programming". Conversely, one teacher who favors product criteria says of the same statement, "That is the starting point. If the program is not working the task is not complete". Thus, these statements show that the teachers' view of assessment also says something about their views on learning and knowledge in programming. This is something that is also visible in the teachers' comments on the criterion *The student has a holistic view, good system thinking*. In relation to this statement, one teacher writes: "If I am forced to choose only one thing, it is the whole."

In summary, the results show that teachers tend to focus more on either the process or the product of programming when it comes to assessment. Both groups place high value on the student's ability to explain the code used in the program, which can be interpreted as a combination of product and process, where the student uses the knowledge they have acquired and turns it into their own knowledge. The teachers express this in their answers to the open questions: "If the student can't explain the code, it is clear proof that he doesn't understand the program. If you can't explain the code, you don't know what you have done". Another excerpt makes it clear that both product and process are needed: "It is essential to give insight into the process, provide good opportunities to discuss various choices and methods regarding the code, and prove that the code is written by the student himself". Further, the teachers also place high value on the student's ability to take a holistic approach: "It is extremely important to be able to see the big picture when programming, how your own code relates to a larger whole".

5 Discussion

Assessing the product is a common way of grading students' work, but the process criteria seem to be more useful for formative purposes. Based on the results regarding product and process criteria, we found strong similarities with the work of Lars Lindström (2006), who designed an instrument to assess creativity in Art education.

Creativity is closely related to computer science and has a central role in fostering motivation and interest in this field of study. Studies have found a bi-directional connection between creativity and computer science. On the one hand, creativity may serve as a catalyst for solving algorithmic problems, creating computational artifacts, and developing new knowledge. Scores from a standardised creativity test predicted creativity in problem solving in computer programming among undergraduate students (Liu & Lu, 2002). Perez-Poch et al. (2016) also found a strong correlation between results from a creativity test and students' gradings in a computer science course. Many other studies have connected programming skills and computational thinking with the concept of creativity (Hershkovitz et al., 2019). Teaching fundamentals of programming is a complex task that involves student acquisition of diverse knowledge and skills. It is also well known that programming often requires a certain degree of creativity.

Creativity is one of the most important cognitive skills in our fast-changing world, and various techniques have been designed to maximise the generation of creative ideas over the past few decades. However, before creative ideas can be implemented, the most creative ideas must be selected from a larger pool of ideas. Although idea generation and idea selection are tightly linked in creativity theories, research on idea evaluation and idea selection lags far behind research on idea generation. This is unfortunate, as history is replete with cases where creative ideas were initially rejected. For example, concepts such as air travel, personal computers, and online shopping were at first viewed as crazy, but eventually became great successes that changed our world. In addition, the scarce research on idea selection has shown that people perform poorly at selecting creative ideas: they tend to select mainstream ideas at the expense of creative ones (Zhu et al., 2017). Another interesting finding from this area of research is that creativity is mostly referred to as product- or person-related. Consequently, it is measured mostly by self-report or product rating measures. On the other hand, computational thinking (CT) is rarely measured by assessing artifacts produced by learners. This may be explained by the existence of well-established product-related measures for creativity and the lack of such measures for CT. On top of that, it is important to note that planning and carrying out a learning scenario that includes building an artifact using CT, either physical or virtual, is challenging for both teachers and learners. Importantly, engaging learners with an actual building process is one of the cornerstones of CT, as the concept originally emerged from the constructionist approach to learning (Harel & Papert, 1991). More efforts should be made to align CT and creativity, mainly by incorporating product-related learning processes in the context of CT (Israel-Fishelson & Hershkovitz, 2022). There are creative elements in programming, just as there are in technology education and in several other school subjects.

Art is a subject where creativity is central, and previous research on assessment in that field is of interest when exploring assessment in programming. In a large study on creative development in the context of Art education, Lindström et al. aimed to identify criteria or descriptions of general abilities used by teachers to assess creative work (Lindström et al., 1999). Using observational data, a literature review and interviews, a list of important factors/criteria for the development of creativity was formulated and tested. Lindström and colleagues corroborated their work with an American research group that designed a similar instrument, "studio habits of mind" (Hetland et al., 2007). Lindström's criteria (2006) are divided into three groups:

Product criteria:
1. Visibility of the intention/Goal fulfillment
2. Colour, form, and composition/Visual qualities
3. Craftsmanship/Technical skill

Process criteria:
4. Investigative work/Persistence in the pursuit
5. Inventiveness/Imagination and risk taking
6. Ability to use models/Being able to learn from others
7. Capacity for self-assessment/ Knowing one's strengths/weaknesses

Others:
8. Overall judgment

In previous studies, Lindström's criteria have been transferred and used in technology education contexts, e.g., Björklund (2009). Using our results, we have adapted and applied Lindström's instrument to programming:

Product criteria:
1. Visibility of the function/Goal fulfillment
2. Use of programming language, knowledge of syntax
3. Program quality, coding skill, and documentation

Process criteria:
4. Investigative work/Persistence in the pursuit
5. Inventiveness/Imagination and risk taking
6. Ability to use models/Being able to learn from others
7. Capacity for self-assessment/Knowing one's strengths/weaknesses

Our results showed two types of criteria used by the teachers: product criteria assessing the program, and process criteria assessing the ongoing learning process and the student's "studio habits of mind". This somewhat contradicts results from Simon (2012) showing that examiners emphasised product criteria when they assessed programming skills. We compared the criteria used in our study with Lindström's instrument for measuring the development of creativity. Our preliminary results lead us to believe that some kind of process criteria could be used (and indeed seemed to be used by teachers) to assess students struggling with programming tasks.

5.1 *Proposed Criteria to Assess the Process of Programming—Implications*

In this section, we propose a way to use the results of our study for formative assessment. The rubrics in each proposed criterion could be used by the teacher in a formative way to lead the student forward in their development of programming skills. A tentative description of the process criteria could be useful in programming education. Some examples of rubrics follow below.

5.1.1 Investigative Work/Persistence in the Pursuit

The student tackles and completes the work with tenacity and patience. Demonstrates an ability to develop and stick to a theme or problem over time.

1. The student gives up easily, does not complete any ideas and only does what the teacher asks for.
2. The student shows some patience and tries out her/his own solutions and approaches but does not develop them further.
3. The student puts in great effort, approaches themes and problems from several perspectives, and develops the program through a sequence of drafts, sketches and try-outs (trial and error).

5.1.2 Inventiveness/Imagination and Risk-Taking

The student takes risks and tries out their own solutions and ideas.

1. The student does not formulate problems on her/his own and does not experiment with sensors or actuators or try different algorithms to reach the goal.
2. The student sometimes formulates problems of her/his own or slightly reformulates some of the problems stated by the teacher. Shows tendencies to experiment and try out her/his own ideas.
3. The student often states her/his own problems or reformulates the problems stated by the teacher. The student moves on, experiments and is willing to take risks. The student often finds unexpected solutions.

5.1.3 Ability to Use Models/Being Able to Learn from Others

The student searches for existing models, and finds links between these models and their own work.

1. The student does not show interest in solutions made by others and is unable to take advantage of them even when the teacher helps to find them.

2. The student makes some active efforts to find models for her/his own work, and shows some interest in other people's solutions, but confines her/himself to copying them.
3. The student is active in finding different kinds of examples of programs and uses them in a diverse, independent, and well-integrated way in their own work.

5.1.4 Capacity for Self-Assessment/Knowing One's Strengths/ Weaknesses

A capacity for self-assessment is not innate; it is something that students can develop and refine. The student has the ability to evaluate their own work, and can point out other works or parts of works that are either successful or that require further work.

1. The student is unable to identify merits and shortcomings in their own work, or to distinguish between good and bad programs. The student has no opinions of their classmates' programming work.
2. The student can, with some assistance, identify merits and shortcomings in their own work and distinguish between good and bad programs. The student begins to produce qualified judgements of peers' work.
3. The student sees merits and shortcomings in their own work, and can identify examples of programs and others' work that reflect their own development. The student can motivate their choices and explain why a particular result was produced. The student can give nuanced reviews of classmates' programs and contribute with constructive feedback.

As an example, let us return to the assignment on traffic lights that teacher Jill was wondering about at the beginning of this chapter. We can look, for example, at the criterion *Ability to use models/Being able to learn from others.* If one of Jill's students has problems finding a solution to the problem, she can motivate her student to try and see how others have solved this or similar problems to do with sequencing, showing examples of how to control external lamps, pausing the program, and using a loop. Further, she can also motivate her student to find other solutions by herself:

> Now Jill knows more about how other teachers think about assessing programming and how she can use formative assessment to help her student to learn. She understands that it is important that teachers reflect on what they want their students to learn. Also, she realises that it is common for teachers to focus mainly on the parts of programming that are related to how programming works and how the code is constructed.

> She reminds herself to also teach the students about how programming affects our everyday lives and to discuss with them how, for example, a self-driving car might involve ethical issues.

In this study on teachers' views regarding the assessment of programming, criteria related to the product and the process of programming were in focus. In technology education in Sweden, aspects of programming related to how it affects us as humans, and how it relates to society and the environment, are also relevant. Since our results are based on teachers' spontaneous statements about the assessment of programming, the results might indicate that teachers do not include those aspects of programming in their assessment. Another explanation is that at upper secondary and university level, the actual programming is the focus. The social aspects of programming in assessment would be interesting to examine further in future studies.

References

Barr, V., & Stephenson, C. (2011). Bringing computational thinking to k-12: What is involved and what is the role of the computer science education community? *ACM Inroads*, 2(1), 48–54.

Björklund, L. (2009). The forming and assessment of creative skills, from a neurocognitive point of view. In *Patt-22 Conference – Strengthening Technology Education in the School Curriculum, Delft, the Netherlands.*

Björklund, L., & Nordlöf, C. (2021). Teacher's assessment in programming: Comparing teachers' individual judgement criteria in a programming course. *Techne Series: Research in Sloyd Education and Crafts Science, 28*(2), 188–195.

Dagienė, V., & Stupurienė, G. (2016). Bebras – a sustainable community building model for the concept based learning of informatics and computational thinking. *Informatics in Education, 15*(1), 25–44.

Daly, C., & Waldron, J. (2004). Assessing the assessment of programming ability. *ACM SIGCSE Bulletin, 36*(1), 210–213.

Harland, J., D'Souza, D., & Hamilton, M. (2013). A comparative analysis of results on programming exams. In *Proceedings of the Fifteenth Australasian Computing Education Conference-Volume 136* (pp. 117–126).

Hershkovitz, A., Sitman, R., Israel-Fishelson, R., Eguíluz, A., Garaizar, P., & Guenaga, M. (2019). Creativity in the acquisition of computational thinking. *Interactive Learning Environments, 27*(5–6), 628–644. https://doi.org/10.1080/10494820.2019.1610451

Israel-Fishelson, R., & Hershkovitz, A. (2022). Studying interrelations of computational thinking and creativity: A scoping review (2011–2020). *Computers & Education, 176*, 1–22.

Jurczyk, J., & Ramlo, S. (2004). A new approach to performing course evaluations: Using Q methodology to better understand student attitudes. In *International Q Methodology Conference, Athens, GA.*

Lindström, L. (2006). Creativity: What is it? Can you assess it? Can it be taught? *International Journal of Art & Design Education, 25*(1), 53–66.

Lindström, L., Ulriksson, L., & Elsner, C. (1999). *Portföljvärdering av elevers skapande i bild [Portfolio Assessment of Students' Creative Skills in the Visual Arts]* (99:488). Skolverket.

Liu, M. C., & Lu, H. F. (2002). A study on the creative problem-solving process in computer programming. In *International Conference on Engineering Education, Manchester, UK.*

Moreno-León, J., Robles, G., & Román-González, M. (2016). Examining the relationship between socialization and improved software development skills in the scratch code learning environment. *Journal of Universal Computer Science, 22*(12), 1533–1557.

Pérez Poch, A., Olmedo Torre, N., Sánchez Carracedo, F., Salán Ballesteros, M. N., & López Álvarez, D. (2016). On the influence of creativity in basic programming learning at a first-year Engineering course. *International Journal of Engineering Education, 32*(5), 2302–2309.

Simon, D., Chinn, M., de Raadt, A., Philpott, J., Sheard, M. J., Laakso, D., D'Souza, J., Skene, A., Carbone, T., Lister, R., & Warburton, G. (2012). Introductory programming: examining the exams. In *Proceedings of the Fourteenth Australasian Computing Education Conference, Melbourne, Australia.*

Skolverket. (2022). *Läroplan för grundskolan, förskoleklassen och fritidshemmet 2022.* Skolverket.

Soloway, E. (1986). Learning to program = learning to construct mechanisms and explanations. *Communications of the ACM, 29*(9), 850–858.

Stephenson, W. (1953). *The study of behavior: Q-technique and its methodology.* University of Chicago Press.

Straw, S., Bamford, S., & Styles, B. (2017). *Randomised controlled trial and process evaluation of code clubs.* National Foundation for Educational Research and the Raspberry Pi Foundation.

Wing, J. M. (2006). Computational thinking. *Communication of the ACM, 49*(3), 33–35.

Zhu, Y., Ritter, S. M., Müller, B. C. N., & Dijksterhuis, A. (2017). Creativity: Intuitive processing outperforms deliberative processing in creative idea selection. *Journal of Experimental Social Psychology, 73*, 180–188.

Index

www.ingramcontent.com/pod-product-compliance
Lightning Source LLC
Chambersburg PA
CBHW071421230926
51846CB00001B/100
9789004687899